OXFORD MASTER SERIES IN PARTICLE PHYSICS, ASTROPHYSICS, AND COSMOLOGY

OXFORD MASTER SERIES IN PHYSICS

The Oxford Master Series is designed for final year undergraduate and beginning graduate students in physics and related disciplines. It has been driven by a perceived gap in the literature today. While basic undergraduate physics texts often show little or no connection with the huge explosion of research over the last two decades, more advanced and specialized texts tend to be rather daunting for students. In this series, all topics and their consequences are treated at a simple level, while pointers to recent developments are provided at various stages. The emphasis in on clear physical principles like symmetry, quantum mechanics, and electromagnetism which underlie the whole of physics. At the same time, the subjects are related to real measurements and to the experimental techniques and devices currently used by physicists in academe and industry. Books in this series are written as course books, and include ample tutorial material, examples, illustrations, revision points, and problem sets. They can likewise be used as preparation for students starting a doctorate in physics and related fields, or for recent graduates starting research in one of these fields in industry.

CONDENSED MATTER PHYSICS
1. M. T. Dove: *Structure and dynamics: an atomic view of materials*
2. J. Singleton: *Band theory and electronic properties of solids*
3. A. M. Fox: *Optical properties of solids*
4. S. J. Blundell: *Magnetism in condensed matter*
5. J. F. Annett: *Superconductivity*
6. R. A. L. Jones: *Soft condensed matter*

ATOMIC, OPTICAL, AND LASER PHYSICS
7. C. J. Foot: *Atomic physics*
8. G. A. Brooker: *Modern classical optics*
9. S. M. Hooker, C. E. Webb: *Laser physics*

PARTICLE PHYSICS, ASTROPHYSICS, AND COSMOLOGY
10. D. H. Perkins: *Particle astrophysics*
11. Ta-Pei Cheng: *Relativity and Cosmology*

Particle Astrophysics

D.H. PERKINS

Particle and Astrophysics Department
Oxford University

OXFORD
UNIVERSITY PRESS

OXFORD
UNIVERSITY PRESS

Great Clarendon Street, Oxford OX2 6DP

Oxford University Press is a department of the University of Oxford.
It furthers the University's objective of excellence in research, scholarship,
and education by publishing worldwide in

Oxford New York

Auckland Bangkok Buenos Aires Cape Town Chennai
Dar es Salaam Delhi Hong Kong Istanbul Karachi Kolkata
Kuala Lumpur Madrid Melbourne Mexico City Mumbai Nairobi
São Paulo Shanghai Taipei Tokyo Toronto

Oxford is a registered trade mark of Oxford University Press
in the UK and in certain other countries

Published in the United States
by Oxford University Press Inc., New York

A catalogue record for this title is available from the British Library

Library of Congress Cataloging in Publication Data
(Data available)
ISBN 0 19 850951 0 (Hbk)
ISBN 0 19 850952 9 (Pbk)

10 9 8 7 6 5 4 3 2

Typeset by Newgen Imaging Systems (P) Ltd., Chennai, India
Printed in Great Britain on acid-free paper by
Antony Rowe Ltd, Chippenham, Wiltshire

Preface

The main object of this text is to present the twin fields of elementary particle physics and astrophysics at a level suitable for undergraduate students in physics. During the last two or three decades in particular, the subjects of elementary particles, astrophysics, and cosmology have become inextricably woven together and interdependent, and this symbiosis has led to tremendous advances in our understanding of the universe at the very largest and the very smallest scales of both time and distance. Undoubtedly, the main features of the cosmos today depend on the nature of the elementary particles and of their interactions, which together were the main players in the very earliest stages of the universe.

The merging together of the subjects of elementary particles and astrophysics in a single text also seemed appropriate at this time. For the last fifty years, an intensive programme of research at accelerator laboratories worldwide has led to an impressively exact account of the properties and interactions of the fundamental quarks and leptons that constitute our own world. Both experimental precision and theoretical predictions in this field have reached unprecedented levels of accuracy in verifying the so-called 'Standard Model', reaching a level of the order of 10^{-6} in the case of quantum electrodynamics. However, the days are gone when particle physicists might claim that experiments at accelerators were elucidating the fundamental structure of the entire material universe. Indeed, to an increasing extent over the last decades, physicists with a background in elementary particle physics at accelerators have moved over to the astrophysics field, for example in the study of solar and atmospheric neutrinos, in the observations of high redshift supernovae, and in the search for astronomical point sources of very high energy γ-rays and neutrinos.

According to our present picture, the known elementary particles account for only about 5% of the total energy density of the universe, the remainder being accounted for by dark matter and dark energy, the nature and origins of which are so far unknown. For simple economic reasons, if none other, it will never be possible to build accelerators taking us to the highest energies, which we believe were involved in the birth pangs of the universe. What exactly happened in those early stages, what particles and interactions played the leading role, will probably only be answered from astronomical observations. In the field of astrophysics and cosmology, enormous strides have been made over the last few decades, and a rather exact picture of the fundamental cosmological parameters—the 'Standard Model' of cosmology—is slowly beginning to emerge. Certainly, the famous old criticism of Landau, describing cosmologists as "people who are frequently wrong, but never in doubt", no longer applies.

A full discussion of the development of the universe and its cosmological history requires as a basis, first, the general theory of relativity, and second a discussion of the fundamental particles and their interactions via relativistic quantum mechanics and quantum field theory. Both these subjects are outside the scope of the average physics undergraduate course. Nevertheless, I felt that the subject matter of particle astrophysics was important enough to warrant a more elementary approach, since it will introduce the student to the big questions and challenges at one of the main frontiers of physics, and perhaps lead some of them on to more advanced and specialized texts.

The material in this text is presented assuming that the reader has some elementary knowledge of quantum mechanics and of special relativity. A résumé of the latter has been included in an appendix. An attempt has been made to make the various chapters self-contained as far as possible. For example, a shorter course could be obtained by omitting Chapter 3 on symmetries and Chapters 6 and 7 on cosmic rays and particle physics in stars. Sets of problems, mostly numerical, appear at the ends of chapters. Answers, and solutions for the harder problems, are given at the end of the book. Worked examples also appear inside the chapter material and these, plus the questions in the problem sets, range from easy to challenging. However, none should be beyond the abilities of the average student. They are an integral part of the book in the sense that, as Kelvin stated long ago, it is only by working through numerical problems that the meanings of the various ideas, concepts, and experimental projects can be fully appreciated.

Acknowledgements

For permission to reproduce photographs, figures and diagrams I am indebted to the authors cited in the text and to the following individuals, laboratories, journals, and publishers:

> Addison-Wesley Publishers for Figures 1.11, 1.13, 3.1, 3.15, 6.8 and 6.9, reproduced from '*Introduction to High Energy Physics*' 3rd edition (1987) by D.H. Perkins; and for Figure 7.1, reproduced from '*Introduction to Nuclear Physics*' (1972) by H.A. Enge.
> American Institute of Physics, publishers of *Physical Review* for Figure 1.2
> *Astronomy and Astrophysics* for Figure 4.2
> *Astrophysical Journal* for Figure 2.2
> *Annual Reviews of Nuclear and Particle Science* for Figure 6.3
> Cambridge University Press for Figures 2.12, 5.3, 7.6, 10.2, 10.3 and 10.4, reproduced from '*Introduction to High Energy Physics*' 4th edition (2000) by D.H. Perkins.
> CERN Information Services, CERN, Geneva for Figure 1.1 and Figure 1.6
> DESY Laboratory, Hamburg for Figure 1.5
> Elsevier Science for Figure 6.13, reproduced from '*Physics Reports*' Vol. 305, 93 (1998) by R.A. Ong
> European Southern Observatory for Figure 7.6
> Fermilab Visual Media Services, Fermilab, Chicago for Figure 1.10
> *Nature* for Figure 4.4
> *Nuclear Physics B* (Elsevier Science) for Figure 7.3
> Prof. Chris Carilli, NRAO for Figure 6.15
> Prof. Y. Totsuka of the Superkamiokande collaboration for Figure 3.13
> Prof. Trevor Weekes of the Whipple Observatory, Arizona for Figure 6.11

Oxford University   *D.H.P.*

To my family
For their patience and encouragement

Contents

Quarks and leptons and their interactions

1

High energy particle physics is concerned with the study of the fundamental constituents of matter and the interactions between them. Experiments in this field have been carried out with giant accelerators and their associated detection equipment, which have probed the structure of matter down to very small scales, of order 10^{-17} m, which is about one hundredth of the radius of a proton. In contrast, astrophysics is concerned with the structure and evolution of the universe in the large, including the study of the behaviour of matter and radiation on enormous scales, up to about 10^{26} m. Experimental observations have been made with telescopes on the Earth or on satellites, covering the visible, infrared, and ultraviolet regions of the spectrum, as well as with detectors of radio waves, X-rays, γ-rays, and neutrinos. These have revealed an astonishing range of extra-terrestrial phenomena from the most distant regions of the cosmos. The object of this text is to show how studies of particles on a laboratory scale have helped in our understanding of the development of the universe, and conversely, how celestial observations have, in turn, shed light on our understanding of particle interactions. Although, in this chapter, we discuss the constitution of matter as determined by accelerator experiments on Earth, it is becoming clear that on cosmological scales, other quite different forms of matter and energy may be important or even dominant, as discussed in Chapter 4.

1.1 Preamble

First, we should note the units employed in the study of the fundamental quark and lepton constituents of matter. The unit of length is the femtometre (1 fm = 10^{-15} m), an appropriate unit because, for example, the charge radius of a composite particle such as a proton is 0.8 fm. The typical energy scale is the giga-electronvolt (1 GeV = 10^9 eV); for example, the mass energy equivalent of a proton is $M_p c^2 = 0.938$ GeV. Table 1.1 lists the units employed in high energy physics, together with their equivalents in SI units. A list of appropriate physical constants is given in Appendix A.

In the description of particle interactions at the quantum level, the quantities $\hbar$ and c frequently occur, and it is often convenient to employ the so-called natural units, which set $\hbar = c = 1$. Having chosen these two units, we are free to specify one more unit, which is taken as that of energy, the GeV. The unit of mass is then $Mc^2/c^2 = 1$ GeV, of length $\hbar c/Mc^2 = 1$ GeV$^{-1} = 0.1975$ fm, and of time, $\hbar c/Mc^3 = 1$ GeV$^{-1} = 6.59 \times 10^{-25}$ s.

Table 1.1 Units in high energy physics

Quantity	High energy unit	Value in SI units
Length	1 fm	10^{-15} m
Energy	1 GeV	1.602×10^{-10} J
Mass, E/c^2	1 GeV/c^2	1.78×10^{-27} kg
$\hbar = h/2\pi$	6.588×10^{-25} GeV s	1.055×10^{-34} J s
c	2.998×10^{23} fm s^{-1}	2.998×10^{8} m s^{-1}
$\hbar c$	0.1975 GeV fm	3.162×10^{-26} J m

1.2   Quarks and leptons

In the so-called Standard Model of particle physics, which is strongly supported by extensive laboratory experiments and is further discussed in Chapter 3, the material universe is assumed to be built from a small number of fundamental constituents, the **quarks** and the **leptons**. The names of these, together with their electrical charges are given in Table 1.2. All these particles are **fermions**, that is, they have half-integral intrinsic angular momentum or spin, $\frac{1}{2}\hbar$ For each of the particles in the table there is an **antiparticle**, with the opposite value of electric charge and magnetic moment, but with identical mass and lifetime to those of the particle. For example, the positron (see Fig. 1.2) e$^+$ is the antiparticle of the electron, e$^-$. In contrast to the proton and neutron, which are extremely small but nevertheless extended objects, the quarks and leptons are considered to be **pointlike**: as far as we know today, they are truly elementary and are not composed of other, even more fundamental entities.

Let us first consider the leptons, of which the electron is familiar to everyone. The muon μ and the tauon τ are heavier, highly unstable versions of the electron, with mean lifetimes of 2.2×10^{-6} s and 2.9×10^{-13} s, respectively. Associated with each charged lepton is a neutral lepton, called a neutrino, denoted by the generic symbol, ν. A different neutrino ν_e, ν_μ, or ν_τ is associated with each different type or **flavour** of charged lepton. For example, in nuclear beta decay, a (bound) proton in a nucleus transforms to a neutron together with a positron e$^+$, which is emitted together with an electron-type neutrino, that is, $p \rightarrow n + e^+ + \nu_e$. In a subsequent interaction, this neutrino may transform into an electron, that is, $\nu_e + n \rightarrow e^- + p$, but not into a charged muon or tauon (see Fig. 1.1 for examples of such transformations).

All the particles (and their antiparticles) in Table 1.2, with the exception of the neutrinos, are fermions with two spin substates each; relative to the momentum (z-)axis, the spin components are $s_z = \pm\frac{1}{2}\hbar$. However, a neutrino has only one spin state, $s_z = -\frac{1}{2}\hbar$, while an antineutrino has $s_z = +\frac{1}{2}\hbar$ only. This fact, and the violation of parity in weak interactions is discussed in Sections 3.4 and 3.5.

The quarks in Table 1.2 have fractional electric charges, of $+2|e|/3$ and $-|e|/3$ where $|e|$ is the numerical value of the electron charge. As for the leptons, the masses increase as we go down the table (see also Tables 1.3 and 1.4). Apart from charge and spin, the quarks, like the leptons, have an extra internal degree of freedom, again called the **flavour**. The odd names for the various quark flavours—'up', 'down', 'charm' etc.—have arisen historically. Just as for the leptons, the six flavours of quark are arranged in three doublets, the components of which differ by one unit of electric charge.

Table 1.2 Quark and lepton flavours

Symbol	Name	$Q/|e|$	Symbol	Name	$Q/|e|$
u	up	$+\frac{2}{3}$	e	electron	-1
d	down	$-\frac{1}{3}$	ν_e	e-neutrino	0
c	charm	$+\frac{2}{3}$	μ	muon	-1
s	strange	$-\frac{1}{3}$	ν_μ	μ-neutrino	0
t	top	$+\frac{2}{3}$	τ	tauon	-1
b	bottom	$-\frac{1}{3}$	ν_τ	τ-neutrino	0

(a)

v beam →

(b)

v beam →

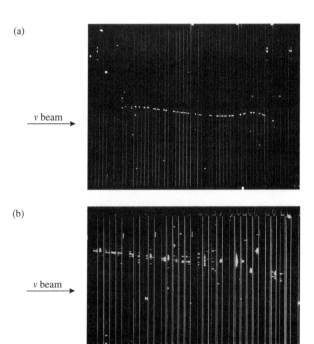

Fig. 1.1 Interaction of neutrino beam, from left, of about 1 GeV energy, in a CERN experiment employing a spark chamber detector. This consists of an array of parallel, vertical metal plates maintained at high voltages. A charged particle will ionize the gas between the plates, and this leads to a complete breakdown of the gas in a spark (Geiger) discharge. Thus, charged particle trajectories appear as rows of sparks. The event at the top is attributed to a muon-type neutrino. Upon interaction in the plate it transforms to a muon, which traverses many plates before coming to rest. The event at the bottom is due to an electron-type neutrino, transforming to an electron. The latter generates scattered sparks characteristic of an electron–photon shower, as described in Chapter 6, quite distinct from the rectilinear muon track. In both cases, the reactions are 'elastic', of the form $\nu_l + n \rightarrow l + p$, where $l = \mu$ or e, and the recoiling proton is stopped inside the plate. (Courtesy CERN Information Services.)

Table 1.3 Lepton masses in energy units, mc^2

Flavour	Charged lepton mass [MeV]	Neutral lepton mass
e	0.511 MeV	ν_e: <2.5 eV
μ	105.66 MeV	ν_μ: <0.16 MeV
τ	1777 MeV	ν_τ: <18 MeV

Table 1.4 Constituent quark masses

Flavour	Quantum number	Approximate rest-mass, GeV/c^2
Up or down	—	0.31
Strange	$S = -1$	0.50
Charm	$C = +1$	1.6
Bottom	$B = -1$	4.6
Top	$T = +1$	175

While the leptons exist as free particles, the quarks do not. It is a peculiarity of the strong force between the quarks that they are always found associated in quark composites called **hadrons**. These are of two types: **baryons** consist of three quarks, QQQ, while **mesons** consist of a quark–antiquark pair, Q$\bar{\text{Q}}$. For example,

$$\text{Proton} = \text{u u d} \quad \text{Neutron} = \text{d d u}$$

$$\text{Pion: } \pi^+ = \text{u}\bar{\text{d}}, \quad \pi^- = \bar{\text{u}}\text{d}$$

The common material of the world today is built from u and d quarks, forming the protons and neutrons of atomic nuclei, which together with the electrons e$^-$ form atoms and molecules. The heavier quarks c, s, t, and b are also observed to form baryon composites such as sud, sdc,..., and mesons such as b$\bar{\text{b}}$, c$\bar{\text{c}}$, c$\bar{\text{b}}$,..., but these heavy hadrons are all highly unstable and decay rapidly to states containing u and d quarks only. Likewise, the heavier charged leptons μ and τ decay to electrons and neutrinos. These heavy quarks and leptons can be produced in collisions at laboratory accelerators, or naturally in the atmosphere as a result of collisions of high energy cosmic rays. However, they appear to play practically no role in today's relatively cold universe. For example, while several hundred high energy muons (coming down to earth as secondary components of the cosmic rays) pass through everyone each minute, this is a trivially small number compared with the human tally of electrons, of order 10^{28}. Of course, we believe that these heavier flavours of quarks and leptons would have been as prolific as the light ones at a very early, intensely hot stage of the Big Bang, when the temperature was such that the mean thermal energy kT far exceeded the mass energy of these particles. Indeed, it is clear that the type of universe we inhabit today must have depended very much, in its initial evolution, on these heavier fundamental particles.

The masses of the quarks and leptons are given in Tables 1.3 and 1.4. The masses shown for neutrinos in Table 1.3 are upper limits deduced from energy and momentum conservation in decays involving neutrinos (e.g. from the kinematics of the decay $\pi \rightarrow \mu + \nu_\mu$). Indirect evidence, from oscillations of neutrino flavour in very long baseline experiments, described in Sections 6.8 and 6.9, indicates neutrino masses that are very much smaller than the limits given in the table.

As discussed below, the quarks are held together in hadrons by the gluon carriers of the strong force, and the 'constituent' quark masses in Table 1.4 include such quark binding effects. The u and d quarks have nearly equal masses (each of about one-third that of the nucleon) as indicated by the smallness of the neutron–proton mass difference of 1.3 MeV/c^2. Isospin symmetry in nuclear physics results from this coincidence in the light quark masses.

High energy scattering experiments often involve 'close' collisions between the quarks. In this case, the quarks can be temporarily separated from their retinue of gluons, and the so-called 'current' quark masses which then apply are smaller than the constituent masses by about 0.30 GeV/c^2. So the current u and d quark masses are a few MeV/c^2 only.

In the strong interactions between the quarks, the flavour quantum number is conserved, and is denoted by the quark symbol in capitals. For example, a strange s quark has a strangeness quantum number $S = -1$, while a strange antiquark $\bar{\text{s}}$ has $S = +1$. Thus, in a collision between hadrons containing u and

Table 1.5 The fundamental interactions ($Mc^2 = 1$ GeV)

	Gravitational	Electromagnetic	Weak	Strong
Field boson	graviton	photon	W, Z	gluon
Spin/Parity	2^+	1^-	$1^+, 1^-$	1^-
Mass	0	0	$M_W = 80.2$ Gev $M_Z = 91.2$ Gev	0
Source	mass	electric charge	weak charge	colour charge
Range, m	∞	∞	10^{-18}	$<10^{-15}$
Coupling constant	$GM^2/4\pi\hbar c$ $= 5 \times 10^{-40}$	$\alpha = e^2/4\pi\hbar c$ $= 1/137$	$G_F(Mc^2)^2/(\hbar c)^3$ $= 1.17 \times 10^{-5}$	$\alpha_s \leq 1$

d quarks only, heavier quarks can be produced, but only as quark–antiquark pairs, so that the net flavour is conserved. In weak interactions, on the contrary, the quark flavour may change—for example, one can have transitions of the form $\Delta S = \pm 1$, $\Delta C = \pm 1$ etc. As an example, a baryon called the lambda hyperon of $S = -1$ decays to a proton and a pion, $\Lambda \rightarrow p + \pi^-$, with a mean lifetime of 2.6×10^{-10} s, typical of a weak interaction of $\Delta S = +1$. In quark nomenclature, this decay would be expressed as sud $\rightarrow$ uud + dū.

A few words are appropriate here about the practical attainment in the laboratory of high mass scales in high energy particle physics. The completion of Table 1.2 of fermions and Table 1.5 of bosons took over forty years of the twentieth century, as bigger and more energetic particle accelerators were able to excite production of more and more massive fundamental states. The first evidence for the existence of the lighter quarks u,d,s appeared in the 1960s, from experiments at the CERN PS (Geneva) and Brookhaven AGS (Long Island) proton synchrotrons, with beam energies of 25–30 GeV, as well as the 25 GeV electron linear accelerator at SLAC, Stanford. The weak bosons W and Z, with masses of 80 and 90 GeV/c^2 were first observed in 1983 at the CERN proton–antiproton collider with oppositely circulating beams of energy 270 GeV (see Fig. 1.6). The most massive particle so far produced, the top quark of mass 175 GeV/c^2, was first observed in 1995 at the Fermilab proton–antiproton collider (Chicago), with 900 GeV energy in each beam. (see Fig. 1.10)

1.3   Fermions and bosons: the spin-statistics theorem: supersymmetry

As stated above, the fundamental particles consist of half-integer spin fermions, the quarks and leptons, the interactions of which are mediated, as described below, by integer spin **bosons**. The distinction between the two types is underlined by the **spin-statistics theorem**. This specifies the behaviour of an ensemble of identical particles, described by some wave function ψ, when any two particles, say 1 and 2, are interchanged. The probability $|\psi|^2$ cannot be altered by the interchange, since the particles are indistinguishable, so under the operation, $\psi \rightarrow \pm\psi$. The rule is as follows:

Identical bosons: under interchange $\psi \rightarrow +\psi$ Symmetric

Identical fermions: under interchange $\psi \rightarrow -\psi$ Antisymmetric

Suppose, for example, that it were possible to put two identical fermions in the **same** quantum state. Then, under interchange, ψ would not change sign, since the particles are indistinguishable. However, according to the above rule ψ **must** change sign. Hence two identical fermions cannot exist in the same quantum state—the famous Pauli Principle. On the other hand, there are no restrictions on the number of identical bosons in the same quantum state, an example of this being the laser.

One exciting development in connection with theories unifying the fundamental interactions at very high mass scales, has been the postulate of a fermion–boson symmetry called **supersymmetry**. For every known fermion state there is assigned a boson partner, and for every boson a fermion partner. The reasons for this postulate are discussed in Chapter 3, and a list of proposed supersymmetric particles given in Table 3.2. At this point we content ourselves with the remark that supersymmetric particles created in the early universe could be prime candidates for the mysterious **dark matter** which, as we shall see in Chapter 4, constitutes the bulk of the material universe. However, at the present time there is no direct experimental evidence for the existence of supersymmetric particles.

1.4   Antiparticles

In 1931, Dirac wrote down a wave equation describing an electron, which had four solutions. Two of these were for the electron, and corresponded to the two possible **spin substates** with projections $s_z = +\frac{1}{2}\hbar$ and $-\frac{1}{2}\hbar$ along the quantization axis. The other two solutions were attributed to the **antiparticle**, with properties similar to that of the electron except for the opposite value of the electric charge. These predictions followed from the two great conceptual advances in twentieth century physics, namely the classical theory of relativity and the quantum mechanical description of atomic and subatomic phenomena.

The relativistic relation connecting energy E, momentum p, and rest mass m is

$$E^2 = p^2c^2 + m^2c^4 \tag{1.1}$$

(The reader is referred to Appendix B for a summary of relativistic transformations.) From this equation we see that the total energy can, in principle, assume both negative and positive values:

$$E = \pm\sqrt{p^2c^2 + m^2c^4} \tag{1.2}$$

While in classical mechanics negative energies appear to be meaningless, in quantum mechanics we represent a stream of electrons travelling along the positive x-axis by the plane wavefunction

$$\psi = A\exp\left[-i\frac{Et - px}{\hbar}\right] \tag{1.3}$$

where the angular frequency is $\omega = E/\hbar$, the wavenumber is $k = p/\hbar$, and A is a normalization constant. As t increases, the phase $(Et - px)$ advances in the direction of positive x. However, (1.3) can equally well represent a particle

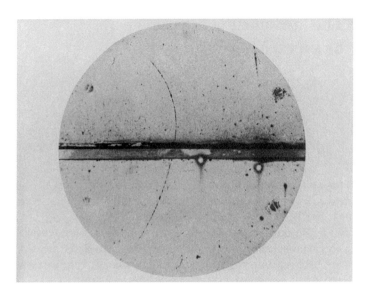

Fig. 1.2 The discovery of the positron by Anderson in 1932, in a cloud chamber operated at a mountain station to investigate the cosmic rays. The cloud chamber consists of a glass-fronted cylindrical tank of gas saturated with water vapour. Upon applying an expansion by means of a piston at the rear of the chamber, the gas cools adiabatically, it becomes supersaturated and water condenses as droplets, especially on charged ions created by the passage of a charged particle. A magnetic field applied normal to the chamber plane allows the measurement of particle momentum from the track curvature. Note that this curvature is larger in the top half of the chamber, because the particle loses momentum in traversing the central metal plate. Hence, it was established that the particle was positive and travelling upwards, and that its mass was very much less than that of a proton, and consistent with that of the electron. Since Anderson's experiment, many other types of antiparticles have been observed, including the antiproton in 1956 and the anti-hydrogen atom in 1995.

of energy $-E$ and momentum $-p$ travelling in the negative x direction and **backwards in time**, that is, replacing Et by $(-E)(-t)$ and px by $(-p)(-x)$:

$$E > 0 \qquad\qquad E < 0$$

$$t_1 \rightarrow t_2 \qquad\qquad t_1 \leftarrow t_2 \qquad (t_2 > t_1)$$

A stream of negatively charged electrons flowing backwards in time is equivalent to a positive charge flowing forwards, thus with $E > 0$. Hence, the negative energies are formally connected with the existence of a positive energy antiparticle e^+, the positron. This particle was first observed by Anderson in 1932, quite independently of Dirac's prediction (see Fig. 1.2). The existence of antiparticles is a general property of both fermions and bosons, but for fermions alone there is a conservation rule. One can define a fermion number, $+1$ for a fermion and -1 for an antifermion and postulate that the total fermion number is conserved. Thus, fermions can only be created or destroyed in particle–antiparticle pairs, such as e^+e^- or $Q\bar{Q}$. For example, a γ-ray, if it has energy $E > 2mc^2$ where m is the electron mass, can create a pair (in the presence of an atom to conserve momentum), and an e^+e^- pair can annihilate to γ-rays. As another example, in massive stars reaching the supernova phase, fermion number conserving reactions such as $e^+ + e^- \rightarrow \nu + \bar{\nu}$ are expected to be commonplace.

1.5   The fundamental interactions: boson exchange

The elementary fermions—the quarks and leptons—are postulated to interact via the **exchange of boson mediators**, the boson carrying momentum from one fermion to the other. The rate at which momentum is exchanged in this way provides the force between the interacting particles. There are four known types of interactions, each with its characteristic boson exchange particle. They are as follows:

- The **electromagnetic** interaction occurs between all types of charged particles, and is brought about by the exchange of a **photon**, with spin 1 and zero mass.
- The **strong** interactions occur between the quarks, via exchange of the **gluon**, a particle again with spin 1 and zero mass. Such interactions are responsible not only for the binding of quarks in hadrons but also for the force holding neutrons and protons together in atomic nuclei.
- The **weak** interactions take place between all types of quarks and leptons. They are mediated by the exchange of **weak bosons**, $W^{\pm}$ and Z^0. These particles are also of spin 1 but have masses of 80 GeV and 91 GeV, respectively. Weak interactions are responsible for radioactive beta decay of nuclei.
- The **gravitational** interactions take place between all forms of matter or radiation. They are mediated by the exchange of **gravitons**, of zero mass but spin 2.

For orientation on the magnitudes involved, the relative strengths of the different forces between two protons when just in contact are approximately

$$
\begin{array}{lll}
\text{Strong} & 1 & \\
\text{Electromagnetic} & 10^{-2} & \\
\text{Weak} & 10^{-7} & \quad(1.4)\\
\text{Gravitational} & 10^{-39} &
\end{array}
$$

It turns out that the electromagnetic and weak interactions are in fact different aspects of a single **electroweak** interaction, as described below.

Figure 1.3 shows diagrams depicting the above exchange processes, and Table 1.5 lists some of the properties of the interactions. In these diagrams (called in a more sophisticated form, Feynman diagrams after their inventor) solid lines entering or leaving the boundaries represent real particles—usually—quarks or leptons—with time flowing from left to right. The arrows along these lines indicate the direction of fermion number flow. An arrow indicating an electron flowing backwards in time is equivalent to a positron (i.e. anti-fermion) moving forwards: the convention is to use such time-reversed arrows for antiparticles.

Wavy, curly, or broken lines run between the vertices where the exchange interactions take place, and they represent the mediating bosons, which are **virtual** particles; that is, they carry energy and momentum such that the mass does not correspond to that of the free particle. To understand this, consider for example an electron of total energy E, momentum p, and mass m being

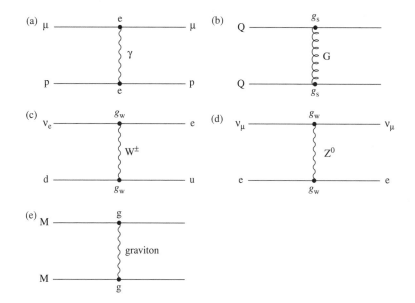

Fig. 1.3 Diagrams representing examples of single quantum exchange processes in electromagnetic, strong, weak, and gravitational interactions. (a) The electromagnetic interaction between a muon μ and proton p, via photon (γ) exchange with coupling e. (b) The strong interaction between quarks Q via gluon (G) exchange with coupling g_s. (c) The weak interaction involving charged W boson exchange, transforming an electron-neutrino v_e to an electron e, and a neutron (quark composition ddu) to a proton (duu). (d) The weak interaction involving neutral Z boson exchange, showing a muon–neutrino $v_μ$ scattering from an electron, e. In both (c) and (d), the couplings have been denoted g_w, but there are different numerical coefficients (of order unity) associated with the W and Z exchanges, as described in Chapter 3. (e) Gravitational interaction between two masses M, mediated by graviton (g) exchange. For macroscopic masses, multiple graviton exchanges will be involved.

scattered as another electron absorbs the exchanged photon. The relativistic relation between E, p, and m is given in (1.1), which in units $c = 1$ is

$$E^2 - p^2 = m^2$$

If the electron emits a photon of energy ΔE and momentum Δp then

$$E\Delta E - p\Delta p = 0$$

so that the mass of the exchanged photon is

$$\Delta m^2 = \Delta E^2 - \Delta p^2 = \frac{-m^2 \Delta p^2}{E^2} < 0 \tag{1.5}$$

Thus, as we know from common sense, a free electron cannot spontaneously emit a real photon, and the exchanged photon mass is imaginary—hence, the term virtual. The energy ΔE has been 'borrowed' by the photon, and this is permitted for a time Δt limited by the Uncertainty Principle: $\Delta E \Delta t \sim \hbar$. However, if the virtual photon is absorbed by the second electron within the time Δt, energy and momentum balance can be satisfied. The quantity Δm^2 in (1.5) is defined in the rest frame of the exchanged particle and is, therefore, a relativistically invariant quantity. It is usually called q^2, the square of the four-momentum transferred between the electrons.

Note that, if ΔE is large, Δt is correspondingly small and the range of the interaction $\Delta r \approx c\Delta t$ is correspondingly short. In 1935 Yukawa showed that the interaction potential $V(r)$ due to a spinless exchange boson of mass M had the form (as shown in Appendix D)

$$V(r) \propto \left(\frac{1}{r}\right) \exp\left(\frac{-r}{r_0}\right) \tag{1.6a}$$

where $r_0 = \hbar/Mc$ is the effective range of the interaction. The W- and Z-bosons have large masses and so the range r_0 is very short (of order 0.0025 fm). This is the reason why weak interactions are so much feebler than electromagnetic. The free photon associated with electromagnetism has rest mass $M = 0$ and $r_0 = \infty$ so that the value of ΔE of the virtual photon can be arbitrarily small and the range of the interaction can, therefore, be arbitrarily large.

Instead of discussing the range of a static interaction, these features can be taken into account in a scattering process by defining a so-called **propagator**, measuring the amplitude for scattering with a momentum transfer q. Neglecting spin, this has the general form, following from the Yukawa potential (see Appendix D),

$$F(q^2) = \frac{1}{-q^2 + M^2} \tag{1.6b}$$

where $q^2 = \Delta E^2 - \Delta p^2$ is the (negative) four-momentum transfer squared in (1.5) and M is the (free particle) rest-mass of the exchanged boson. So, for photons $M = 0$, for a weak boson $M = M_W$ and so on. The square of $F(q^2)$ enters into the cross section describing the probability of the interaction, as discussed below. As an example, for Coulomb scattering $M = 0$ and the differential cross section $d\sigma/dq^2$ varies as $1/q^4$, as given by the famous Rutherford scattering formula.

1.6   The boson couplings to fermions

1.6.1   Electromagnetic interactions

Apart from the effect of the boson propagator term, the strength of a particular interaction is determined by the coupling strength of the fermion (quark or lepton) to the mediating boson. For electromagnetic interactions, shown in Fig. 1.3(a), the coupling of the photon to the fermion is denoted by the electric charge $|e|$ (or a fraction of it, in the case of a quark), and the product of the couplings of the two fermions to each other is

$$e^2 = 4\pi\alpha\hbar c$$

more usually written as $4\pi\alpha$ in natural units with $\hbar = c = 1$. Here, $\alpha \approx 1/137$ is the dimensionless fine structure constant. The cross section or rate of a particular interaction is proportional to the square of the transition amplitude. This amplitude is proportional to the product of the vertex factors and the propagator term, that is $\alpha/|q^2|$ for the electromagnetic interaction, corresponding to a factor α^2/q^4 for the rate.

1.6.2   Strong colour interactions

For strong interactions, as in Fig. 1.3(b), the coupling of the quark to the gluon is denoted by g_s with $g_s^2 = 4\pi\alpha_s$. Typically, α_s is a number of order unity. While in the electromagnetic interactions there are just two types of electric charge, denoted by the symbols $+$ and $-$, as in Fig. 1.4(a), the interquark interactions involve six types of strong charge. This internal degree of freedom is called **colour**. Quarks can carry one of three colours, say red, blue or green, while antiquarks carry the anticolour. Gluons, unlike photons, carry a charge, consisting of one colour and one anticolour. As an example, Fig. 1.4(b) shows

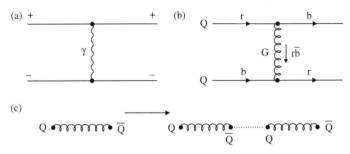

Fig. 1.4 The electromagnetic interaction in (a) involves two types of electric charge, + and − and is mediated by an uncharged photon. In (b) the strong interquark force involves six types of colour charge. The diagram depicts the interaction of a red quark with a blue quark via the exchange of a red–antiblue gluon. Diagram (c) depicts two quarks, connected by a gluon 'string', being pulled apart. Because of the confinement term in equation (1.7) the potential energy in the string grows linearly with the distance, and eventually it requires less energy to create a fresh quark–antiquark pair, involving two short strings, rather than one long one. Hadrons are of course colourless, for example a baryon is made up of one blue, one green and one red quark, while a meson consists of say, a blue quark plus an antiblue antiquark.

a red quark r interacting with a blue quark b via the exchange of an $r\bar{b}$ gluon. The potential between two quarks due to the colour force is usually taken to be of the form

$$V(\text{colour}) = -\frac{4}{3}\frac{\alpha_s}{r} + kr \qquad (1.7)$$

where r is the interquark separation, to be compared with the Coulomb potential between two unit charges of

$$V(\text{Coulomb}) = -\frac{\alpha}{r} \qquad (1.8)$$

The factor $\frac{4}{3}$ in (1.7) is a colour factor. Basically, this comes about because there are eight possible colour–anticolour combinations of gluon ($3^2 = 9$, minus 1 which is a colourless singlet combination) to be divided between the six colours and anticolours of quark and antiquark. Both potentials have a $1/r$ dependence at small distances, corresponding to the fact that both photons and gluons are massless. However, at larger distances the second term in (1.7) is dominant and is responsible for **quark confinement**. The value of k is about 0.85 GeV fm^{-1}.

> **Example 1.1** *Calculate the force in tonnes weight between a pair of quarks separated by a few fm.*
> In equation (1.7) the attractive force at large r is $dV/dr = k = 0.85$ GeV fm^{-1} or 1.36×10^5 J m^{-1}. Inserting the acceleration due to gravity, $g = 9.81$ m s^{-2}, a mass of 1000 kg exerts a force of 1 tonne weight $= 9.8 \times 10^3$ J m^{-1}. Dividing, one finds $k = 13.9$ tonnes weight—a great deal for those tiny quarks, each weighing less than 10^{-24} g.

Because gluons carry a colour charge (unlike photons, which are uncharged) there is a strong gluon–gluon interaction. Thus, the 'lines of colour force' between a pair of quarks, analogous to the lines of electric field between a pair of charges, are pulled into a tube or string. In Fig. 1.4(c) such a gluon 'string' is depicted connecting a quark–antiquark pair. If one tries to pull apart the two quarks, the energy required to do so grows linearly with the string length as in (1.7), and eventually it requires less energy to produce another quark–antiquark pair, thus involving two short strings instead of one long one. Thus even the most violent efforts to separate quarks just result in the production of many quark–antiquark pairs (mesons).

We may note at this point a peculiar property of the quark and lepton quantum numbers. Each of the three 'families' consists of a doublet of quarks of charges $+\frac{2}{3}|e|$ and $-\frac{1}{3}|e|$, respectively, and a pair of leptons with charges $-1|e|$ and 0. If allowance is made for the colour degree of freedom, the total charge of the quarks is $3 \times (\frac{2}{3} - \frac{1}{3})|e| = +1|e|$ per family, while for the leptons it is $(-1 + 0)|e| = -1|e|$. This is true for each family, so the total electric charge of the fermions is zero. It turns out that this is a crucial property, in making the theory free of so-called 'triangle anomalies' and ensuring that it is renormalizable. However, the fundamental reason for the existence of three families which are, so to speak 'carbon copies' of one another, is at present unknown. It may, however, be of interest to remark that, as discussed in Chapters 2 and 3, the observed matter–antimatter asymmetry of the universe is connected with violation of symmetry under the *CP* operation (charge conjugation followed by spatial inversion). In the unlikely event that the universe could be completely described in terms of the present Standard Model of particle physics, such *CP* violation would imply an arbitrary phase in the quark wavefunctions, which could only exist if there are at least three generations.

The confining force associated with the interquark potential (1.7) has dramatic effects on the process of high energy electron–positron annihilation to hadrons. A first stage of the process is annihilation to a quark–antiquark pair, which in a second stage transforms into hadrons, $e^+ e^- \rightarrow Q\bar{Q} \rightarrow$ hadrons. The transverse momentum of a hadron is of order 0.3 GeV/c that is, $\hbar/a$ where $a \sim 1$ fm is the force range, while the typical longitudinal momentum of a hadron from a high energy collision is much larger; hence, the usual appearance of two oppositely directed 'jets' of secondary particles. Such 'jets' of hadrons are the nearest that one ever gets to 'seeing' an actual quark. It may be remarked here that the observed cross section for this process, compared with that for $e^+ e^- \rightarrow \mu^+ \mu^-$ via photon exchange at the same energy, gave the first convincing evidence for the colour degree of freedom (see Fig. 1.9). Both are reactions proceeding via photon exchange, with a rate proportional to the square of the electric charges of the particles involved. The observed two-jet event rate was consistent with that expected, provided a factor 3 enhancement was included, to take account of the fact that the quark–antiquark pair could be emitted in three colours ($r\bar{r}$ or $b\bar{b}$, or $g\bar{g}$).

Figure 1.5 shows an event containing three, rather than two, jets of particles. In this case one of the quarks has radiated a high energy gluon at wide angle, $e^+ e^- \rightarrow Q + \bar{Q} + G$. The ratio of three-jet to two-jet events clearly gives a measurement of the strong coupling α_s.

1.6.3   Weak interactions

Figure 1.3(c) and (d) show the W and Z exchanges of the weak interactions, with couplings that we can generically denote by the symbol g_w. Thus, the product of the propagator term and the coupling would give an amplitude in this case of

$$\underset{q^2 \to 0}{\text{Lt}} \frac{g_w^2}{-q^2 + M_W^2} \equiv G_F \tag{1.9}$$

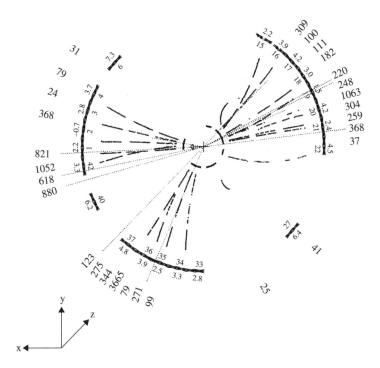

Fig. 1.5 Example of hadron production following e^+e^- annihilation observed in the JADE detector at the PETRA collider at DESY, Hamburg. The total centre-of-momentum energy is 30 GeV. Trajectories of charged pions are shown as crosses, and of γ-rays from decay of neutral pions as dotted lines. The γ-rays are detected when they produce electron–photon showers in lead glass counters. Note the collimation of the hadrons into three distinct 'jets'. (Courtesy DESY laboratory.)

We have made the identification here with a point coupling G_F between the fermions involved, which Fermi had postulated in the earliest days of nuclear beta decay (1934). In those processes $|q^2| \ll M_W^2$ so that $G_F = g_w^2/M_W^2$. The process of $W^\pm$ exchange in Fig. 1.3(c) involves a change in the charge of the lepton or quark, and is therefore, sometimes referred to as a 'charged current' weak interaction, while the Z^0 exchange in Fig.1.3(d) does not affect the charges of the particles and is termed a 'neutral current' weak interaction.

1.6.4   Electroweak interactions

As stated above, the electromagnetic and weak interactions are unified. The electroweak model was developed in the 1960s, particularly by Glashow (1961), Salam (1967), and Weinberg (1967). Basically, what this means is that the couplings of the W- and Z-bosons to the fermions are the same as that of the photon, that is $g_w = e$. Here, for simplicity, we have omitted certain numerical factors of order unity. If one inserts this equality in (1.9) for the limit of low q^2 one obtains the expected large masses for the weak bosons:

$$M_{W,Z} \sim \frac{e}{\sqrt{G_F}} = \sqrt{\frac{4\pi\alpha}{G_F}} \sim 100 \text{ GeV} \qquad (1.10)$$

where the value of $G_F = 1.17 \times 10^{-5} \text{ GeV}^{-2}$ has been inserted from the measured rate for muon decay (see also Table 1.5). So what (1.9) and (1.10) are telling us is that, although the photons and the weak bosons have the same couplings to leptons (to within a constant), the effective strength of the weak interaction is much less than that of the electromagnetic interaction because

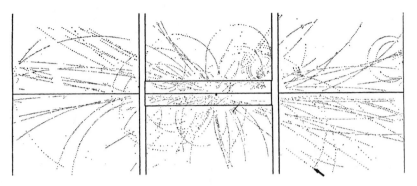

Fig. 1.6 One of the first examples of the production of a W boson at the CERN proton–antiproton collider in 1983. This reconstruction is of signals from drift chamber detectors surrounding the horizontal vacuum pipe: 270 GeV protons coming from the right collided with 270 GeV antiprotons from the left. Among the 66 tracks of secondary particles, one, shown by the arrow is an energetic (42 GeV) positron identified in a surrounding electromagnetic calorimeter. This positron has transverse momentum of 26 GeV/c, while the missing transverse momentum in the whole event is 24 GeV/c in the opposite sense, consistent with that of a neutrino, produced in the decay $W^{\pm} \rightarrow e^{+} + \nu_{e}$ (from Arnison *et al.* 1983).

the larger mediating boson mass implies a much shorter range for the interaction. Because of this difference in boson masses, from zero for the photon to 80–90 GeV for the weak bosons, the electroweak symmetry is a broken symmetry. Figure 1.6 shows an example of the first observation of a W particle, in proton–antiproton collisions in 1983. Electroweak interactions are described in detail in Chapter 3.

1.6.5   Gravitational interactions

Finally, Fig. 1.3(e) depicts the gravitational interaction between two masses, via graviton exchange. The force between two equal point masses M is given by GM^2/r^2 where r is the separation and G is the Newtonian gravitational constant. Comparing with the electrostatic force between two charges $|e|$ of e^2/r^2, the quantity $GM^2/\hbar c$ is seen to be dimensionless. If we take as the unit of mass $Mc^2 = 1$ GeV, then,

$$\frac{GM^2}{4\pi\hbar c} = 5.3 \times 10^{-40} \tag{1.11a}$$

to be compared with

$$\frac{e^2}{4\pi\hbar c} = \frac{1}{137.036} \tag{1.11b}$$

Thus, for the energy or mass scales of GeV common in high energy physics experiments, the gravitational coupling is negligible. Of course, on a macroscopic scale, gravity is important and indeed dominant, because it is cumulative, since all particles with energy and momentum are attracted by their mutual gravitation. Thus, the gravitational force on a charged particle on the Earth's surface is the sum of the attractive effects of all the matter in the Earth. Since the Earth is electrically neutral, however, the enormously larger electrical force due to all the protons in the Earth is exactly cancelled by the opposing force due to the electrons.

However, even on subatomic scales the gravitational coupling can become strong for hypothetical elementary particles of mass equal to the **Planck mass,** defined as

$$M_{\text{PL}} = \left(\frac{\hbar c}{G}\right)^{1/2} = 1.2 \times 10^{19} \text{ GeV}/c^2 \tag{1.12}$$

The **Planck length** is defined as $L_{\text{PL}} = \hbar/M_{\text{PL}}c$, that is, the Compton wavelength of a particle of Planck mass. Two pointlike particles each of the Planck mass and separated by the Planck length would, therefore, have a gravitational potential energy equal to their rest masses, so quantum gravitational effects can become important at the Planck scale.

We should emphasize here that, although we have drawn a parallel between the inverse square law of force between point charges and point masses in (1.11) and (1.12), there is a fundamental difference between the two. Due to the attractive force between two masses, the latter can acquire momentum and kinetic energy (at the cost of potential energy), which is equivalent to an increase in the effective mass through the Einstein relation $E = mc^2$, and thence in the gravitational force. For close enough encounters, therefore, the force will increase faster than $1/r^2$. Indeed, one gets non-linear effects, which is one of the problems in formulating a quantum field theory of gravity. The effects of gravitational fields (including the non-linear behaviour) are enshrined in the Einstein field equations of general relativity, which interpret these effects in terms of the curvature of space. In this book, we do not discuss the general theory of relativity, which would require a text on its own, although we shall where necesssary quote predictions from it.

While for the strong, electromagnetic and weak interactions, there is direct laboratory evidence for the existence of the mediating bosons—gluons, photons, and weak bosons, respectively—so far, the direct detection of gravitational waves (gravitons) has escaped us. Even the most violent events in the universe are expected to produce only incredibly small (10^{-22}) fractional deviations in detecting apparatus on Earth, caused by the compressing and extending effects of gravitational radiation. However, indirect evidence from the slow-down rate of binary pulsars discussed in Section 6.13, shows that gravitational radiation does indeed exist, and at exactly the rate predicted by general relativity.

At this point, we may note in passing that the gravitational, electromagnetic and strong interactions can all give rise to (non-relativistic) **bound states.** A planetary system is an example of gravitational binding. Atoms and molecules are examples of binding due to electromagnetic interactions, while the strong interquark forces lead to three-quark (baryon) states as well as quark–antiquark bound states—for example, the ϕ meson $s\bar{s}$ and the J/ψ meson $c\bar{c}$, appearing as resonances in electron–positron annihilation at the appropriate energy (see Fig. 1.9). Strong interactions are of course also responsible for the binding of atomic nuclei. The weak interactions do not lead to any bound states, because of the rapid decrease of the potential with distance as mentioned above.

Example 1.2   *Calculate at what separation r the weak potential between two electrons falls below their mutual gravitational potential.*
  As indicated in (1.6), the Yukawa formula for the weak potential between two electrons, mediated by a weak boson of mass M and weak coupling g_{w}

to fermions is

$$V_{wk}(r) = \left(\frac{g_w^2}{r}\right) \exp\left(\frac{-r}{r_0}\right) \quad \text{where } r_0 = \frac{\hbar}{Mc}$$

to be compared with the gravitational potential between two electrons of mass m

$$V_{grav}(r) = \frac{Gm^2}{r}$$

In the electroweak theory, $g_w \sim e$ so that $g_w^2/4\pi\hbar c \sim \alpha = 1/137$, while (see Table 1.5) $Gm^2/4\pi\hbar c = 1.6 \times 10^{-46}$. Hence, inserting $M_w = 80$ GeV to obtain $r_0 = 2.46 \times 10^{-3}$ fm, one finds the two potentials are equal when $r \sim 100r_0 = 0.25$ fm.

To summarize this section, the characteristics of the fundamental interactions are listed in Table 1.5 (page 5).

1.7   The quark–gluon plasma

As already stated, laboratory experiments indicate that quarks do not exist as free particles, but rather as three-quark and quark–antiquark bound states, called hadrons. There has long been speculation that this quark confinement mechanism is a low energy phase and that at sufficiently high energy densities, quarks and gluons might undergo a phase transition, to exist in the form of a plasma. An analogy can be made with a gas, in which at sufficiently high temperatures the atoms or molecules become ionized, and the gas transforms to a plasma of electrons and positive ions. If such a quark–gluon plasma is possible, the conditions of temperature and energy density in the very early stages (the first 25 μs) of the Big Bang would have certainly resulted in such a state existing, before the temperature fell as the expansion proceeded and the quark–gluon 'soup' froze out into hadrons.

Attempts have been made over the years to reproduce the quark–gluon plasma in the laboratory, by making head-on collisions of heavy nuclei (e.g. lead on lead) accelerated to relativistic energies. The critical quantity is the energy density of the nuclear matter during the very brief (10^{-23} s) period of the collision. In lead–lead collisions at 0.16 GeV per nucleon in each of the colliding beams, a threefold enhancement has been observed in the frequency of strange particles and antiparticles (from creation of s$\bar{s}$ pairs) as compared with proton–lead collisions at a similar energy per nucleon. Obviously, present and future studies of such plasma effects in the laboratory (specifically at the RHIC heavy ion collider at Brookhaven National Laboratory, and the Large Hadron Collider at CERN), could be very important in shedding light on exactly how the early universe evolved.

1.8   The interaction cross section

The strength of the interaction between two particles, for example in the two-body → two-body reaction

$$a + b \rightarrow c + d$$

is specified by the **interaction cross section** σ defined as follows. Suppose the particles a are in a parallel beam, incident normally on a target of thickness dx containing n_b particles of type b per unit volume (see Fig. 1.7). If the density of incident particles is n_a per unit volume, the flux—the number of particles per unit area and per unit time—through the target will be

$$\phi_i = n_a v_i \qquad (1.13)$$

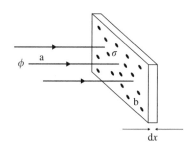

Fig. 1.7 Diagram indicating a beam of particles of type a incident on a target containing particles of type b.

where v_i is the relative velocity of beam and target. If each target particle has an effective cross section of σ, then the fraction of the target area obscured, and the probability of collision, will be $\sigma n_b \, dx$. The reaction rate per unit time and per unit area of the target will then be $\phi_i \sigma n_b \, dx$. Per target particle the reaction rate will be

$$W = \phi_i \sigma \qquad (1.14)$$

so that the cross section is equal to the reaction rate per target particle and per unit incident flux. Cross sections are measured in units called the **barn**: 1 barn = 1 b = 10^{-28} m^2. This is roughly the geometric area of a nucleus of mass number $A = 100$. Appropriate units in particle physics are the millibarn (1 mb = 10^{-3} b), the microbarn (1 μb = 10^{-6} b), the nanobarn (1 nb = 10^{-9} b), and the picobarn (1 pb = 10^{-12} b).

The quantity W is given by an expression from (non-relativistic) perturbation theory, usually referred to as 'Fermi's Second Golden Rule', and derived in standard texts on atomic physics. It has the form

$$W = \frac{2\pi}{\hbar} |T_{if}|^2 \rho_f \qquad (1.15)$$

where the transition amplitude or matrix element T_{if} between initial and final states is effectively an overlap integral $\int \psi_f^* U \psi_i \, dV$ between initial and final state wavefunctions brought about by the interaction potential U. The quantity $\rho_f = dN/dE_f$ is the energy density of final states, that is, the number of states in phase space available to the product particles per unit interval of the final state energy E_f. The number of states in phase space available to a particle in the momentum interval $p \to p + dp$ and directed into the solid angle dΩ and enclosed in a volume V is

$$dN = Vp^2 \frac{dp \, d\Omega}{(2\pi\hbar)^3} \qquad (1.16)$$

In the reaction a+b $\to$ c+d, the final state wavefunction ψ_f will be the product wavefunction $\psi_c \psi_d$, so that to ensure that we end up with just one particle of each type when we integrate over the volume in the transition matrix, a $V^{-1/2}$ normalization factor is needed for the wavefunction of each final state particle. When T_{if} is squared, the resulting $1/V$ factor cancels with the V factor in the phase space in (1.16), for each particle in the final state. Similarly, the normalization factors for the wavefunctions of the particles in the initial state cancel with the factors proportional to V for the incident flux and for the number of target particles. Hence, the arbitrary normalization volume V cancels out, as indeed it must.

It is most convenient to express the cross section in terms of quantities defined in the **centre-of-momentum frame** (CMS) of the collision, that is, in a reference frame in which the vector sum of the momenta of the colliding particles is zero (see Appendix B for definitions and a discussion). Then, from the above formulae one obtains (with $n_a = 1$ per normalization volume)

$$\frac{d\sigma}{d\Omega} = \frac{W}{\phi_i} = \frac{W}{v_i} = \left(\frac{|T_{if}|^2}{v_i}\right) p_f^2 \left(\frac{dp_f}{dE_f}\right) \left(\frac{1}{4\pi^2 \hbar^4}\right) \tag{1.17}$$

where p_f is the numerical value of the oppositely directed momenta of c and d in the CMS, and $E_f = E_c + E_d$ is the total energy in the CMS. Energy conservation gives

$$\sqrt{p_f^2 + m_c^2} + \sqrt{p_f^2 + m_d^2} = E_f$$

and thus

$$\frac{dp_f}{dE_f} = \frac{E_c E_d}{E_f p_f} = \frac{1}{v_f}$$

where v_f is the relative velocity of c and d. Then,

$$\frac{d\sigma}{d\Omega}(a + b \rightarrow c + d) = \frac{1}{4\pi^2 \hbar^4}|T_{if}|^2 \frac{p_f^2}{v_i v_f} \tag{1.18}$$

We have so far neglected the spins s_a, s_b, s_c, and s_d of the particles involved. If a and b are unpolarized, that is, their spin substates are chosen at random, the number of possible substates for the final state particles is $g_f = (2s_c+1)(2s_d+1)$ and the cross section has to include the factor g_f. For the initial state, the factor is $g_i = (2s_a + 1)(2s_b + 1)$. Since a given reaction has to proceed through a particular spin configuration, one must average the transition probability over all possible initial states, all equally probable, and sum over all final states. This implies that the cross section has to be multiplied by the factor g_f/g_i.

The reduction in the number of incident particles $n = n_a$ after passage through a thickness dx of absorber in Fig. 1.7 is

$$dn = -n\sigma\rho\,dx$$

where $\rho = n_b$ is the density of target particles. Integrating, we obtain

$$n(x) = n(0)\exp(-\sigma\rho x) \tag{1.19}$$

Thus, the proportion of incident particles that survive without interaction falls to 1/e in a distance

$$\lambda = \frac{1}{\sigma\rho} \tag{1.20}$$

The quantity λ is called the **mean free path** for interaction. It is left as an exercise to show that, from the distribution (1.19) and the definition (1.20), λ is the mean path length between collisions.

1.9   Examples of elementary particle cross sections

The two-body to two-body reactions described above are important in discussing the role of elementary particle interactions in the early universe, and here we give some examples. A full evaluation of the cross sections would in some cases involve rather lengthy Dirac algebra, so that all that we shall do here is to give approximate expressions which can be justified simply on dimensional grounds, and which will give orientation on the dependences and magnitudes involved. For simplicity we also neglect considerations of spin at this stage. Spin and helicity factors are taken into account later in the text, when exact cross sections are required.

As (1.18) indicates, the differential cross section for an extreme relativistic two-body to two-body elastic collision has the form

$$\frac{d\sigma}{d\Omega} = \frac{|T_{if}|^2 s}{64\pi^2} \quad \text{(extreme relativistic two-body} \rightarrow \text{two-body)} \quad (1.21)$$

where we have used units $\hbar = c = 1$, so that $s = E_f^2$ and $E_f = 2p_f$, since the masses involved are small compared with the energies, and $v_i = v_f = 2$. In this case also, the four-momentum transfer squared is $|q^2| = 2p_f^2(1 - \cos\theta)$, where θ is the angle of emission of the secondary particles in the CMS, so that $dq^2 = p_f^2 \, d\Omega/\pi$ and (1.21) can also be written as

$$\frac{d\sigma}{dq^2} = \frac{|T_{if}|^2}{16\pi} \quad (1.22)$$

Examples of electromagnetic cross sections are as follows:

(a) $\mathbf{e^-\mu^+ \rightarrow e^-\mu^+}$

This is an example of the Coulomb scattering between singly charged leptons, as shown in Fig. 1.8(a). The couplings and photon propagator term together give $|T_{if}| = e^2/|q^2|$ so that from (1.21)

$$\frac{d\sigma}{d\Omega} \sim \frac{\alpha^2 s}{q^4} \quad (1.23)$$

where $\alpha = e^2/4\pi$. This is just the Rutherford formula for pointlike scattering, and if θ is the angular deflection of the incident particle, then $q = 2p\sin(\theta/2)$ and one obtains the famous $\mathrm{cosec}^4(\theta/2)$ dependence on the scattering angle:

(b) $\mathbf{e^+e^- \rightarrow \mu^+\mu^-}$.

The diagram for this process, in Fig. 1.8(b), is just that in (a) rotated through 90° and with outgoing leptons replaced by incoming antileptons and vice versa.

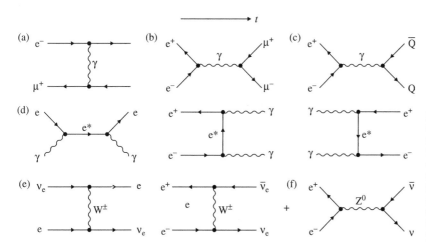

Fig. 1.8 Feynman diagrams for various elementary two-body to two-body reactions. In these diagrams, time flows from left to right. The convention is that right-pointing arrows denote particles, while left-pointing arrows denote antiparticles. Diagrams (a)–(d) refer to electromagnetic interactions, and (e) and (f) refer to weak interactions. (a) $e^-\mu^+ \rightarrow e^-\mu^+$; (b) $e^+e^- \rightarrow \mu^+\mu^-$; (c) $e^+e^- \rightarrow Q\bar{Q} \rightarrow$ hadrons; (d) $e\gamma \rightarrow e\gamma$, $e^+e^- \rightarrow \gamma\gamma$, $\gamma\gamma \rightarrow e^+e^-$; (e) $\nu e \rightarrow \nu e$; (f) $e^+e^- \rightarrow \nu\bar{\nu}$.

These two diagrams are said to be 'crossed' diagrams. In this case $|q^2| = s$, the square of the CMS energy, so that

$$\frac{d\sigma}{d\Omega} \sim \frac{\alpha^2}{s} \tag{1.24}$$

In fact, a full calculation gives for the total cross section

$$\sigma = \frac{4\pi\alpha^2}{3s} \tag{1.25}$$

This is the cross section based upon single photon exchange, as shown by the dashed line in Fig. 1.9. There will also be a contribution from Z^0 exchange, but because of the propagator term in (1.9), this is strongly suppressed at GeV energies. The above result is also expected on dimensional grounds. In units $\hbar = c = 1$, the cross section has dimensions of GeV^{-2} and if the CMS energy dominates over the lepton masses involved, the $1/s$ dependence must follow. Since, from Table 1.1, $1\ GeV^{-1} = 1.975 \times 10^{-16}$ m, the above cross section is readily calculated to be 87/s nb, where s is in GeV^2.

(c) $e^+e^- \rightarrow Q\bar{Q} \rightarrow$ hadrons

The same formula (1.24) applies, again assuming the quarks have relativistic velocities, replacing the unit charges by the fractional quark charges at the right-hand vertex (Fig. 1.8(c)), and multiplying by a factor 3 for the number of quark colours. Of course, one does not observe the actual quarks: they 'fragment' into hadrons by gluon exchanges, but this is a relatively slow and independent second stage process. Except near meson resonances, the cross section is determined entirely by the elementary $e^+e^- \rightarrow Q\bar{Q}$ process.

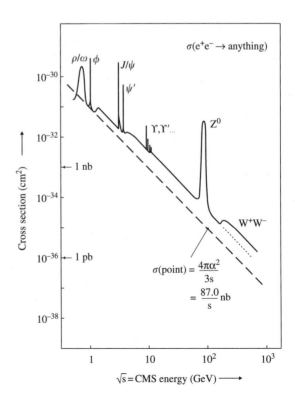

Fig. 1.9 The cross section for the reaction $e^+e^- \rightarrow$ anything, as a function of CMS energy. The prominent peaks are due to various boson resonances, as described in the text. The overall $1/s$ dependence, typical of a pointlike scattering process, is clear. The magnitude of the cross section for $e^+e^- \rightarrow$ anything, compared with that for muon pair production, shown as a dashed line, provides evidence for a factor 3 for the hadronic cross section due to the colour degree of freedom of the quarks. The data are taken from measurements at various electron–positron colliders, the largest of which was the LEP collider at CERN, Geneva, in which 100 GeV electrons collided head-on with 100 GeV positrons.

Figure 1.9 shows a plot of the cross section as a function of energy. The peaks are due to the excitation of bound quark–antiquark states or resonances forming short-lived mesons: for example, the ρ- and ω-mesons formed from u and d quarks and antiquarks; the φ formed from s̄s; the J/ψ formed from cc̄, the Υ from bb̄ and finally the Z^0 resonance. Nevertheless the general $1/s$ dependence of the cross section, aside from these resonance effects, is quite clear.

Example 1.3 *Calculate the ratio R of the cross section for* $e^+e^- \rightarrow$ *QQ̄ → hadrons, to that for* $e^+e^- \rightarrow \mu^+\mu^-$, *via photon exchange, as a function of CMS energy (see also Fig. 1.9).*

The quark–antiquark cross section is proportional to the square of the quark charges and carries a factor 3 for colour. Thus, for uū, dd̄, and s̄s quarks, the factor is

$$R = 3 \times \left[\left(\tfrac{2}{3}\right)^2 + \left(\tfrac{1}{3}\right)^2 + \left(\tfrac{1}{3}\right)^2 \right] = 2$$

and this is shown as the first 'step' above the pointlike cross section for a muon pair in Fig. 1.9. Above the charmed (cc̄) threshold appears a second step with $R = \tfrac{10}{3}$, and above the bb̄ threshold at CMS energy of 10 GeV, $R = \tfrac{11}{3}$.

Example 1.4 *Estimate an approximate value for the cross section for the production of the top quark, of mass* $m_t = 175$ GeV$/c^2$, *in proton–antiproton collisions at centre-of-mass energies large compared with the quark mass.*

The top quark was discovered in 1995 at the Fermilab proton–antiproton collider, with CMS energy 1.8 TeV. The principal process is that of production of a tt̄ pair in the collision of a u or d quark from the proton with an antiquark from the antiproton, via gluon exchange, that is, $u + \bar{u} \rightarrow t + \bar{t}$, as in Fig. 1.8(c) but with incident e^+e^- replaced by uū (or dd̄) and gluon exchange with coupling α_s replacing photon exchange with coupling α. Then, from (1.24) we expect the cross section to be $\sigma \sim F\alpha_s^2/s$, where $s \sim (2m_t)^2$, that is, assuming an incident QQ̄ centre-of-mass energy just above threshold, with F representing the probability that the colliding quarks are above threshold. If we set $\alpha_s = 0.1$ as a typical value, $m_t = 175$ GeV, then $\sigma \sim 30F$ pb. The value of F depends on the momentum distribution of quarks in the nucleon. A detailed calculation ends up with a cross section of 7 pb, in agreement with observation.

Since the total proton–antiproton collision cross section is 80 mb, this meant that the top quark was produced in only about 1 in 10^{10} collisions. Despite this huge background, top quarks could be detected because of the very distinguishing features of their decays, for example, $t \rightarrow W^+ + b$ and $\bar{t} \rightarrow W^- + \bar{b}$. The W-bosons decay to give muons at wide angle and neutrinos, which manifest themselves as 'missing' energy and momentum, while the b quarks produce hadronic jets, slightly displaced from the main vertex because of the finite lifetime of the B-mesons. The signal is so characteristic and specific that the background could be reduced below the 10% level. One of the detectors used in discovering the top quark is shown in Fig. 1.10.

(d) $e^+e^- \rightarrow \gamma\gamma$; $\gamma\gamma \rightarrow e^+e^-$; $\gamma e \rightarrow \gamma e$

These are important processes in astrophysics. They can all be represented by the same Feynman diagram, as in Fig. 1.8(d). Because in this case a virtual

Fig. 1.10 Photograph of the CDF detector employed in the discovery of the top quark in 1995, in collisions of 0.9 TeV protons with 0.9 TeV antiprotons at the proton–antiproton collider at Fermilab, near Chicago. In the centre of the picture is the central tracking detector, which records the trajectories of individual secondary particles in drift chambers, and measures their momenta from track curvature in the applied magnetic field. Inside this are precision solid state (silicon strip) detectors that can record tracks and secondary vertices very close to the main interaction vertex. Surrounding the central tracker are calorimeters that measure the total energy in charged and neutral particles. They are built in two arches, which are withdrawn in the photograph. Outside the calorimeter modules and magnet yoke are further chambers to record penetrating muons from the annihilation reactions. (Courtesy Fermilab Visual Media Services.)

electron, rather than a photon, operates between the vertices, the formula (1.24) is modified by a logarithmic term. Of course, although represented by the same diagram, the three processes have different dynamics (thresholds). In the limit of high energy, that is, for CMS energy squared $s \gg m^2$ where m is the electron mass, the cross sections have the asymptotic forms

$$\sigma(e^+e^- \to \gamma\gamma) = \left(\frac{2\pi\alpha^2}{s}\right)\left[\ln\left(\frac{s}{m^2}\right) - 1\right] \tag{1.26a}$$

$$\sigma(\gamma\gamma \to e^+e^-) = \left(\frac{4\pi\alpha^2}{s}\right)\left[\ln\left(\frac{s}{m^2}\right) - 1\right] \tag{1.26b}$$

$$\sigma(\gamma e \to \gamma e) = \left(\frac{2\pi\alpha^2}{s}\right)\left[\ln\left(\frac{s}{m^2}\right) + \frac{1}{2}\right] \tag{1.26c}$$

The fact that the cross section (1.26a) is half that of (1.26b) arises because in the first process there are two indistinguishable particles in the final state, so that the phase space volume is halved. At collision energies large compared with the W mass, a formula similar to (1.26a) would apply to the reaction $e^+e^- \to W^+W^-$. The last process (1.26c) is known as Compton scattering. The same formula applies when the roles of incident and target particles are reversed, that is γ-rays are accelerated to higher energies following collision with incident electrons. This inverse Compton effect is believed to be important in the production of high energy γ-rays from point stellar sources.

At the other extreme of low CMS energy, the electron mass will dominate the energy scale and s is replaced by m^2, so that $\sigma \sim \alpha^2/m^2$. In fact, the classical Thomson cross section, which applies for Compton scattering as $E_\gamma \to 0$, has the value

$$\sigma(\gamma e \to \gamma e)_{\text{Thomson}} = \frac{8\pi\alpha^2}{3m^2} = 0.666 \text{ b} \qquad (1.26)$$

For the process $\gamma\gamma \to e^+e^-$, the threshold energy is $s_{\text{th}} = 4m^2$ and $\sigma \sim \beta\alpha^2/m^2$, where $\beta = (1 - 4m^2/s)^{1/2}$ is the CMS velocity of the electron or positron. Thus, the cross section at first increases with energy, and reaches a maximum of about $0.25\sigma_{\text{Thomson}}$ at $s \sim 8m^2$, before falling off at higher s values.

In summary, the above processes involve massless or almost massless photon or electron propagators and their effect is to introduce a $1/s$ dependence to the cross sections. We now discuss weak processes that involve the massive W and Z propagators.

(e) $\nu_e e \to \nu_e e$; $e^+e^- \to \nu_e\bar{\nu}_e$

Figure 1.8(e) shows the diagram for neutrino–electron scattering via W exchange. There is also a contribution from Z exchange, but we consider here only the former. The propagator gives a term $1/(|q^2| + M_W^2)$ in $|T_{\text{if}}|$. Because of the large value of the W mass (80 GeV), $|q^2| \ll M_W^2$ at normal (<1 TeV) neutrino energies and, therefore,

$$\sigma \sim \frac{g_w^4 s}{M_W^4} \sim G_F^2 s \qquad (1.27a)$$

where G_F is the Fermi constant defined in (1.9). An exact calculation gives

$$\sigma(\nu_e e \to \nu_e e) = \frac{G_F^2 s}{\pi} \qquad (1.27b)$$

again assuming that lepton masses can be neglected in comparison with the collision energy. For the 'crossed' reaction $e^+e^- \to \nu_e\bar{\nu}_e$ shown in Fig. 1.8(f), the cross section is (see Section 3.6)

$$\sigma(e^+e^- \to \nu_e\bar{\nu}_e) = \frac{G_F^2 s}{6\pi} \qquad (1.28)$$

In addition to $W^\pm$ exchange, this reaction can also proceed through Z^0 exchange, and indeed this is the only possibility for the reactions $e^+e^- \to \nu_\mu\bar{\nu}_\mu$ or $\nu_\tau\bar{\nu}_\tau$. These processes are also discussed in Section 3.6. They are of astrophysical significance, both in the very early stages of the universe, and in the later supernova stages of giant stars.

1.10   Decays and resonances

As indicated in (1.15), an unstable state has a decay rate W, usually quoted as a **width** Γ in energy units, which corresponds to the fact that a non-stationary state with a finite lifetime must have a spread in energy, in accord with the Uncertainty Principle, that is, $\Gamma = \hbar W = \hbar/\tau$ where $\tau = 1/W$ is the mean lifetime of the state.

As an example of a decaying state, let us consider muon decay, $\mu^+ \to e^+ + \nu_e + \bar{\nu}_\mu$. Clearly the transition amplitude for this weak decay $|T_{if}| \propto G_F$, the Fermi coupling constant, which has dimensions (energy)$^{-2}$—see Table 1.5. Hence, the square of the transition amplitude has dimensions (energy)$^{-4}$ while the decay rate or width has dimensions of energy. Thus, by dimensional arguments the phase space factor in (1.15) must vary as (energy)^5, and since the largest energy or mass involved is $m_\mu c^2$ it follows that the muon decay rate $\Gamma \sim G_F^2 m_\mu^5$. In fact a full (and quite lengthy) calculation gives

$$\Gamma(\mu^+ \to e^+ \nu_e \bar{\nu}_\mu) = \frac{G_F^2 m_\mu^5}{192\pi^3} \tag{1.29}$$

It is left as an exercise to verify the value of the Fermi constant in Table 1.5 from the measured lifetime $\tau_\mu = \hbar/\Gamma = 2.197\ \mu\text{s}$ and mass $m_\mu c^2 = 105.66$ MeV.

One can understand that a state with a measurable lifetime usually has an unmeasurably small width, while one with a broad and measurable width, for example, one decaying through the strong interactions, usually has an unmeasurably short lifetime, and is referred to as a **resonance**. Such resonances can readily be formed in collisions between the particles into which they decay. The exponential nature of the time distribution of decays determines the form of the line shape of the resonance.

Denoting the central angular frequency of the resonant state by ω_R, the wavefunction describing this state can be written as

$$\psi(t) = \psi(0) \exp(-i\omega_R t) \exp\left(\frac{-t}{2\tau}\right) = \psi(0) \exp\left\{\frac{-t(iE_R + \Gamma/2)}{\hbar}\right\} \tag{1.30}$$

where the central energy is $E_R = \hbar\omega_R$ and the width $\Gamma = \hbar/\tau$. The intensity $I(t) = \psi^*(t)\psi(t)$ obeys the usual radioactive decay law

$$\frac{I(t)}{I(0)} = \exp(-\Gamma t/\hbar) \tag{1.31}$$

The energy dependence of the cross section for forming the resonance is the Fourier transform of the time pulse in the same way that, in wave optics, the angular distribution of the beam diffracted by a slit system is the Fourier transform of the slit profile. The Fourier transform of (1.30) is

$$g(\omega) = \int \psi(t) \exp(i\omega t)\,dt$$

With $E = \hbar\omega$, the amplitude as a function of E is then (in units $\hbar = c = 1$)

$$A(E) = \psi(0) \int \exp\left\{-t\left[\left(\frac{\Gamma}{2}\right) + i(E_R - E)\right]\right\} dt = \frac{K}{E - E_R - i\Gamma/2} \tag{1.32}$$

where K is some constant. The cross section $\sigma(E)$, measuring the probability of two particles a and b forming a resonant state c will be proportional to A^*A, that is,

$$\sigma(E) = \sigma_{\max} \frac{\Gamma^2/4}{(E - E_R)^2 + \Gamma^2/4} \tag{1.33}$$

which is called the **Breit–Wigner resonance formula**. The shape of the resonance is shown in Fig. 1.11, from which we note that the cross section falls to half its peak value for $E - E_R = \pm\Gamma/2$. The value of the peak cross section in (1.33) can be evaluated as follows. An incident particle beam of momentum p will be described by a plane wave, which can be decomposed into a superposition of spherical waves of different angular momenta l with respect to the scattering centre, where $l\hbar = pb$ and b is the 'impact parameter'. Particles of angular momentum in the interval $l \rightarrow l + 1$, therefore, impinge on an annular ring of cross sectional area

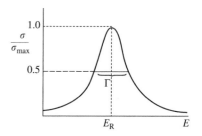

Fig. 1.11 The Breit–Wigner resonance curve.

$$\sigma = \pi(b_{(l+1)}^2 - b_l^2) = \pi\lambda^2(2l + 1) \tag{1.34}$$

where $\lambda = \hbar/p$. If the scattering centre is totally absorbing, σ in (1.34) will be the reaction or absorption cross section. More generally, we can write the radial dependence of the outgoing amplitude (for the lth partial wave) in the form

$$r\psi(r) = \exp[ikr]$$

so that the total flux $4\pi r^2|\psi(r)|^2$ through radius r is independent of r. This is for the case of no scattering centre, while if the scattering centre is present

$$r\psi(r) = \eta\exp[i(kr + \delta)]$$

Here $0 < \eta < 1$ and δ is a phase shift. By conservation of probability, the reaction cross section σ_r will be given by the difference of intensities with and without the scattering:

$$\sigma_r = \sigma(1 - \eta^2)$$

The scattered amplitude will clearly be

$$A = \exp[ikr] - \eta\exp[i(kr + \delta)]$$

giving for the scattered intensity

$$A^*A = 1 + \eta^2 - 2\eta\cos\delta$$

Thus, for $\eta = 0$ (total absorption) both elastic and reaction cross sections are equal to σ in (1.34). The elastic cross section in this case corresponds to the elastically diffracted beam from the absorbing obstacle. The other extreme case is that of pure scattering ($\eta = 1$) without absorption but just a shift in phase. Then,

$$\sigma_{el} = 4\sigma\sin^2\left(\frac{\delta}{2}\right)$$

The maximum effect is for a phase shift of π radians, leading to a scattering amplitude equal to twice that for total absorption, or a cross section

$$\sigma_{el}(\max) = 4\pi\lambda^2(2l + 1) \tag{1.35}$$

So far, we have omitted the effects of particle spin. The appropriate spin multiplicity factors were given in Section 1.6. Putting all these things together, the complete Breit–Wigner formula becomes

$$\sigma(E) = \frac{4\pi \lambdabar^2 (2J + 1)\Gamma^2/4}{(2s_a + 1)(2s_b + 1)[(E - E_R)^2 + \Gamma^2/4]} \tag{1.36}$$

where s_a and s_b are the spins of the incident particles and J is the spin of the resonant state (all in units of $\hbar$). Usually, the resonance from the reaction a + b → c can decay in a number of modes, each one with a **partial width** Γ_i for the ith mode, so that the fractional probability of decaying through that mode is Γ_i/Γ. In general, the resonance is formed through channel i and decays through channel j, and the cross section is then given by multiplying (1.36) by the ratio $\Gamma_i \Gamma_j/\Gamma^2$.

1.11   Examples of resonances

We cite here three examples of resonances, of importance both in particle physics and in astrophysics. Figure 1.12 shows the cross-section for the process $e^+e^- \to$ anything, in the neighbourhood of the Z^0 resonance (the Z^0 being the mediator of the neutral current weak interaction). The central mass $E_R = 91$ GeV and the total width $\Gamma = 2.5$ GeV. This resonance has many possible decay modes; into hadrons via pairs of u, d, s, c, or b quarks and antiquarks, into pairs of charged leptons e^+e^-, $\mu^+\mu^-$, or $\tau^+\tau^-$, or into neutrino pairs $\nu_e \bar{\nu}_e$, $\nu_\mu \bar{\nu}_\mu$, or $\nu_\tau \bar{\nu}_\tau$. At the time that this resonance was first investigated, there was some question about the total number of families of quarks and leptons (and the top quark had not yet been discovered). Could there be more than three types or flavours of neutrino? The curves in Fig. 1.12 show the effect on the width of assuming two, three, or four flavours of neutrino, based on the couplings of the Z to quarks and leptons as prescribed by the Standard Model. Clearly the observed width bears out the Standard Model assumption of three families. The number of flavours of neutrino, as discussed in the next chapter, has an effect on the primordial helium/hydrogen ratio and thence on the subsequent evolution of the stars.

> **Example 1.5**   *Calculate the peak cross section for the production of the Z^0 resonance in the reaction $e^+ + e^- \to Z^0$, where the partial width is given by $\Gamma_{ee}/\Gamma_{total} = 0.033$. Compare the answer with the result in Fig. 1.12.*
>
> Inserting $\lambdabar = \hbar c/pc$ where the CMS momentum $pc = M_Z/2 = 45.5$ GeV, and with $J = 1$, $s = \frac{1}{2}$, and $\Gamma_{ee}/\Gamma_{tot} = 0.033$, one obtains from (1.36), $\sigma(\text{peak}) = [12\pi/M_Z^2] \times 0.033 \, (\hbar c)^2 = 58$ nb. The actual cross section (Fig. 1.12) is about half of this. The difference is due to radiative corrections to the electron and positron in the initial state, which smear the energy distribution and depress the peak cross section.

Figure 1.13 shows the first resonance ever to be discovered in high energy physics, namely the Δ (1232) pion–proton resonance observed in 1952. It has central mass 1232 MeV/c^2 and width $\Gamma = 120$ MeV. This observation was followed by that of many other meson–meson and meson–baryon resonances in the 1950s and 1960s. These resonant states were important in providing the essential clues that led to the development of the quark model by Gell-Mann and by Zweig in 1964. The Δ resonance is also of present astrophysical significance

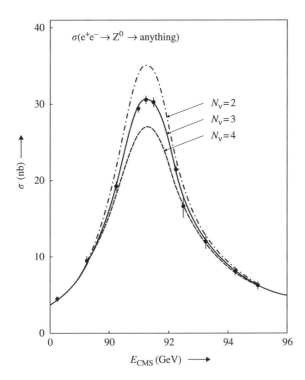

Fig. 1.12 The electron–positron annihilation cross section as a function of energy near the Z^0 resonance. The observed values are averages from four experiments at the LEP electron–positron collider at CERN. The three curves are the Standard Model predictions for two, three, or four flavours of neutrino. In this case the Breit–Wigner curve is asymmetric, because the nominal beam energy is modified by synchrotron radiation losses.

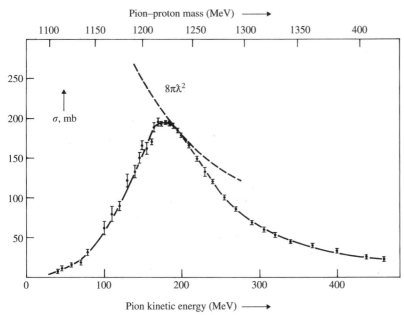

Fig. 1.13 The pion–proton resonance Δ(1232) first observed by Anderson *et al.* in 1952.

in connection with the very highest energy cosmic rays, since it can be excited in collisions of protons above 10^{20} eV energy with the cosmic microwave background, with quantum energy of the order of 0.25 meV (milli-electronvolt). Known as the GKZ effect after Greisen, Kuzmin, and Zatsepin who suggested it, the resonance could lead to a cut-off in the cosmic ray spectrum above 10^{20} eV (see Section 6.6).

From the viewpoint of the human race, possibly the most important resonance is that of the 0^+ excited state of the ^{12}C nucleus at an excitation energy of 7.654 MeV. The width is about 10 eV only. The production of carbon in helium-burning red giant stars, discussed in Chapter 7, is achieved through the so-called triple alpha process, $3\alpha \rightarrow {}^{12}$C. First, two alpha particles combine to form ^{8}Be in its ground state, which is unstable with a lifetime of only 10^{-16} s. It may, nevertheless, capture a third alpha particle to form carbon, which however usually decays back into beryllium plus an alpha particle, but can with a small (10^{-3}) probability decay by gamma emission to the ground state of carbon. The rate of ^{12}C production depends crucially on the existence of this resonance level, occuring just 400 keV above the threshold energy, to enhance the triple alpha cross section. Indeed, Hoyle had predicted the need for such a resonance and its properties in 1953, before it was finally found in laboratory experiments. Without the existence of this resonance, it is almost certain that carbon-based biological evolution in the universe could not have taken place.

1.12 New particles

The Standard Model of the fundamental quarks and leptons, and their interactions described here and in Chapter 3, is able to account for essentially all laboratory experiments at accelerators to date, and describes with great accuracy the physics of the fundamental particles, at least in our particular corner of the universe. However, it apparently does not describe the building blocks of the universe on large scales: indeed on such scales it may account for only about 5% of the total energy density! As described in detail in Chapter 4, the study of the kinematics of large-scale cosmic structures—galaxies, galaxy clusters, and superclusters—indicates that the bulk of the matter in the universe is invisible (i.e. non-luminous) **dark matter**. The nature of this dark matter is presently unknown. There is so far no direct experimental evidence for the detection of individual dark matter particles. It is possible that the proposed massive supersymmetric particles mentioned in Section 1.3 could be candidates.

Present experiments in astrophysics, to be described in the following chapters, also suggest that, although in the past the universal expansion following the Big Bang was indeed slowing down (on account of the restraining pull of gravity), the expansion is now accelerating, this being ascribed to **dark energy**, which actually exerts a gravitational repulsion. This dark energy appears to exceed the energy density in all other matter and radiation. Again, the actual nature and origin of this dark energy is presently unknown.

The possibilities of new particles and new interactions also follow from the very successes of the Standard Model, in being able to unify electromagnetic interactions with weak interactions in the electroweak theory. This suggests that it might also be possible to unify strong and electroweak interactions in a so-called grand unified scheme. There is, however, essentially no experimental support for such higher levels of unification at the present time. The reasons for these schemes, and their consequences for the experimental situation are discussed in Chapters 3 and 4. The evidence for higher mass scales, well above anything attainable in the laboratory, may also be suggested by inflationary models of the early universe, described in Chapter 5, and by the evidence for finite neutrino masses, described in Chapter 6.

1.13 Summary

- Matter is built from elementary fermion constituents, the quarks and leptons, occurring in three 'families'. Each family consists of a quark with charge $+2|e|/3$ and one of $-|e|/3$, a charged lepton with charge $-|e|$ and a neutral lepton (neutrino). The antiparticles of these states have electric charges of opposite sign but are otherwise identical to the particles.
- The observed strongly interacting particles (hadrons) consist of bound quark combinations. Baryons consist of three quarks, and mesons of a quark–antiquark pair.
- Quarks and leptons interact via exchange of fundamental bosons, characteristic of four fundamental interactions: strong, electromagnetic, weak, and gravitational. The exchanged bosons are virtual particles.
- Strong interactions occur between quarks and are mediated by gluon exchange; electromagnetic interactions are between all charged particles and are mediated by photon exchange; weak interactions are mediated by $W^{\pm}$ and Z^0 exchange; gravitational interactions via graviton exchange. Photons, gluons, and gravitons are massless. W and Z bosons have masses of 80 GeV/c^2 and 91 GeV/c^2, respectively.
- The basic boson exchange process can be visualized by diagrams called Feynman diagrams, depicting the exchange of a virtual boson between two interacting fermions. The amplitude for this process is the product of the couplings g_1 and g_2 of the fermions to the exchanged boson, multiplied by a propagator term which depends on the (free) boson mass M and the momentum transfer q, of the form $1/(M^2 - q^2)$, where q^2 is a negative quantity. The cross section for the interaction is the product of the square of the above amplitude and a phase space factor. It is numerically equal to the reaction rate per target particle and per unit incident flux.
- The strength of an interaction is also measured by the decay rate or width of unstable hadronic or leptonic states. If the width is large enough to be measurable, the state is referred to as a resonance.
- Under normal conditions, quarks are confined as combinations in baryons (QQQ) or mesons (Q$\bar{\text{Q}}$). At sufficiently high temperatures, with $kT > 0.3$ GeV, it is expected that quarks would no longer be confined, and that hadrons would undergo a phase transition to a quark–gluon plasma.

Problems

A table of physical constants can be found in Appendix A. More challenging problems are marked with an asterisk.

(1.1) Find the fractional change in total energy (including rest energy and gravitational potential energy) when two equal and isolated point masses M are brought from infinity to a separation R. Calculate what this is when (a) $M = 1 M_{\odot}$ and $R = 1$ pc, and (b) when $M = M_{PL}$ and $R = 1$ fm.

*(1.2) The bombardment of a proton target by a pion beam of energy 1 GeV results in the reaction $\pi^- + p \rightarrow \Lambda + K^0$ with a cross section of about 1 mb. The K^0 particle is a meson of strangeness $S = +1$, while the Λ is a baryon

of strangeness $S = -1$. Write down the above reaction in terms of quark constituents.

Both of the product particles are unstable. One undergoes decay in the mode $\Lambda \rightarrow p + \pi^-$ with a mean lifetime of 10^{-10} s, while the other decays in the mode $K^0 \rightarrow \pi^+ + \pi^-$, also with a lifetime of 10^{-10} s. Both decay rates and interaction cross sections are proportional to the squares of the coupling constants associated with the interactions responsible. Explain qualitatively how the long lifetimes for the above decays can be reconciled with the large production cross section.

(1.3) Calculate the energy carried off by the neutrino in the decay of a pion at rest, $\pi^+ \rightarrow \mu^+ + \nu_\mu (m_\pi c^2 = 139$ MeV, $m_\mu c^2 = 106$ MeV, $m_\nu \sim 0$). Pions in a beam of energy 10 GeV decay in flight. What are the maximum and minimum energies of the muons from these decays? (See Appendix B for formulae on relativistic transformations.)

(1.4) The Δ^{++} resonance shown in Fig. 1.13 has a full width of $\Gamma = 120$ MeV. What is the mean proper lifetime of this state? How far on average would such a particle, of energy 100 GeV, travel before decaying?

(1.5) The Ω^- is a baryon of mass 1672 MeV/c^2 and strangeness $S = -3$. It decays principally to a Λ baryon of mass 1116 MeV/c^2 and $S = -1$ and a K^- meson of mass 450 MeV/c^2 and $S = -1$. Express the decay process in terms of quark constituents in a Feynman diagram.

State which of the following decay modes are possible for the Ω particle.

(a) $\Omega^- \rightarrow \Xi^0 + \pi^-$ $(m_\Xi = 1315$ MeV/c^2, $S = -2)$

(b) $\Omega^- \rightarrow \Sigma^- + \pi^0$ $(m_\Sigma = 1197$ MeV/c^2, $S = -1)$

(c) $\Omega^- \rightarrow \Lambda^0 + \pi^-$ $(m_\Lambda = 1116$ MeV/c^2, $S = -1)$

(d) $\Omega^- \rightarrow \Sigma^+ + K^- + K^-$

*(1.6) The following decays are all ascribed to the weak interaction, resulting in three final-state particles. For each process, the available energy Q in the decay is given together with the decay rate W:

(a) $\tau^+ \rightarrow e^+ + \nu_e + \bar{\nu}_\tau$ $Q = 1775$ MeV

$$W = 6.1 \times 10^{11} \text{ s}^{-1}$$

(b) $\mu^+ \rightarrow e^+ + \nu_e + \bar{\nu}_\mu$ $Q = 105$ MeV

$$W = 4.6 \times 10^5 \text{ s}^{-1}$$

(c) $\pi^+ \rightarrow \pi^0 + e^+ + \nu_e$ $Q = 4.1$ MeV

$$W = 0.39 \text{ s}^{-1}$$

(d) $^{14}O \rightarrow {}^{14}N^* + e^+ + \nu_e$ $Q = 1.8$ MeV

$$W = 5.1 \times 10^{-3} \text{ s}^{-1}$$

(e) $n \rightarrow p + e^- + \bar{\nu}_e$ $Q = 0.78$ MeV

$$W = 1.13 \times 10^{-3} \text{ s}^{-1}$$

Using dimensional analysis show that within one or two orders of magnitude, the Q values and decay rates are compatible with the same weak coupling. Comment on any trends with the values of Q.

(1.7) The weak decays in Problem 1.6 are mediated by the boson W, with a mass of 80 GeV/c^2. If proton decay was mediated by a boson of mass M_X with exactly the same weak coupling, give a rough estimate of the value of M_X which would result in a proton lifetime of 10^{32} years (yr), given the quoted lifetime and mass of the muon in the previous questions (1.3) and (1.6).

*(1.8) The cross section for neutrino–electron scattering via W exchange is given in (1.27b) as $\sigma(\nu_e + e \rightarrow \nu_e + e) = G_F^2 s/\pi$, where s is the square of the CMS energy. At high energies, deep inelastic neutrino–nucleon scattering can be treated as elastic scattering of a neutrino by a quasi-free quark constituent of the nucleon, the scattered quark then 'fragmenting' into secondary hadrons. Assuming that the struck quark carries on average 25% of the mass of the nucleon, calculate the neutrino–nucleon cross section in cm^2 as a function of the laboratory neutrino energy in GeV.

*(1.9) In the previous problem, the cross section formula assumes a pointlike interaction specified by the Fermi constant G_F. However, at very high energies, the effect of the finite W-boson mass in the propagator term (1.6) must be taken into account. Write down an expression for the differential cross section $d\sigma/dq^2$ for neutrino–electron scattering in this case, based on equation (1.27b) and the fact that the maximum value of momentum transfer squared is $q^2(\text{max}) = s$. Show that, as $q^2(\text{max}) \rightarrow \infty$, the neutrino–electron cross section tends to a constant, and find its value. At what neutrino energy does the cross section reach half of its asymptotic value?

(1.10) When electron–antineutrinos $\bar{\nu}_e$ traverse interstellar matter, they can interact with electrons to form the W-boson resonance: $\bar{\nu}_e + e^- \rightarrow W^-$. Assuming the target electrons are at rest, calculate the 'resonant' energy required for these antineutrinos.

(1.11) The charmed meson D^+ (mass 1.87 GeV/c^2) undergoes weak $\Delta C = 1$ decay in the mode $D^+ \rightarrow K^0 + l^+ + \nu_l$ where $l = e$ or μ, with a 15% branching ratio. The quark constitution of the charmed meson is $D^+ = c\bar{d}$ and of the kaon is $K^0 = s\bar{d}$, so the decay can be written as the

Baryon	Quark structure	Q (MeV)	Decay mode	Lifetime or width
$\Sigma^0(1192)$	u ds	74	$\Lambda\gamma$	$\tau = 7.4 \times 10^{-20}$ s
$\Sigma^+(1189)$	u us	187	$p\pi^0, n\pi^+$	$\tau = 8 \times 10^{-11}$ s
$\Sigma^0(1385)$	u ds	208	$\Lambda\pi^0$	$\Gamma = 36$ MeV

transformation of a charmed to a strange quark (with a $\bar{\text{d}}$-quark as 'spectator'):

$$c \rightarrow s + l^+ + \nu_l$$

in close analogy with muon decay in Problem 1.6. Draw the Feynman diagram for c quark decay, and assuming a mass of the c quark of 1.6 GeV/c^2, and neglecting the mass of the decay products, estimate the lifetime of the D-meson from that of the muon as given in Problem 1.6.

(1.12) The Σ-baryons are combinations of s, u, and d quarks, with strangeness $S = -1$. The first two entries in the Table at the top of this page are for the ground-state combination of spin $J = \frac{1}{2}$, while the third is an excited state of $J = \frac{3}{2}$. The rest-masses in MeV/c^2 are given in brackets, followed by the quark constitution, the Q value, principal decay mode and lifetime or width. All the decay products have $S = 0$ except for the Λ-baryon, which has $S = -1$.

State which of the fundamental interactions are responsible for the above decays, and from the decay rates estimate the relative values of their coupling strengths.

(1.13) Draw the Feynman diagram representing electron–electron scattering to first order in the coupling constant. If you carefully label the incoming and outgoing electron states, in fact, you will find that **two** diagrams are possible. Draw some second-order diagrams involving exchanges of photons and/or electron–positron pairs. Compare the interaction rates with those for the first-order process.

(1.14) The following transitions have Q values and mean lifetimes as indicated:

Transition	Q value (MeV)	Lifetime (s)
(a) $\mu^+ \rightarrow e^+ + \nu_e + \bar{\nu}_\mu$	105	2.2×10^{-6}
(b) $\mu^- + {}^{12}\text{C} \rightarrow {}^{12}\text{B} + \nu_\mu$	93	2×10^{-6}
(c) $\pi^0 \rightarrow 2\gamma$	135	10^{-17}
(d) $\Delta^{++} \rightarrow p + \pi^+$	120	10^{-23}

State which interactions are responsible in each case, and estimate the relative coupling strengths from the given quantities.

*(1.15) Calculate the ratio R of the cross section for $e^+e^- \rightarrow Q\bar{Q} \rightarrow$ hadrons to that for the reaction $e^+e^- \rightarrow \mu^+\mu^-$ in (1.25), as a function of increasing CMS energy up to 20 GeV. Assume the quark masses given in Table 1.4.

At a certain energy, the process $e^+e^- \rightarrow \pi^+ + \pi^- + \pi^0$ is observed. Draw a Feynman diagram to illustrate such an event.

*(1.16) The Δ pion–nucleon resonance (see Fig. 1.13) has a central mass of 1232 MeV/c^2 and spin $J = \frac{3}{2}$. It decays predominantly into a pion of $J = 0$ plus a nucleon of $J = \frac{1}{2}$, but also decays in the mode $\Delta \rightarrow n + \gamma$ with a branching ratio of 0.55%. Using equation (1.33), calculate the peak cross section for the process $\gamma + p \rightarrow \Delta^+$. The cosmic microwave background consists of photons with a temperature of $T = 2.73$ K and density of 400 cm^{-3}. Estimate the energy that primary cosmic ray protons would require in order to excite the Δ resonance in collisions with the microwave background. Assume a mean photon energy of $2.7kT$ and head-on collisions. What is the mean free path for collision of such protons?

2 The expanding universe

2.1 The Hubble expansion

Everyone is familiar with the fact that the universe is populated by stars and that these occur in huge assemblies called galaxies. A typical galaxy such as our own Milky Way will contain of the order of 10^{11} stars, together with clouds of gas and dust. Various forms of galaxies are observed, the most common being the elliptical and spiral galaxies. In spiral galaxies, the older, population II stars are located in a central spherical hub, which is surrounded by a flattened structure or disc in the form of a spiral, associated with the formation of younger, population I stars moving in roughly circular orbits and concentrated in spiral arms. Figure 2.1(a) shows a picture of the spiral galaxy M31, which is similar in structure to our own Milky Way, sketched in Fig. 2.1(b), and which, together with our nearest neighbour galaxy, the Large Magellanic Cloud, forms part of the Local Group of around thirty galaxies. The total number of observable galaxies is enormous, of the order of 10^{11}. They occur in clusters—see Fig. 2.2(a) for the Coma cluster—and superclusters separated by enormous voids as in Fig. 2.2(b). In other words, the material of the universe is not distributed at random, but there is structure on the very largest scales. Typical sizes and masses are given in Table 2.1.

From the last line of Table 2.1 we note that the (negative) gravitational potential energy GM^2/R and the mass energy Mc^2 of the universe are about equal at 10^{70} J, so that the total energy is near zero. As indicated later, it turns out that the measured value of the curvature parameter on very large scales is consistent with it being exactly zero.

In 1929 Hubble, observing the spectral lines from distant galaxies with the new 100-inch Mount Wilson telescope, noted that the lines were shifted towards the red end of the spectrum, the amount of shift depending on the apparent brightness of the galaxy and hence on the distance. He measured the velocity of recession of a galaxy, v, interpreting the redshift as due to the Doppler effect. The wavelength in this case is increased from λ to λ' so that

$$\lambda' = \lambda \sqrt{\frac{1+\beta}{1-\beta}} = \lambda(1+z) \qquad (2.1)$$

where $\beta = v/c$ and the redshift $z = \Delta\lambda/\lambda$. Hubble discovered a linear relation between v and the distance r:

$$v = H_0 r \qquad (2.2a)$$

(a)

(b)

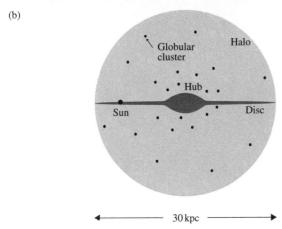

Globular cluster

Halo

Hub

Sun

Disc

30 kpc

Fig. 2.1 (a) The spiral galaxy M31 in Andromeda, believed to be very similar in form to our own galaxy, the Milky Way. Two dwarf elliptical galaxies appear in the same picture. (b) Sketch of edge-on view of the Milky Way. As well as stars and dust, the spiral arms of the disc contain gas clouds, predominantly of hydrogen, detected from the 21 cm wavelength emission line due to flipover of the electron spin relative to that of the proton. The Milky Way contains at least 150 globular clusters (see Fig. 7.3), each containing of the order of 100 000 very old stars of similar age. The halo region may contain dark matter as described in Chapter 4, and the central hub, a massive black hole of order 3×10^6 solar masses.

where H is called the Hubble constant. In Hubble's early measurements, its value was vastly overestimated. The accepted value today is approximately

$$H_0 = 70 \pm 10 \text{ km s}^{-1} \text{ Mpc}^{-1} \tag{2.2b}$$

where the megaparsec has the value 1 Mpc $= 3.09 \times 10^{19}$ km. The subscript '0' to H is to signify that this is the value measured today. Clearly if the universal expansion implied above is accelerating or decelerating, H will be a function of time.

The interpretation of the redshift in terms of the Doppler effect is permissible for the small redshifts of $z < 0.003$ observed by Hubble. For such nearby galaxies, Newtonian concepts of space and time are applicable. Expanding (2.1) for small values of v/c we get

$$\lambda' \approx \lambda(1 + \beta)$$

and hence

$$z = v/c$$

However, for distant galaxies and large redshifts $z \sim 1$, the Doppler formula is certainly inapplicable, since at such distances additional, gravitational redshifts could then become important (see Appendix B). The empirical relation observed is, therefore, of a linear dependence of the redshift on the distance of the

(a)

(b)

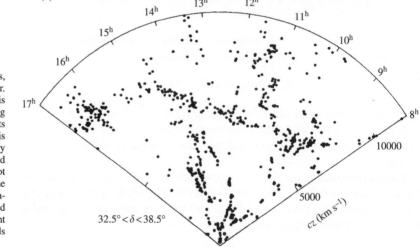

Fig. 2.2 (a) The Coma cluster of galaxies, in which both spirals and ellipticals appear. The space between galaxies in clusters is usually filled with very hot, X-ray emitting gas, which includes ions of heavy elements like iron, indicating that much of the gas is debris expelled from early generations of very massive stars that have long since disappeared from view (Palomar Observatory). (b) A plot showing the distribution of a sample of some 700 galaxies over a small range of declination angle δ. The redshift velocity cz is plotted radially, the angular coordinate being the right ascension. The existence of clusters and voids is apparent (from de Lapparent *et al.* 1986).

galaxy. The distance is estimated from the apparent brightness or luminosity, and is therefore called the **luminosity distance** D_L. It is determined from the (supposedly known) intrinsic luminosity L or total power radiated by the star, and the measured energy flux F at the Earth:

$$F = \frac{L}{(4\pi D_L^2)}$$

Table 2.1 Approximate sizes and masses in the universe
(1 parsec = 1 pc = 3.09×10^{16} m = 3.26 lightyears)

	Radius	Mass
Sun	7×10^8 m	2×10^{30} kg $= M_\odot$
Galaxy	15 kpc	$10^{11} M_\odot$
Cluster	5 Mpc	$10^{14} M_\odot$
Supercluster	50 Mpc	$10^{15} M_\odot$
Universe	4500 Mpc	$10^{23} M_\odot$

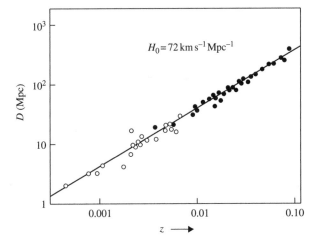

Fig. 2.3 Log–log plot of distance versus redshift, for small redshifts, $z < 0.1$. The points for $z < 0.01$ are from Cepheid variables (open circles), and those of higher z (full circles) include results from Type Ia and Type II supernovae. The straight line is that for the Hubble parameter $H_0 = 72$ km s^{-1} Mpc^{-1} (after Freedman *et al.* 2001). The absolute distance scale has been established by parallax measurements of nearby sources, working out to larger distances, eventually reaching the nearest Cepheids.

In fact astronomers use a logarithmic scale of luminosity, called **magnitude**, running (perversely) from small values of magnitude for the brightest stars to large values for the faintest. The defining relation between the apparent magnitude $m(z)$ at redshift z, the so-called absolute magnitude M (equal to the value that m would have at $D_L = 10$ pc) and the distance D_L in Mpc, is given by the **distance modulus**

$$m(z) - M = 5 \log_{10} D_L(z) + 25 \qquad (2.3)$$

In the Hubble diagram, $(m - M)$ or $\log_{10} D_L$ is plotted against $\log_{10} z$.

A modern version of the Hubble plot at small redshifts is shown in Fig. 2.3, for events of $z < 0.1$. The various sources in this plot include, for example, Cepheid variable stars for $z < 0.01$, and Type Ia and Type II supernovae for higher redshifts. Cepheid variables can be used as 'standard candles', since they vary in luminosity due to oscillations of the envelope, the period τ being determined by the time for sound waves to cross the stellar material—$\tau \propto L^{0.8}$. Supernovae, discussed in Chapter 7, signal the death throes of stars in the final stages of evolution, and when they occur, their light output for a time—typically weeks or even months—can completely dominate that from the local galaxy. So, in principle, they are useful for probing out to large distances and redshifts, or equivalently, back to earlier times. The absolute distance scale to the thirty or so nearest spiral galaxies where a few Type Ia or Type II supernovae have occurred, has been established by observations on Cepheid variables, and this provides a means of calibrating supernova luminosity.

The data in Fig. 2.3 is seen to be consistent with a very constant and uniform Hubble flow, and this particular sample leads to $H_0 = 72$ km s^{-1} Mpc^{-1}. As discussed later in Section 4.11, at very high redshifts (z up to and beyond unity, that is, looking back into the distant past) the data seem to indicate that H is not in fact constant with time, that it was smaller in the past and that the universe is now **accelerating**. However, the evidence for and implications of all this are deferred to Chapter 4.

The Hubble relation (2.2) implies a uniform and homogeneous expansion of the universe with time. If H were independent of time, it would imply an increase in the size of the universe by a factor e in the **Hubble time**

$$t(\text{Hubble}) = \frac{1}{H_0} = 14 \text{ Gyr} \quad (1.4 \times 10^{10} \text{ years}) \tag{2.4}$$

where H_0 is the current value of the Hubble parameter. The actual or **physical coordinate distance** from the Earth, say, to some distant galaxy at time t is written as the product

$$r(t) = R(t)r_0 \tag{2.5}$$

where $R(t)$ is the value of the **scale parameter** and r_0 is the **comoving coordinate distance** measured in a reference frame that is comoving with the expansion (or equivalently the true distance measured today if we define the scale parameter at the present time as $R(0) = 1$). The quantity r_0 is a time-independent constant (for the distance to a particular galaxy), while according to the cosmological principle discussed below, the expansion parameter $R(t)$ is the same over all space and depends **only** on time, in a way determined by the exact geometry (curvature) of the universe, as indicated in Fig. 2.4. Its value at time t, as compared with the value today, is of course just equal to the reciprocal of the redshift factor in (2.1):

$$R(t) = \frac{R(0)}{1+z} \tag{2.6}$$

Substituting (2.5) in (2.2) it is seen that the Hubble law is then a statement about the rate of change of the scale parameter:

$$\dot{R}(t) = HR(t) \tag{2.7}$$

where $\dot{R} = \mathrm{d}R/\mathrm{d}t$. The expansion is often compared with the stretching of the surface of a balloon under inflation in the two-dimensional case. However, it must be emphasized that the expansion applies only to truly cosmological distances, that is, to those between galaxies or galaxy clusters. It does not, for example, represent the dependence of the size of the solar system or of a hydrogen atom on time. In the balloon analogy, the galaxies should be represented by dots on the balloon surface. The pattern of galaxies does not change in shape, but just expands in size.

The expansion of the universe is usually referred to as the Big Bang, suggesting that a sudden explosion occurred at a singular point in space-time. Obviously, referred to such an origin, this could reproduce the Hubble relation (2.2), since the particles of largest velocity will have travelled the farthest from the origin. However, the accepted view of the early universe is based on the **cosmological principle**, namely, that on large scales the universe was both isotropic and homogeneous, so that no direction or location was to be preferred over

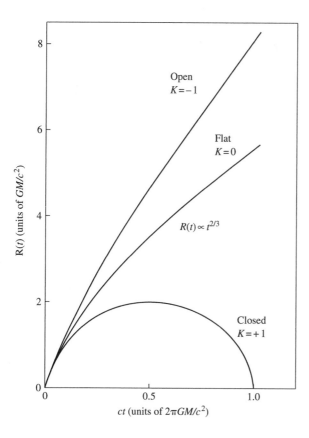

Fig. 2.4 The scale factor R as a function of time in a matter-dominated universe, for three different values of the curvature parameter K. Present evidence is that the value of $K \approx 0$, corresponding to a flat universe. For a vacuum dominated universe, on the contrary, the scale factor would increase exponentially with time (see Table 2.2). In the distant past, when the universe was less than half of its present age, it seems that it was indeed matter dominated—see Fig. 4.11—but that at the present time, the contributions to the total energy density from vacuum and matter are comparable (see Section 2.5).

any other, and thus it must appear the same to all observers no matter where they are. Observationally, the universe is indeed found to be isotropic on large enough scales, and this also implies homogeneity, since a non-homogeneous universe would appear isotropic only to a favoured observer stationed at its centre, if it had spherical symmetry. So the 'Big Bangs' occur everywhere at once and the expansion appears the same to all observers irrespective of their location.

2.2 Olbers' paradox

In the nineteenth century, Olbers asked the question 'Why is the sky dark at night?' He supposed the universe to be unlimited in extent and filled uniformly with sources of light (stars). The light flux reaching us from a star at distance r varies as r^{-2}, while the number of stars in the spherical shell $r \rightarrow r + dr$ varies as $r^2\,dr$. Hence, the total light flux will increase as $r(\text{max})$, which is infinite in the model.

There are several reasons why Olbers' arguments are invalid. First, we believe the observable universe is not infinite but has a finite age, and began at a time t_0 in the past with the Big Bang, which started off the Hubble expansion. This means that light can only reach us from a maximum horizon distance ct_0 and the flux must be finite. A second point is that the light sources (stars) are finite in size, so that nearby sources will block out light from more distant sources. Their light is absorbed exponentially with distance by the intervening stars and dust. Third, stars only emit light for a finite time t, and the flux from the most distant stars will therefore be reduced by a factor t/t_0. Finally, the expansion of the universe results in an attenuation at large enough redshifts of light of any particular frequency; for example, red light will disappear into the infrared and the flow of light energy will fall off. However, we may remark that, as indicated below, the 2.7 K microwave background radiation, which is the cooled and redshifted remnant of the original expanding fireball of the Big Bang, although invisible to the eye, is just as intense at night time as during the day. So in this sense Olbers was right!

2.3   The Friedmann equation

The evolution of the universe is described theoretically by the solution of Einstein's field equations of general relativity. The 'Standard Model' of present day cosmology is the solution proposed by Friedmann–Lemaitre–Robertson–Walker (FLRW for short), which assumes a completely isotropic and homogeneous distribution of matter and radiation, behaving like a perfect frictionless fluid, and these assumptions and that of energy conservation are enough to fix the form of the evolution equation. Of course, certain constants are needed to specify the evolution quantitatively, and these are found by reference to general relativity and to Newton's law of gravity.

The FLRW assumption of isotropy and homogeneity is a statement of the cosmological principle mentioned above. However, it is obvious that, whatever the case at early times in the universe, matter today shows enormous fluctuations in density in the form of galaxies and larger structures. But the average separation between galaxies is of the order of 100 times their diameter, and the overall expansion of the universe of billions of galaxies on large enough scales, that is orders of magnitude larger than the intergalactic separations, still appears to be well described by the FLRW model. Thus, the universe is homogeneous in the same sense as is a volume of gas on a scale large compared with the intermolecular separation. The best evidence for isotropy and homogeneity in fact comes from observations of the cosmic microwave background discussed in Section 2.7, which reflects the distribution of matter and radiation when the universe was only 300 000 yr old (as compared to 14 Gyr today) and long before either stars or larger structures had started to form.

The solution for the temporal development of the universe predicted by this model was first found by Friedmann (1922) and has the form (see also (2.17))

$$H^2 = \left(\frac{\dot{R}}{R}\right)^2 = \frac{8\pi G \rho_{\text{tot}}}{3} - \frac{Kc^2}{R^2} \tag{2.8}$$

where $R = R(t)$ is the expansion parameter in (2.4) and (2.5), ρ_{tot} is the total density of matter, radiation and vacuum energy, as described below, and G is Newton's gravitational constant. The term Kc^2/R^2 is the so-called **curvature term**. One of the consequences of general relativity is that the 'flat' Euclidean space of special relativity is replaced by curved space. Light beams travel along paths of least time called **geodesics**, which are straight lines in Euclidean space, but, in the presence of gravitational fields, the paths are curved (just as in the two-dimensional case, the shortest path between two points on a spherical surface is along a great circle). In the language of particle physics, one could say that space appears curved because photons are deflected by gravitational fields, mediated by graviton exchange. The curvature parameter K can in principle assume positive, zero, or negative values, corresponding to positive, zero, and negative curvature, respectively. The two-dimensional analogy for positive or convex curvature is the surface of a sphere, while that for negative or concave curvature is as in a saddle.

Equation (2.8) has been quoted without derivation, but as pointed out by Milne (1934), we can understand in terms of Newtonian mechanics what it implies in the special case where the energy density is dominated by non-relativistic matter. For, consider a point mass m being accelerated by gravity at the surface of a sphere of radius r, density ρ, and mass $M = 4\pi r^3 \rho / 3$. According to Newtonian mechanics, the assumed spherically symmetric and homogeneous distribution of matter outside the sphere can make no contribution to the force. It turns out that this is also true in general relativity (by a theorem due to Birkhoff). So, the force equation is simply

$$m\ddot{r} = -\frac{mMG}{r^2} \tag{2.9}$$

In this equation, if we express M in terms of r and ρ, factors of r_0 from (2.5) cancel out, and, for brevity in what follows, we choose units such that $r_0 = 1$ (but must remember that all cosmological distances are the product $R(t)r_0$). After integrating (2.9) we then get

$$\frac{m\dot{R}^2}{2} - \frac{mMG}{R} = \text{constant} = -\frac{mKc^2}{2} \tag{2.10}$$

If we multiply through by $2/mR^2$, we obtain an equation in agreement with (2.8), after setting the constant of integration equal to the value given by general relativity. We note that the terms on the left-hand side of (2.10) correspond to the kinetic and potential energies of the mass m, and, therefore, the so-called curvature term on the right simply represents the total energy. $K < 0$ corresponds to negative curvature and positive energy, that is to say an **open** universe expanding without limit. For $K = -1$, $\dot{R}(t) \equiv r_0 \dot{R}(t) \to c$ at large enough values of R. A value $K > 0$ corresponds to a **closed** universe with negative total energy and positive curvature, which reaches a maximum radius and then collapses. $K = 0$ is the simplest case, where the kinetic and potential energies just balance and the total energy and curvature are both zero. The universe expands forever but the velocity tends asymptotically to zero at large t. This case is called the **flat** universe. The three cases for K values of $+1, -1$, and 0 are illustrated in Fig. 2.4.

Present data indicate that on large scales the universe is extremely close to being flat, with $K \approx 0$. In that case, the constant of integration on the right-hand side of (2.10) representing the total energy, potential plus kinetic, is practically zero.

> **Example 2.1** *Show that for a curvature term with $K = +1$, the Big Bang would be followed by a Big Crunch at time $t = 2\pi GM/c^3$, where M is the (assumed conserved) mass of the universe.*
>
> For $K = +1$, the Friedmann equation becomes $(\dot{R}/R)^2 = 2GM/R^3 - c^2/R^2$. From this expression it is clear that $\dot{R} = 0$ when $R = 2GM/c^2$, which is the maximum radius. The element of time is then given by $dt = dR/(2GM/R - c^2)^{1/2}$. Substituting $2GM/R - c^2 = c^2 \tan^2 \theta$ the

total time to the maximum is $(4GM/c^3) \int \cos^2 \theta \, d\theta = GM\pi/c^3$. By symmetry, the time to the subsequent crunch is just twice this. The value of M from Table 2.1 gives $t \sim 10$ Gyr.

Upon integrating (2.8) for the case $K = 0$ and a universe dominated by non-relativistic matter of conserved mass M one finds

$$R(t) = \left(\frac{9GM}{2}\right)^{1/3} t^{2/3} \tag{2.11}$$

so that the Hubble time (2.4) is $1/H_0 = R/\dot{R} = 3t_0/2$, and the age of the universe is then

$$t_0 = \frac{2}{3H_0} = 8\text{--}11 \text{ Gyr} \tag{2.12}$$

Other estimates of the age of the universe give larger values. They come, for example, from the study of the luminosity–colour relation (Herzsprung–Russell diagram) in the oldest star populations, the globular clusters (see the caption to Fig. 7.3); from cooling rates of white dwarf stars; and from dating using isotopic ratios of radioactive elements in the Earth's crust and in very old stars. All these estimates straddle the usually accepted range for the age of the universe:

$$t_0 = 14 \pm 2 \text{ Gyr} \tag{2.13}$$

The discrepancy between this figure and that for a flat, matter-dominated universe (2.12) could be due, in principle, either to curvature ($K \neq 0$) or to the existence of a cosmological constant (see (2.19)). However, measurements to be described later indicate that $K \approx 0$. In fact, as shown in Example 2.3, when account is taken of the effect of the vacuum energy/cosmological constant, the age of the universe estimated from the Hubble parameter is in excellent agreement with the result (2.13). Indeed, it is quite remarkable that completely independent estimates of the age come out in agreement to within 20% or so.

Example 2.2   *Find solutions of the Friedmann equation for the case of a matter-dominated universe of total mass M, and values of K > 0 and K < 0.*

The Friedmann equation (2.8) in this case takes the form $\dot{R}^2 = 2MG/R - Kc^2$. For $K > 0$, the solution for R as a function of t has the parametric form of a cycloid curve (that is, the curve traced out by a point on the circumference of a circular disc rolling along a plane):

$$R = a(1 - \cos\theta)$$
$$t = b(\theta - \sin\theta) \tag{2.14}$$

as can be verified by substitution. Here, the constants $a = MG/Kc^2$ and $b = MG/(Kc^2)^{3/2}$, and the parameter θ is the angle of rotation of the cycloid. For the case $K < 0$, the corresponding solution is

$$R = a(\cosh\theta - 1)$$
$$t = b(\sinh\theta - \theta) \tag{2.15}$$

with the above values of a and b, and K replaced by $|K|$. The curves in Fig. 2.4 were plotted from these expressions, and the solutions for the maxima and minima in Example 2.1 are found by setting $\theta = \pi$ and 2π in (2.14).

By expanding the above circular functions for small values of θ, it is straightforward to show that for either $K > 0$ or $K < 0$, the expansion parameter $R \propto t^{2/3}$, which is the same as for the case $K = 0$ in (2.11).

2.4   The sources of energy density

The conservation of energy E in a volume element dV of our perfect cosmic fluid can be expressed as

$$dE = -P\,dV$$

where P is the pressure. Then, with ρc^2 as the energy density, this becomes

$$d(\rho c^2 R^3) = -P\,d(R^3)$$

which leads to

$$\dot{\rho} = -3\left(\frac{\dot{R}}{Rc^2}\right)(P + \rho c^2) \tag{2.16}$$

Differentiating (2.8) and substituting for $\dot{\rho}$ we get the differential form of the Friedmann equation:

$$\ddot{R} = -\left(\frac{4\pi GR}{3}\right)\left(\rho + \frac{3P}{c^2}\right) \tag{2.17}$$

which is the same as (2.9) for the case $P \approx 0$ for non-relativistic matter. Generally, the quantities ρ and P will be connected by an equation of state (see Table 2.2).

The overall density ρ in the Friedmann equation will be made up of three components, corresponding to the contributions from matter, radiation, and the vacuum state:

$$\rho_{\text{tot}} = \rho_{\text{m}} + \rho_{\text{r}} + \rho_{\text{v}} \tag{2.18}$$

The quantity ρ_{v} (also denoted ρ_Λ in some texts) can be incorporated in the Friedmann equation as a **cosmological constant** Λ, such that

$$\rho_{\text{v}} = \frac{\Lambda}{8\pi G} \tag{2.19}$$

The quantity Λ had originally been introduced by Einstein, before the advent of the Big Bang hypothesis, in an attempt to achieve a static (non-expanding and non-contracting) universe. Clearly, if a term $\Lambda/3$ is added to the right-hand side of (2.8), then at large enough $R(t)$, this term will dominate and the expansion will become **exponential**, that is, $R(t) \propto \exp(\alpha t)$ where $\alpha = (\Lambda/3)^{1/2}$. Present evidence, discussed in Chapters 4 and 5, indicates a finite value of Λ, with ρ_{v} comparable with and indeed somewhat larger than ρ_{m}.

Table 2.2 Energy density and scale parameters for different regimes

Dominant regime	Equation of state	Energy density	Scale parameter
Radiation	$P = \rho c^2/3$	$\rho \propto R^{-4}$	$R \propto t^{1/2}$
Matter	$P = \frac{2}{3}\rho c^2 \times (v^2/c^2)$	$\rho \propto R^{-3}$	$R \propto t^{2/3}$
Vacuum	$P = -\rho c^2$	$\rho = \text{const.}$	$R \propto \exp(\alpha t)$

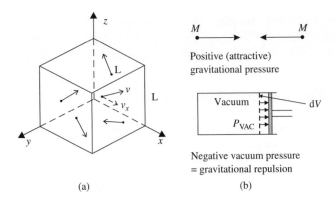

(a)   (b)

Fig. 2.5

In Table 2.2, the dependences of ρ on $R(t)$ and of $R(t)$ on t are given for different possible regimes, namely a radiation-dominated, matter-dominated or a vacuum-dominated universe. The equations of state for radiation and for non-relativistic matter are found as follows. Suppose we have an ideal gas consisting of particles of mass m, velocity v, and momentum mv, confined within a cubical box of side L with walls with which the particle collides elastically (see Fig. 2.5(a)). A particle with x-component of velocity v_x will strike a particular face normal to the x-axis at a rate of $v_x/2L$ collisions per unit time. As the component of momentum $p_x = mv_x$ is reversed at each collision, the rate of change of momentum and, therefore, the force exerted by the particle will be $2mv_x(v_x/2L)$. The pressure exerted by the particle on the face of the box, which has area $A = L^2$, is therefore mv_x^2/L^3, where $V = L^3$ is the volume. If there are n particles per unit volume, it follows that the pressure they exert will be $mn\langle v_x^2 \rangle$ where $\langle v_x^2 \rangle$ is a mean square value. Since the gas is isotropic, the mean square values of the x-, y-, and z-components of velocity will be equal and the pressure will be

$$P = \tfrac{1}{3}\, mn\langle v^2 \rangle = \tfrac{1}{3}n\langle pv \rangle \tag{2.20}$$

Let us first assume that the gas consists of **non-relativistic** particles. Then, the values of the kinetic energy density ε and of the pressure are

$$\varepsilon = \tfrac{1}{2}\, mn\langle v^2 \rangle$$

$$P_{\text{non-rel}} = \frac{2}{3}\varepsilon = \frac{2}{3}\rho c^2 \times \left(\frac{v^2}{c^2}\right) \tag{2.21}$$

where ρc^2 is the total energy density of matter, including the mass energy. Since, for cosmic matter in general, $v^2 \ll c^2$, the pressure it exerts is very small.

If the gas particles have **extreme relativistic** velocities $v \approx c$, then the energy density and the pressure, usually called the **radiation pressure**, have the values

$$\rho c^2 = n\, mc^2 = n\langle pc \rangle$$

$$P_{\text{rel}} = \frac{\rho c^2}{3} \tag{2.22}$$

That the vacuum may contain an energy density and exert a pressure equivalent to a **gravitational repulsion** may seem strange, since in classical physics, a vacuum supposedly contains absolutely nothing. However, in quantum field theory, as will be discussed for the electroweak model in Chapter 3, the Uncertainty Principle actually requires that the vacuum contains virtual particle–antiparticle pairs that spontaneously appear and disappear, and the vacuum itself is defined, not as nothing but as the state of lowest possible energy of the system. Because the virtual particles carry energy and momentum, if only on a temporary basis, general relativity implies that they must be coupled to gravitation. Indeed, the measurable effect of such vacuum energy is through its gravitational influence. Speculations on the magnitude of the vacuum energy density are discussed in Chapter 4.

The relation $P = -\rho c^2$ for this lowest energy vacuum state can be formally shown to follow from Lorentz invariance, that is, the requirement that the state must look the same to all observers, implying also that the energy density must have the same constant value everywhere and for all time. A plausibility argument for the pressure–density relation is as follows. Assume that we have a piston enclosing an isolated cylinder filled with the vacuum state of energy density ρc^2 (see Fig. 2.5(b)). If the piston is withdrawn adiabatically by an element of volume dV, the extra vacuum energy created will be $\rho c^2 \, dV$, and this must be supplied by the work done by the vacuum pressure, $P \, dV$. Hence, by energy conservation $P = -\rho c^2$ and from (2.16), $\rho = $ constant.

Note that in (2.17), the deceleration $\ddot{R}$ is due to the gravitational attraction associated with the density ρ plus the pressure P. An increase in pressure due to relativistic particles is proportional to an increase in their energy density and hence in their gravitational potential, through the Einstein relation $E = mc^2$. Thus, a negative pressure will correspond to a gravitational repulsion and the exponential expansion indicated in Table 2.2.

2.5   Observed energy densities: the age of the universe

For the case $K = 0$, (2.8) gives a value for the **critical density** that (today) would just close the universe:

$$\rho_c = \frac{3}{8\pi G} H_0^2 = 9.2 \times 10^{-27} \text{ kg m}^{-3}$$

$$= 5.1 \text{ GeV m}^{-3} \tag{2.23}$$

taking H_0 from (2.2). In the second line, we have quoted a critical energy density, $\rho_c c^2$. The ratio of the actual density to the critical density is called the **closure parameter** Ω, which at the present time from (2.8) and for arbitrary K is

$$\Omega = \frac{\rho}{\rho_c} = 1 + \frac{Kc^2}{[H_0 R(0)]^2} \tag{2.24}$$

One sees that a flat universe with $K = 0$ will have $\Omega = 1$ for all values of t. The different contributions to the total value of Ω are then, in parallel with (2.18), for radiation, non-relativistic matter and vacuum densities, respectively:

$$\Omega = \Omega_r + \Omega_m + \Omega_v \tag{2.25}$$

If $K \neq 0$, one can express the curvature term as $\Omega_k = \rho_k/\rho_c = -Kc^2/[H_0^2 R(0)^2]$, when from (2.24) one obtains

$$\Omega + \Omega_k = \Omega_r + \Omega_m + \Omega_v + \Omega_k = 1 \tag{2.26}$$

At the present time, as described in the next section, the density of radiation corresponds to $\Omega_r = 5 \times 10^{-5}$ as in (2.43), and is completely negligible in comparison with that of matter. To anticipate later results, the overall density is made up of several components as follows:

1. For **luminous baryonic matter** (i.e. visible protons, neutrons, and nuclei) in the form of stars, gas, and dust it is found that

$$\rho_{lum} \approx 9 \times 10^{-29} \text{ kg m}^{-3}$$

or

$$\Omega_{lum} \approx 0.01 \tag{2.27}$$

2. The **total density of baryons**, visible or invisible, as inferred from the model of nucleosynthesis described below, is about 0.25 baryons m^{-3}, or an energy density

$$\rho_b \approx 4.5 \times 10^{-28} \text{ kg m}^{-3}$$

and

$$\Omega_b \approx 0.05 \tag{2.28}$$

3. The **total matter density**, as inferred from the gravitational potential energy deduced from galactic rotation curves (see Section 4.1) and the kinematics of large-scale structures in the universe (see Section 5.7) is found to be

$$\rho_m \approx 3 \times 10^{-27} \text{ kg m}^{-3}$$

and

$$\Omega_m \approx 0.30 \tag{2.29}$$

4. The **vacuum energy density** can be estimated from the evidence for an upward curvature in the Hubble plot, obtained from the study of Type Ia supernovae at large redshifts, discussed in detail in Section 4.11, and also from the observation of 'acoustic peaks' in the temperature fluctuations of the microwave background radiation at small angular scales, discussed in Section 5.9. These suggest values of

$$\Omega_v \approx 0.70 \tag{2.30}$$

to bring the value of the total density parameter Ω in (2.25) close to unity. Such a flat universe with $K = 0$ is predicted by the inflationary model described in Section 5.2.

There are several important conclusions from equations (2.25)—(2.30). First, most of the baryonic matter is non-luminous. Second, baryons account for only a small fraction, of the order of 15%, of the total matter. The bulk is ascribed to dark matter, as discussed in detail in Chapter 4. The nature of such dark matter is presently unknown. Lastly, it appears that at the present epoch, the bulk of the energy density is in the form of dark vacuum energy. Here, as in Chapter 4, we identify this dark energy with vacuum energy, but other possibilities have been proposed, such as a fifth type of fundamental interaction, with the dark energy density being a function of time. Like dark matter, the source of the dark energy is unknown at the present time.

It is also perhaps worth stating here that as a result of recent measurements, the postulated partition of energy among the various components has changed dramatically over the last decade. It was thought originally that vacuum energy would make only a minor contribution, and that the value $\Omega \sim 1$ was made up largely of dark matter. It should also be emphasized that all the numbers quoted above are still very approximate, with uncertainties that could well be of the order of 30%.

Finally, we note that, after allowing for the contribution from the vacuum energy density, it appears somewhat miraculous that, of all the conceivable values of Ω, the one observed today appears to be quite close to the value of unity expected for a flat universe with zero total energy, exactly as predicted by inflationary models (discussed in detail in Chapter 5).

An estimate of the age of the universe, including all the sources of energy density, can be made as follows. From (2.8) and (2.23)–(2.26) the Hubble parameter at time t is given by

$$
\begin{aligned}
H(t)^2 &= \frac{8\pi G}{3}[\rho_m(t) + \rho_r(t) + \rho_v(t) + \rho_k(t)] \\
&= H_0^2[\Omega_m(t) + \Omega_r(t) + \Omega_v(t) + \Omega_k(t)] \\
&= H_0^2[\Omega_m(0)(1+z)^3 + \Omega_r(0)(1+z)^4 + \Omega_v(0) + \Omega_k(0)(1+z)^2]
\end{aligned}
$$
(2.31)

where we have used the fact that $R(0)/R(t) = (1+z)$ from (2.6), and that matter, radiation, and curvature terms vary as $1/R^3, 1/R^4$ and $1/R^2$ (see Table 2.2 and Section 2.7). The vacuum energy, is independent of z, while $\Omega_k(0) = -Kc^2/(R(0)H_0)^2$ as in (2.24). Also,

$$
H = \frac{1}{R}\frac{dR}{dt} = -\left(\frac{dz/dt}{(1+z)}\right)
$$

and hence

$$
dt = -\frac{dz}{(1+z)H}.
$$
(2.32)

We integrate to obtain the interval from the time t when the redshift was z, to the present time, t_0, when $z = 0$:

$$
t_0 - t = \frac{1}{H_0}
$$

$$
\times \int_0^Z \frac{dz}{(1+z)[\Omega_m(0)(1+z)^3 + \Omega_r(1+z)^4 + \Omega_v(0) + \Omega_k(0)(1+z)^2]^{1/2}}
$$
(2.33a)

The age is found by setting the upper limit as $z = \infty$ at $t = 0$. In the general case this integral has to be evaluated numerically, but there are a few cases

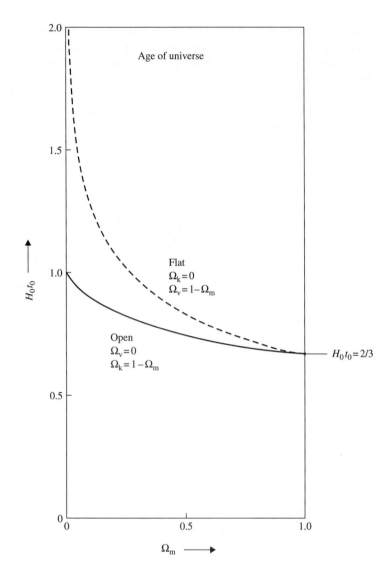

Fig. 2.6 Plot of the age of the universe versus the parameter Ω_m, the ratio of the matter density to the critical density. The solid curve is for an open universe, in which the curvature term $\Omega_k = 1 - \Omega_m$, and radiation and vacuum energy terms are assumed to be zero. The dashed curve is for a flat universe ($\Omega_k = 0$), in which radiation energy is neglected and the vacuum energy $\Omega_v = 1 - \Omega_m$. The present best estimate relates to the flat universe with $\Omega_m \sim 0.30$. The curves have been calculated from the analytical expressions in Example 2.3 and in Problem 2.14.

where analytical solutions are possible, for example, when the radiation term can be neglected and either $\Omega_v = 0$ or $\Omega_k = 0$, as shown in Example 2.3 and in Problem (2.14), and illustrated in Fig. 2.6. The result (2.12) obviously follows when radiation, vacuum, and curvature terms are all zero and the universe is flat and matter-dominated.

Example 2.3   *Estimate the age of a flat universe ($K = 0$) if radiation is neglected and it is presently made up of matter with $\Omega_m = 0.30$ and vacuum energy with $\Omega_v = 0.70$.*

In this case the above integral becomes

$$H_0 t_0 = \int_0^\infty \frac{dz}{(1+z)[\Omega(1+z)^3 + (1-\Omega)]^{1/2}} \tag{2.33b}$$

where $\Omega \equiv \Omega_m(0)$ and $\Omega_v(0) = (1 - \Omega)$. The integral is readily evaluated with the substitution $\Omega(1 + z)^3/(1 - \Omega) = \tan^2 \theta$, when it transforms to the integral $\int d\theta/\sin\theta = \ln[\tan(\theta/2)]$. Finally, one obtains

$$H_0 t_0 = \left(\frac{1}{3A}\right) \ln \left(\frac{1+A}{1-A}\right)$$

where $A = (1 - \Omega)^{1/2}$. For $\Omega = 0.30, (1 - \Omega) = 0.70$, one finds $H_0 t_0 = 0.964$, so that $t_0 \approx 0.96/H_0 = 13.5(\pm 2)$ Gyr. The vacuum energy/cosmological constant term has thus increased the age over the value (2.12).

2.6   The deceleration parameter: the effects of vacuum energy/cosmological constant

One can express the time dependence of the expansion parameter as a Taylor series

$$R(t) = R(0) + \dot{R}(0)(t - t_0) + \frac{1}{2}\ddot{R}(0)(t - t_0)^2 + \cdots .$$

or

$$\frac{R(t)}{R(0)} = 1 + H_0(t - t_0) - \frac{1}{2}q_0 H_0^2(t - t_0)^2 + \cdots$$

where the **deceleration parameter**, which can be time dependent, is defined as

$$q = -\frac{\ddot{R}R}{\dot{R}^2}$$

$$= \left(\frac{4\pi G}{3c^2 H^2}\right)[\rho c^2 + 3P] \tag{2.34}$$

from (2.17). Inserting the values of ρ and P for the components in Table 2.2, it is straightforward to show that this dimensionless parameter has the value

$$q = \Omega_m/2 + \Omega_r - \Omega_v \tag{2.35}$$

Today, $\Omega_m \gg \Omega_r$, so that if Ω_v could be neglected, a flat universe would have $\Omega = \Omega_m = 1$ and $q = 0.5$, that is, the universal expansion must be decelerating, because of the retarding effects of the gravitational attraction of matter. In fact, early attempts to measure q seemed to give results consistent with this value (within large errors). We may note that if Ω_v is large enough, however, $q < 0$ and the expansion would be **accelerating**, the vacuum energy having the same effect as a gravitational repulsion. At least two independent surveys on Type Ia supernovae at high redshifts, treating them as 'standard candles', appear to indicate that q is indeed negative, as discussed in Chapter 4 and shown in Fig. 4.11. This plot suggests that several billion years ago, that is, for redshifts $z > 1$, the universe **was** indeed decelerating, but that more recently this has been replaced by an acceleration. We note here from (2.35) that an **empty universe**, that is, one with $\Omega_m = \Omega_v = \Omega_r = 0$, and hence $\Omega_k = 1$, is neither accelerating nor decelerating, with H independent of time (see (2.8)). Thus, an empty universe is the yardstick against which, in Chapter 4, we judge that a particular model results in acceleration or retardation.

2.7   Cosmic microwave radiation

One of the major discoveries in astrophysics was made in 1965 by Penzias and Wilson. While searching for cosmic sources of radio waves at a wavelength of approximately 7 cm, they discovered an isotropic background of microwave radiation. Although they were unaware of it, this had been predicted by Gamow many years before, as a relic of the Big Bang, a photon fireball cooled by expansion to a temperature of a few degrees kelvin. Figure 2.7 shows data on the spectral distribution of radiation recorded by the COBE satellite. Satellite and balloon-borne detectors as well as ground-level interferometers have mapped out the spectrum over an enormous range of wavelengths, from 0.05 to 75 cm. Recent data shows precise agreement with the spectrum expected from a black body at a temperature of 2.725 ± 0.001 K; indeed, the cosmic microwave spectrum is **the** black body spectrum **par excellence**. It proves, among other things, that at the time the radiation last interacted significantly with matter, it was in thermal equilibrium with it.

Assuming matter to have been conserved, the matter density of the universe can be expected to vary as $\rho_m \propto R^{-3}$. On the other hand, the density of radiation, assuming it to be in thermal equilibrium, will vary with temperature as $\rho_r \propto T^4$ (Stefan's Law). Since there is no absolute scale of distance, the wavelength of the radiation on the truly cosmic scale associated with the Hubble expansion can only be proportional to the expansion factor R. Thus, the frequency and mean energy per photon will both be proportional to R^{-1}. While the number of photons varies as R^{-3}, the energy density of the radiation will vary as R^{-4}, as indicated already in Table 2.2. The extra factor of $1/R$ in the energy density, as compared with non-relativistic matter, simply arises from the redshift, which will in fact apply to any relativistic particles and not just to photons, provided, of course, that those particles are distributed uniformly on the same cosmological scale as the microwave photons. At the early times we are discussing here, the vacuum energy, which is by assumption independent of R, would have been totally negligible and we can just ignore it.

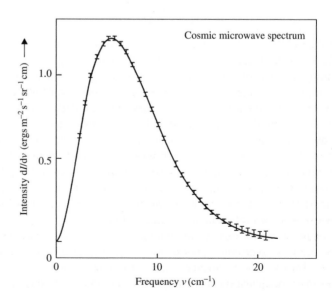

Fig. 2.7 Data on the spectral distribution of the cosmic microwave radiation obtained from the COBE satellite. When satellite data and those from balloon-borne experiments are combined, a very exact fit to a black body spectrum is obtained with $T = 2.725 \pm 0.001$ K and $kT = 0.235$ meV as shown by the curve.

Thus, while the matter density of the universe dominates over radiation today, at early enough times and small enough values of R, radiation must have been dominant. In that case, the second term on the right-hand side of (2.8) can be neglected in comparison with the first, varying as $1/R^4$. Then,

$$\dot{R}^2 = \left(\frac{8\pi G}{3}\right)\rho_{\mathrm{r}}R^2$$

Furthermore, since $\rho_{\mathrm{r}} \propto R^{-4}$,

$$\frac{\dot{\rho}_{\mathrm{r}}}{\rho_{\mathrm{r}}} = -\frac{4\dot{R}}{R} = -4\left(\frac{8\pi G\rho_{\mathrm{r}}}{3}\right)^{1/2}$$

which upon integration gives for the energy density

$$\rho_{\mathrm{r}}c^2 = \frac{3c^2/32\pi G}{t^2} \tag{2.36}$$

For a photon gas in thermal equilibrium

$$\rho_{\mathrm{r}}c^2 = \frac{4\sigma T^4}{c} = \pi^4(kT)^4 \times \left(\frac{g_\gamma}{2}\right)\Big/(15\pi^2\hbar^3c^3) \tag{2.37}$$

where k is the Boltzmann constant, σ the Stefan–Boltzmann constant and $g_\gamma = 2$ is the number of spin substates of the photon. From these last two equations we obtain a relation between the temperature of the radiation and the time of expansion:

$$kT = \left(\frac{45\hbar^3c^5}{32\pi^3G}\right)^{1/4} \times \left(\frac{2}{g_\gamma}\right)^{1/4} \times \frac{1}{t^{1/2}} \approx \frac{1.31\,\mathrm{MeV}}{t^{1/2}} \tag{2.38}$$

where t is in seconds. The corresponding value of the temperature itself is $T \approx 10^{10}\,\mathrm{K}/t^{1/2}$. Since T falls as $1/R$, R increases as $t^{1/2}$ while the temperature falls as $t^{-1/2}$. Hence, the universe started out as a hot Big Bang.

From (2.38) we may roughly estimate the energy of the radiation today, that is, for $t_0 \sim 14\,\mathrm{Gyr} \sim 10^{18}\,\mathrm{s}$. It is $kT \sim 1\,\mathrm{meV}$ (milli-electronvolt), corresponding to a temperature of a few kelvins. This will in fact be an over-estimate since the radiation has cooled more quickly, as $t^{-2/3}$, during the later, matter-dominated era.

Observation of microwave molecular absorption bands in distant gas clouds has made it possible to estimate the temperature of the background radiation at earlier times, when the wavelength would have been reduced, and the temperature increased, by the redshift factor $(1 + z)$. This dependence on redshift has been experimentally verified up to values of $z \approx 3$.

Let us now compare the observed and expected energy densities of radiation. The spectrum of black body photons of energy $E = pc = h\nu$ is given by the Bose–Einstein (BE) distribution, describing the number of photons per unit volume in the momentum interval $p \to p+\mathrm{d}p$. Including $g_\gamma = 2$ as the number of spin substates of the photon, this is

$$N(p)\,dp = \frac{p^2\,dp}{\pi^2\hbar^3\{\exp(E/kT) - 1\}}\left(\frac{g_\gamma}{2}\right) \tag{2.39}$$

In discussing the BE distribution, and later, the Fermi–Dirac (FD) distribution, it will be useful to note the following integrals, from $x = 0$ to $x = \infty$:

$$\int \frac{x^3\,dx}{(e^x - 1)} = \frac{\pi^4}{15}; \quad \int \frac{x^2\,dx}{(e^x - 1)} = 2.404 \qquad\qquad \text{BE}$$

$$\int \frac{x^3\,dx}{(e^x + 1)} = \frac{7}{8}\times\frac{\pi^4}{15}; \quad \int \frac{x^2\,dx}{(e^x + 1)} = \frac{3}{4}\times 2.404 \quad \text{FD}$$

$$\tag{2.40}$$

The total energy density integrated over the spectrum (2.39) is readily calculated to have the value ρ_r in (2.37). The number of photons per unit volume is

$$N_\gamma = \left(\frac{2.404}{\pi^2}\right)\left(\frac{kT}{\hbar c}\right)^3 = 411\left(\frac{T}{2.725}\right)^3 = 411\ \text{cm}^{-3} \tag{2.41}$$

while the energy density from (2.37) is

$$\rho_r c^2 = 0.261\ \text{MeV m}^{-3} \tag{2.42}$$

the equivalent mass density being

$$\rho_r = 4.65 \times 10^{-31}\ \text{kg m}^{-3}$$

with

$$\Omega_r = 5.05 \times 10^{-5} \tag{2.43}$$

some four orders of magnitude less than the present estimated matter density in (2.29).

The temperature of the microwave radiation shows a small anisotropy, of the order of 10^{-3}, attributed to the 'peculiar velocity' $\mathbf{v} = 370\ \text{km s}^{-1}$ of the Earth with respect to the Hubble flow of the radiation. It is given by the Doppler formula $T(\theta) = T_0[1 + (v/c)\cos\theta]$, where θ is the direction of observation with respect to the velocity $\mathbf{v}$. A polynomial analysis of the distribution shows that, apart from this dipole ($l = 1$) term, there are quadrupole ($l = 2$) and higher terms, up to at least $l = 1000$, involving tiny but highly significant anisotropies at the 10^{-5} level. These turn out to be of fundamental importance, reflecting fluctuations in density and temperature in the early universe, which seeded the large-scale structures observed today. These topics are discussed in detail in Section 5.9.

Example 2.4   *Calculate the mean quantum energy and the corresponding wavelength of the cosmic microwave photons for a temperature of $T = 2.725$ K.*

The original discovery of cosmic microwave radiation was made with receivers tuned to 7.3 cm wavelength. What fraction of the photons would have wavelengths in excess of 7.3 cm?

From (2.40) the mean photon energy is $\pi^4 kT/(15\times 2.404) = 2.701\ kT = 6.34 \times 10^{-4}$ eV. The corresponding wavelength is $\lambda = hc/h\nu = 0.195$ cm.

At large wavelengths the curly bracket in (2.39) can be approximated by E/kT if $E/kT \ll 1$. The fraction of photons with quantum energies below $\varepsilon = E/kT$ is then easily shown to be $F = (\varepsilon/kT)^2/(2 \times 2.404)$, which for wavelengths above 7.3 cm is equal to 1.06×10^{-3}.

2.8   Radiations in the early universe

The relation (2.38) for the temperature of the early universe as a function of time applies for radiation consisting of photons. Relativistic fermions, assuming that they are stable enough, will also contribute to the energy density. For a fermion gas, the FD distribution for the number density analogous to (2.39) is

$$N(p)\,dp = \frac{p^2\,dp}{\pi^2\hbar^3\{\exp(E/kT)+1\}}\frac{g_f}{2} \tag{2.44}$$

where $E^2 = p^2c^2 + m^2c^4$, m is the fermion mass and g_f is the number of spin substates. The total energy density, in comparison with (2.37), in the relativistic limit $kT \gg mc^2$ and $E = pc$ is

$$\rho_f c^2 = \frac{7}{8} \times \frac{\pi^4}{15} \times \frac{(kT)^4}{\pi^2\hbar^3 c^3}\frac{g_f}{2} \tag{2.45}$$

Thus, for a mixture of relativistic bosons b and fermions f, the energy density in (2.37) is found by replacing g_γ by a factor g^* where

$$g^* = \sum g_b + \frac{7}{8}\sum g_f \tag{2.46}$$

and the summation is over all types of relativistic particles and antiparticles which contribute to the energy density of radiation in the early universe.

Of course, at very early times when the temperature from (2.38) was high enough for their creation, all types of elementary quarks, leptons, and bosons, plus their antiparticles, would have been present in the primordial 'soup' in which the various components would have been in thermal equilibrium. On the basis of the fundamental particles we know today, the number of degrees of freedom (charge, spin and colour substates) of the fermions would be 90, and that of the gauge bosons 28.

To understand these rather big numbers, recall that the bosons include the massless photon of spin 1, occuring in two spin states since, according to relativistic invariance, a massless particle of spin J can have only two substates, $J_z = \pm J$; the massless gluon also of spin 1, two spin substates and eight substates of colour; the massive bosons W^+, W^-, and Z^0, again of spin 1 but since they are massive, contribute $2J + 1 = 3$ spin substates each; and finally the Higgs scalar spin 0 boson of the electroweak theory described in Chapter 3, bringing the total to 28. The fermions include the quarks, occuring in six flavour states, three colour, and two spin substates, plus their antiparticles, totalling 72 states altogether; the charged leptons in three flavour and two spin substates, plus their antiparticles, that is a total of 12 states; and finally the neutral leptons (neutrinos) in three flavours but only one spin substate each. Including antiparticles the neutrinos contribute six degrees of freedom, making 90 fermion and antifermion states in total. Of course, in this tally, we have counted only the known fundamental particles. If supersymmetry is valid, for example, the number of states will be approximately doubled.

As the expansion proceeded and the temperature fell, the most massive particles, such as the top quark and the W and Z bosons would have been rapidly lost by decay (in less than 10^{-23} s) and not replenished once $kT \ll Mc^2$

where M is the particle mass. After kT fell below the strong QCD scale parameter ~ 0.3 GeV, the remaining quarks, antiquarks, and gluons would no longer exist as separate components of a plasma but as quark bound states, forming the lighter hadrons such as pions and nucleons . However, all hadrons except protons and neutrons would be too short-lived to exist beyond the first few nanoseconds. Similarly, the charged muon and tauon leptons would decay within the first microsecond or so. Once kT had fallen below about 100 MeV, that is, after the first few milliseconds, most of the nucleons and antinucleons would also have annihilated to radiation, as discussed in Section 2.11. This would leave, apart from the photons, the electrons e^- and the ν_e, ν_μ, and ν_τ neutrinos, plus their antiparticles, giving in (2.46) $\sum g_f = 4 + 2 + 2 + 2$ (recalling two spin states each for electrons and positrons, but only one for the neutrinos or antineutrinos). With $g_b = 2$ for the photon this results in a value $g^* = \frac{43}{4}$. The effect is to multiply the expression for kT on the extreme right-hand side of (2.38) by a factor $(g^*/2)^{-1/4}$, which in this case has the value 0.66.

From the formulae of the last two sections we may also express the Hubble parameter $H(t)$ in (2.2) in terms of the temperature T in the radiation-dominated era of the early universe. Since in this era, $\rho \propto R^{-4}$, it follows from (2.36) that

$$H(t) = \frac{\dot{R}}{R} = -\frac{\dot{\rho}}{4\rho} = \frac{1}{2t}$$

and from (2.38)

$$
\begin{aligned}
H(T) &= \left[\frac{4g^*\pi^3 G}{45\hbar^3 c^5}\right]^{1/2} (kT)^2 \\
&= \frac{(4\pi^3 g^*/45)^{1/2}}{M_{PL}\hbar c^2} \times (kT)^2 \\
&= \frac{1.66 g^{*1/2}(kT)^2}{(M_{PL}\hbar c^2)}
\end{aligned}
\tag{2.47}
$$

where in the second line the Newtonian constant is expressed in terms of the Planck mass, that is, $G = \hbar c / M_{PL}^2$ (see Table 1.5).

Example 2.5   *Estimate the time required for the universe to increase its size by 10% during the radiation era, for values of $kT = 100$ MeV and $g^* = 20$.*

Since $H = (1/R)\,dR/dt$, the time required (assuming that H is constant over a short period) is found by integration to be $t = (\ln 1.1)/H$. From (2.47), with $M_{PL} = 1.22 \times 10^{19}$ GeV/c^2 and kT expressed in GeV,

$$H(T) = 2.07 \times 10^5 g^{*1/2}(kT)^2 \text{ s}^{-1}$$

Substituting for kT and g^* we find $H = 9.25 \times 10^3$ s^{-1} and $t = 10.3$ μs.

2.9   Radiation and matter eras

From the above formulae, for example (2.38), it is apparent that at early times in the universe when the temperature and particle densities were extremely high, the various types of elementary fermions and bosons would have been in thermal equilibrium and present in comparable numbers, provided $kT \gg Mc^2$, so that even the most massive particles could have been created. The condition for thermal equilibrium to apply is that the time between collisions should be much shorter than the age t of the universe. Otherwise, there is just not enough time to have had enough collisions to set up equilibrium ratios. The collision rate of a particle will be $W = \langle Nv\sigma \rangle$ where N is the density of other particles with which it collides, σ is the cross section per collision, and an average is taken over the distribution in relative velocity v. So one requires that $W \gg t^{-1}$.

Eventually, particles may fall out of equilibrium as the universe expands and the temperature decreases. For example, the cross section may depend on energy and become so small at low temperatures that W falls below t^{-1} and those particles, therefore, decouple from the rest. We say that they are 'frozen out'. As indicated in Chapter 4, this is the case for the weak reaction

$$e^+ + e^- \leftrightarrow \nu + \bar{\nu} \tag{2.48}$$

for $kT < 3\,\mathrm{MeV}$, that is, when $t > 10^{-2}\,\mathrm{s}$ (see also Problem 4.2). So, after that time, the neutrino fireball is decoupled from matter and expands independently.

Particles may also decouple if they are massive, even if the production cross section is large. For example, this will happen for the reversible process

$$\gamma + \gamma \leftrightarrow p + \bar{p} \tag{2.49}$$

when $kT \ll M_p c^2$. Nucleons and antinucleons that annihilate are then no longer replaced by fresh production. There will of course be a tiny proportion of photons that are in the extreme tail of the Planck spectrum, and thus kinematically able to make a nucleon–antinucleon pair, but the rate of this production will be insufficient to sustain a large enough value of W and thermal equilibrium will be lost.

For some 10^5 years after the Big Bang, matter, consisting largely of protons, electrons, and hydrogen atoms, was in equilibrium with the photons, via the reversible reaction

$$e^- + p \leftrightarrow H + \gamma \tag{2.50}$$

where, in the forward process, a hydrogen atom is formed in the ground state or in an excited state, and in the reverse process, a hydrogen atom is ionized by the radiation. At thermal equilibrium, the ratio of ionized to un-ionized hydrogen is a constant depending on the temperature T. We are interested in what happens as the temperature falls and $kT < I$, the ionization potential of hydrogen ($I = 13.6\,\mathrm{eV}$). Clearly, the rate for the forward reaction is proportional to the product of the densities N_e and N_p of the electrons and protons, while the back reaction rate will be proportional to the number N_H of hydrogen atoms per unit volume. (The number of photons is enormous by comparison—see (2.72) below—so their number is unaffected by the reaction.) Hence,

$$\frac{N_e N_p}{N_H} = f(T) \tag{2.51}$$

The number of bound states available to an electron will be $g_e g_n$ where $g_e = 2$ is the number of spin substates and $g_n = n^2$ is the number of bound states in a hydrogen atom with principal quantum number n and energy E_n. The probability that an electron is bound in a state of energy E_n is found by multiplying by the Boltzmann factor, so that it is $g_e g_n \exp(-E_n/kT)$. Summed over ground ($n = 1$) and excited states ($n > 1$) of the H atom, the probability of finding an electron in a **bound** state is, therefore,

$$P_{bound} = g_e \sum g_n \exp\left(\frac{-E_n}{kT}\right)$$

If we write $-E_n = -E_1 - (E_n - E_1)$ where $-E_1 = I$, the ionization potential, then,

$$P_{bound} = g_e Q \exp\left(\frac{I}{kT}\right) \tag{2.52}$$

where

$$Q = \sum n^2 \exp\left[-\frac{(E_n - E_1)}{kT}\right]$$

Since $(E_n - E_1)/kT \gg 1$ for all values of $n > 1$, the excited states make little contribution and $Q \approx 1$.

The probability that our electron is in an **unbound** state of kinetic energy $E \rightarrow E + dE$ and momentum $p \rightarrow p + dp$ is

$$P_{unbound} = g_e \frac{4\pi p^2 \, dp}{h^3} \exp\left(\frac{-E}{kT}\right)$$

where $4\pi p^2 \, dp/h^3$ is the number of quantum states per unit volume in the interval $p \rightarrow p + dp$, and $\exp(-E/kT)$ is the probability that any such state will be occupied by an electron of kinetic energy $E = p^2/2m$, where m is the electron mass. Here, we have assumed the electron is non-relativistic and that $E \gg kT$, so that the FD occupation probability in (2.44) reverts to the classical Boltzmann factor $\exp(-E/kT)$. The probability that the electron will be unbound with **any** energy $E > 0$ is found by integrating over E, with the result that

$$P_{unbound} = g_e \left(\frac{2\pi mkT}{h^2}\right)^{3/2} \tag{2.53}$$

(see Problem 2.3). Comparing the relative probabilities in (2.52) and (2.53), and using (2.51), the ratio of unbound (ionized) to bound (un-ionized) states is

$$\frac{N_p}{N_H} = \frac{N_{H+}}{N_H} = \frac{1}{N_e}\left(\frac{2\pi mkT}{h^2}\right)^{3/2} \exp\left(\frac{-I}{kT}\right) \tag{2.54}$$

The total number of baryons per unit volume is $N_B = N_p + N_H$, so that if x represents the fraction of hydrogen atoms which are ionized, then $N_e = N_p = xN_B$ and $N_H = (1 - x)N_B$, whence

$$\frac{x^2}{1 - x} = \frac{1}{N_B}\left(\frac{2\pi mkT}{h^2}\right)^{3/2} \exp\left(\frac{-I}{kT}\right) \tag{2.55}$$

which is called the **Saha equation**. Inserting some typical numbers, the reader can easily demonstrate from this formula that for kT between 0.35 (4000 K) and 0.25 eV (3000 K), x drops catastrophically, and so radiation and matter must decouple at around this temperature. A figure of $kT = 0.30$ eV is in fact a good guess for the decoupling temperature. Comparing this with the value $kT_0 = 2.35 \times 10^{-4}$ eV ($T_0 = 2.73$ K) of the microwave radiation at the present day, the redshift at the time of decoupling will have the value

$$(1 + z)_{\text{dec}} = \frac{R_0}{R_{\text{dec}}} = \frac{kT_{\text{dec}}}{kT_0} \approx 1250 \qquad (2.56)$$

Corrections to this result are needed, since a photon emitted upon recombination of one atom can almost immediately ionize another atom, and the level of ionization could therefore be underestimated. In fact, it turns out that slower, two-step processes involving more than one photon are of importance and the value of the redshift deduced from these more detailed calculations is found to be

$$(1 + z)_{\text{dec}} = 1100 \qquad (2.57)$$

After decoupling, matter becomes transparent to the cosmic microwave background radiation, and the formation of atoms and molecules can begin in earnest. Equally important, some vital properties of this radiation, including the very small but very important spatial variations in temperature that can be observed today, are therefore what they were at the 'epoch of the last scattering'. These measurements of temperature fluctuations in very small angular ranges (of order 1°), described in Section 5.9, give rather direct and accurate information on the parameters of the early universe as summarized in Section 2.5.

The decoupling time for matter and radiation estimated from (2.38) is $t_{\text{dec}} = 10^{13}$ s or 3×10^5 yr. It turns out (see Example 2.6) that the energy density of matter, varying as T^{-3}, became equal to that of the radiation, varying as T^{-4}, at a redshift not very different from that for decoupling. Thereafter, matter started to dominate over radiation as far as energy density is concerned, and has thus done so for over 99.9% of the age of the universe. Figure 2.8 shows the variation of temperature with time through the radiation and matter eras.

Example 2.6   *Estimate the value of the redshift and the time t after the Big Bang at which the density of electromagnetic radiation fell below that of baryonic matter. Also, find the value of the redshift below which the energy density of non-relativistic matter started to exceed that of relativistic particles, that is, the universe became 'matter dominated'.*

Referring to equations (2.28) and (2.43), where the baryon and radiation densities are given as $\Omega_b = 0.05$ and $\Omega_r = 5 \times 10^{-5}$, the time t and redshift z at which the energy density of radiation equalled that of baryonic matter is given by the relation

$$\frac{\Omega_b(t)}{\Omega_r(t)} = \frac{\Omega_b(0)}{\Omega_r(0)} \frac{R(t)}{R(0)} = \frac{\Omega_b(0)}{\Omega_r(0)(1 + z)}$$

so that the equality $\Omega_r(t) = \Omega_b(t)$ occurred when $(1+z) = 0.05/5 \times 10^{-5} = 1000$. Since the universe has been baryon dominated since that epoch, from Table 2.2, we find that $(t/t_0)^{2/3} = 1/(1 + z)$, giving $t \sim 4 \times 10^5$ yr.

In the second part of the example, we take the **total** matter density $\Omega_m(0) = 0.30$ from (2.29). In addition to photons, neutrinos also constitute relativistic remnants of the Big Bang, so that, as shown in Section 4.6, the total energy density of relativistic particles is 1.52 times that of the photons alone, or $\Omega_{\text{rel}}(0) = 7.65 \times 10^{-5}$. Thus, non-relativistic matter of

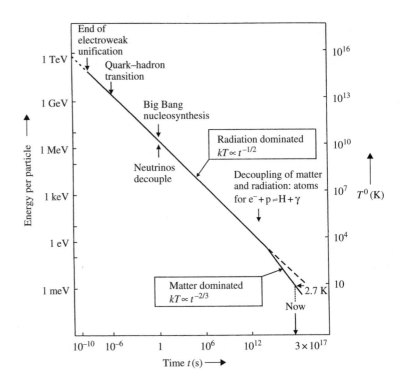

Fig. 2.8 Evolution of the temperature with time in the Big Bang model, with the various eras indicated (see also Fig. 5.2).

all types begins to dominate relativistic particles when $(1 + z)$ falls below $\Omega_m(0)/\Omega_{rel}(0) \sim 4000$, or 6000 if we count only the photons.

2.10   Primordial nucleosynthesis

In this brief initial description of the early universe, we next discuss the synthesis of the nuclei of the light elements—^{4}He, ^{2}H, ^{3}He, and ^{7}Li—since the agreement between the predicted and measured abundances provides very strong support for the Big Bang hypothesis.

As discussed in Section 2.8, once the universe had cooled to a temperature $kT < 100$ MeV, or after a time $t > 10^{-4}$ s, essentially all the hadrons, with the sole exception of neutrons and protons and their antiparticles, would have disappeared by decay. The nucleons and antinucleons would have been present in equal numbers and have nearly, but not quite completely, annihilated to radiation. As described in the next section, a tiny residue of about one billionth of the original numbers of protons and neutrons must have survived to form the constituents of the material universe we inhabit today. The relative numbers of protons and neutrons would have been determined by the weak reactions

$$\nu_e + n \leftrightarrow e^- + p \tag{2.58}$$

$$\bar{\nu}_e + p \leftrightarrow e^+ + n \tag{2.59}$$

$$n \rightarrow p + e^- + \bar{\nu}_e \tag{2.60}$$

Since at the temperatures considered, the nucleons are non-relativistic, then just as in the analysis of the previous section, the equilibrium ratio of neutrons to protons will be governed by the ratio of the Boltzmann factors, so that

$$\frac{N_n}{N_p} = \exp\left(\frac{-Q}{kT}\right); \quad Q = (M_n - M_p)c^2 = 1.293 \text{ MeV} \tag{2.61}$$

The rate or width W for the first two reactions (2.58) and (2.59) must vary as T^5 purely on dimensional grounds. The Fermi constant G_F from (1.9) or Table 1.5 has dimensions E^{-2}, so the cross section (dimension E^{-2}) will vary as $G_F^2 T^2$ and the incident flux as T^3, so the width $W = \sigma\phi$ gets a T^5 factor. On the other hand, the expansion rate of the radiation-dominated universe is $H \sim g^{*1/2}T^2$ from (2.47). Hence, $W/H \sim T^3/(g^*)^{1/2}$ and as the universe expands and the temperature falls, the above reactions will go out of equilibrium when $W/H < 1$. (In fact, as described in Chapter 4, at $kT \sim 3$ MeV neutrinos have already gone out of equilibrium with electrons in the process $e^+ + e^- \leftrightarrow \nu + \bar{\nu}$, since this has an even smaller cross section than (2.58) because of the smaller target mass.) The freeze-out temperature for the reactions (2.58) and (2.59) is calculated to be $kT = 0.80$ MeV, so that the initial value of the neutron–proton ratio will be (see Problem 2.11):

$$\frac{N_n(0)}{N_p(0)} = \exp\left(\frac{-Q}{kT}\right) = 0.20 \tag{2.62}$$

At later times, neutrons will disappear by decay in reaction (2.60). At time t after decoupling, there will then be $N_n(0)\exp(-t/\tau)$ neutrons and $[N_p(0) + N_n(0)\{1 - \exp(-t/\tau)\}]$ protons, with a neutron/proton ratio of

$$\frac{N_n(t)}{N_p(t)} = \frac{0.20\exp(-t/\tau)}{1.20 - 0.20\exp(-t/\tau)} \tag{2.63}$$

where $\tau = 887 \pm 2$ s is the free neutron lifetime. If nothing else were to happen at this juncture, the neutrons would simply die away by decay and the early universe would consist exclusively of protons and electrons. However, as soon as neutrons appear, nucleosynthesis can begin with the formation of deuterons:

$$n + p \leftrightarrow {}^2H + \gamma + Q \tag{2.64}$$

where the binding energy $Q = 2.22$ MeV. This is an electromagnetic process with a cross section of 0.1 mb, very much larger than those of the weak processes (2.58)–(2.60), and consequently it stays in thermal equilibrium for very much longer. Because of the billionfold preponderance of photons over nucleons, the deuterons are not frozen out until the temperature falls to about $Q/40$, that is, $kT = 0.05$ MeV. As soon as the reverse process of photo-disintegration of the deuteron ceases, competing reactions leading to helium production take over:

$${}^2H + n \rightarrow {}^3H + \gamma$$

$${}^3H + p \rightarrow {}^4He + \gamma$$

$${}^2H + p \rightarrow {}^3He + \gamma$$

$${}^3He + n \rightarrow {}^4He + \gamma$$

For $kT = 0.05$ MeV, corresponding to an expansion time from (2.38) of $t \sim 300$ s for $N_\nu = 3$, the neutron–proton ratio from (2.63) becomes

$$r = \frac{N_n}{N_p} = 0.135$$

The helium mass fraction, with the mass of the helium nucleus set equal to four times that of the proton is then given by

$$Y = \frac{4N_{He}}{(4N_{He} + N_H)} = \frac{2r}{(1+r)} = 0.24 \tag{2.65}$$

The mass fraction Y has been measured in a variety of celestial sites, including stellar atmospheres, planetary nebulae, globular clusters, gas clouds etc., with values in the region 0.23–0.24. Problems in evaluating both the predicted and measured values mean that agreement between theory and observation is still uncertain at the 10% level. Nevertheless, this level of agreement was an early and very important success of the Big Bang model. It should be pointed out here that the observed helium mass fraction is far greater than that which could have been produced in hydrogen burning in main sequence stars; their contribution adds only 0.01 to the ratio Y (see Problem 2.5).

An important feature of nucleosynthesis in the Big Bang scenario is that it accounts not only for ^{4}He but also for the light elements ^{2}H, ^{3}He, and ^{7}Li that occur in small but significant amounts, far more in fact than would have survived if they had been produced in thermonuclear interactions in stellar interiors. Figure 2.9 shows the abundances expected from primordial nucleosynthesis, calculated on the basis of the cross sections involved, and plotted in terms of the (present day) baryon density. The results are consistent with a value of the baryon density in the range

$$\rho_{baryon} = (3.0 \pm 1.5) \times 10^{-28} \text{ kg m}^{-3} \tag{2.66}$$

corresponding to a number density of baryons $N_B = 0.18 \pm 0.09$ m^{-3}. Comparing with the number density of microwave photons (2.41), this yields for the baryon/photon ratio

$$\frac{N_B}{N_\gamma} \approx (4 \pm 2) \times 10^{-10} \tag{2.67}$$

So, while in the first nanoseconds of the Big Bang, the relative numbers of baryons, antibaryons, and photons would have been comparable (differing only in spin multiplicity factors), most of the nucleons and antinucleons must later have disappeared by mutual annihilation, leaving a tiny—one part per billion—excess of nucleons as the matter of the everyday world, as discussed in the following section.

After the formation of ^{4}He, there is a bottleneck to further nucleosynthesis, since there are no stable nuclei with $A = 5, 6$ or 8. Formation of ^{12}C via the triple-alpha process, for example, is not possible because of the Coulomb barrier suppression, and this has to await the onset of helium burning in stars at high temperatures. Production of heavier elements in stellar fusion reactions at high temperature is discussed in Chapter 7.

It is of interest to remark here that the helium mass fraction depends on the number of neutrino flavours N_ν since the expansion timescale described

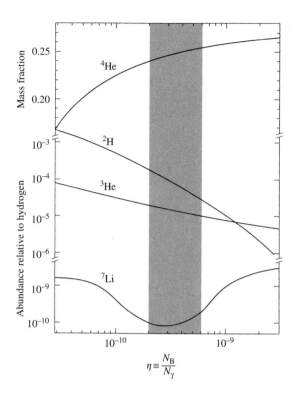

Fig. 2.9 The primordial abundances expected in Big Bang nucleosynthesis of the light elements ^{2}H, ^{3}He, and ^{7}Li, and the mass abundance of ^{4}He, in all cases relative to hydrogen and plotted as a function of the ratio of baryons to photons. The observed values of the abundances are as follows: ^{2}H/H $\sim$ 3×10^{-5}; ^{3}He/H $\sim$ 2×10^{-5}; ^{7}Li/H $\sim$ 1×10^{-10}. The weight abundance of ^{4}He is 0.23–0.24. These results point to a unique value of the baryon to photon ratio given in (2.67) (after Schramm and Turner 1998).

by (2.46) and (2.47) varies inversely as $g^{*1/2}$, the square root of the number of fundamental bosons and fermions. Thus increasing N_ν increases g^* and hence the freeze-out temperature T_F determined by the condition $W/H \sim 1$. This leads, through (2.61) to a higher initial neutron/proton ratio and a higher helium mass fraction. Originally, before the LEP experiments demonstrated that $N_\nu = 3$ (see Fig. 1.12), this argument was used to set a limit on the number of flavours, whereas now it is used to set a better value on the helium mass fraction and the baryon to photon ratio (see also Problem 2.9).

2.11   Baryogenesis and the matter–antimatter asymmetry of the universe

One of the most striking, and as yet unexplained, features of our universe is the absence of antimatter, although the conservation rules described in Chapter 3 seem to indicate an almost exact symmetry between matter and antimatter. We know there is a paucity of antimatter in our own galaxy, because the primary cosmic ray nuclei (see Fig. 6.3), which have been brewed up in stellar reactions over billions of years, and have typically been circulating in the galactic magnetic fields for several million years, are invariably found to be nuclei rather than antinuclei. Furthermore, on a wider scale, there is absolutely no evidence for the intense γ-ray and X-ray emission that would follow annihilation of matter in distant galaxies with clouds of antimatter. Very low fluxes of positrons and antiprotons do exist in the cosmic rays incident on the Earth's atmosphere, but these can be accounted for in terms of the processes of electron–positron or proton–antiproton pair creation resulting from collisions of high energy γ-rays or nuclei with interstellar matter.

In the early stages of the Big Bang, when kT was large compared with the hadron masses, it is expected that many types of hadrons, including protons and neutrons and their antiparticles, would have been in thermal equilibrium with radiation, being created and annihilated in reversible reactions such as

$$p + \bar{p} \leftrightarrow \gamma + \gamma \tag{2.68}$$

Assuming a net initial baryon number of zero, the number density of nucleons and antinucleons at temperature T would be given by (2.44) with $g_f = 2$:

$$N_B = N_{\bar{B}} = \frac{(kT)^3}{\pi^2(\hbar c)^3} \int \frac{(pc/kT)^2 d(pc/kT)}{\exp(E/kT) + 1} \tag{2.69}$$

where p is the three- momentum and E is the total energy given by $E^2 = p^2c^2 + m^2c^4$. This may be compared with the number of photons in (2.41):

$$N_\gamma = \frac{2.404\,(kT)^3}{\pi^2(\hbar c)^3} \tag{2.70}$$

The baryons, antibaryons, and photons are in thermal equilibrium and will stay in equilibrium as long as the rate for the back reaction in (2.68) exceeds the universal expansion rate. Eventually, as the expansion proceeds and the temperature falls, the tiny part of the high energy tail of the photon distribution, with photons above threshold for nucleon–antinucleon pair creation, will become so small that the rate of creation of fresh pairs falls below the expansion rate. Photons cannot produce enough nucleon pairs, nor can nucleons find enough antinucleons with which to annihilate, and the residue of baryons and antibaryons is 'frozen out'. The critical temperature at which this occurs depends on the baryon density (2.69), on the nucleon–antinucleon annihilation cross section and its dependence on velocity, and on the expansion rate. Given these parameters one can solve numerically for the temperature and nucleon density at freeze-out. Here, we just quote the predicted result:

$$kT\text{(critical)} \approx 20\text{ MeV}; \quad \frac{N_B}{N_\gamma} = \frac{N_{\bar{B}}}{N_\gamma} \sim 10^{-18} \tag{2.71}$$

Subsequent to this freeze-out stage, there would be no further nucleon–antinucleon annihilation or creation and the above ratios should hold today. In contrast, the observed values are, using (2.67)

$$\frac{N_B}{N_\gamma} \approx 10^{-9} \quad \frac{N_{\bar{B}}}{N_B} < 10^{-4} \tag{2.72}$$

So, the Big Bang hypothesis gets the baryon–photon ratio wrong by a factor of 10^9 and the antibaryon/baryon ratio wrong by at least a factor of 10^4. Of course, it is possible to avoid this problem by arbitrarily assigning an initial baryon number to the universe, but this would be quite large ($N_B \sim 10^{79}$!) and arbitrary, and it seems more sensible to try to understand the observed values in terms of (hopefully) known physics. This takes us back to a seminal paper in 1967 by Sakharov, who proposed a possible way out.

2.11.1 The Sakharov criteria

Sakharov pointed out the conditions necessary to achieve a baryon–antibaryon asymmetry. Assuming a baryon number $B = 0$ initially, a baryon number asymmetry could obviously only develop as a result of baryon number violating reactions, but requires two other conditions, regarding equilibrium and the charge and space parities C and P. Thus one needs

- B violating interactions;
- non-equilibrium situation;
- CP and C violation.

The first requirement is obvious and its possibility will be discussed in connection with the GUT models and the search for proton decay in Chapter 3. Unfortunately, there is as yet no direct laboratory evidence that baryon number is violated, so we just retain it as an assumption. The second condition follows from the fact that, in thermal equilibrium, the particle density depends only on the particle mass and the temperature. Since particles and antiparticles have identical masses by the CPT theorem (see Section 3.13), no asymmetry could develop. Put another way, at equilibrium, any reaction which destroys baryon number will be exactly counterbalanced by the inverse reaction which creates it. Third, as pointed out in Chapter 3, C and CP violation are necessary if antimatter is to be distinguished unambiguously from matter on a cosmic scale.

The precise conditions required to generate a baryon–antibaryon asymmetry of the observed magnitude are unknown. Such processes are possible in principle via phase transitions in the Standard Model described in Chapter 3, but the baryon number violations associated with grand unified theories (GUTs) are considered more likely sources of the asymmetry. In the SU(5) grand unified model, also discussed in Chapter 3, quarks and leptons are incorporated in the same multiplets, and quark–lepton transitions are therefore possible. For example, as shown in Fig. 3.12, a proton can thus transform into a pion and a positron via X-boson exchange. It is noteworthy that, in this transition, both the baryon number B and the lepton number L have decreased by one unit, so that the difference $(B - L)$ is conserved. This is obviously necessary if one starts out with charge conservation and an electrically neutral universe, so that, irrespective of baryon number violating interactions, the negative charge of the electrons is always matched by the positive charge of the protons.

The bosons X, Y, and their antiparticles of the GUT symmetry are supposedly created in the Big Bang on a 10^{-40} s timescale, and are expected to decay out of thermal equilibrium. The requirements are for two decay channels, say 1 and 2, of different baryon numbers. Suppose that x and $(1 - x)$ are the branching ratios for the decay of X to modes with baryon numbers B_1 and B_2, respectively. For the antiparticle $\bar{\text{X}}$, let the ratios be $\bar{x}$ and $(1 - \bar{x})$, with baryon numbers $-B_1$ and $-B_2$. Since the numbers of X and $\bar{\text{X}}$ particles are equal, by the CPT theorem, the net baryon asymmetry per X$\bar{\text{X}}$ pair will be

$$A = xB_1 - \bar{x}B_1 + (1 - x)B_2 - (1 - \bar{x})B_2 = (x - \bar{x})(B_1 - B_2)$$

B violation ensures that $B_1 \neq B_2$, and CP violation that $\bar{x} \neq x$, so that the asymmetry will be non-zero. It is to be noted that C violation alone, with CP conservation, would give an X decay rate at angle θ equal to the $\bar{\text{X}}$ decay rate at $(\pi - \theta)$, and therefore the same overall rate when integrated over angle. *CP* violation is necessary to ensure different partial decay rates for particles and antiparticles in one particular channel (1 for example).

The required baryon asymmetry must clearly be of the order of the baryon to photon density ratio (2.72). Apart from the fact that the source and degree of *CP* violation observed experimentally, for example, in neutral kaon decay or neutral B-meson decay, is probably not relevant to that connected with the universal baryon asymmetry, it is not clear how processes such as inflation (discussed in Chapter 5) can affect the baryon–antibaryon ratio. In the light of the conservation laws established in experimental particle physics, the baryon asymmetry is one of the most striking and puzzling features of our universe, and one for which we have, at present, no convincing explanation.

2.12 Summary

- The 'Standard Model' of the universe is based on Einstein's general relativity and the cosmological principle, namely that on large scales, the universe is isotropic and homogeneous, and the Big Bang expansion appears the same to all observers.
- Hubble's Law describes the linear relation between the redshift of the light from distant galaxies and the universal expansion parameter R, with $R_0/R_e = \lambda_{observed}/\lambda_{emitted} = (1 + z)$.
- The Friedmann equation relates the Hubble expansion parameter $H = \dot{R}/R$ to the total energy density of the universe and the curvature of space (the parameter K).
- The total energy density is the sum of contributions from matter, radiation and vacuum energy. The vacuum term plays the role of Einstein's cosmological constant.
- The age of the universe is about 14 ± 2 Gyr. Independent estimates, from radioactive isotope ratios, from stellar population analysis and from the most recently measured cosmological parameters ($\Omega_m = 0.30$, $\Omega_v = 0.70$) are in agreement.
- The all-pervading and isotropic microwave background radiation has a black-body spectrum of $T = 2.725$ K and is the cooled remnant of the hot Big Bang.
- The baryonic matter and radiation energy densities were equal at redshift $z \sim 10^3$, when the universe was $\sim$300 000 years old. At about that time, radiation and matter decoupled and atoms (mainly hydrogen) started to form.
- The observed abundances of the light elements deuterium (^{2}H), helium (^{3}He and ^{4}He) and lithium (^{7}Li) can be understood by their creation in nucleosynthesis at temperatures $kT \sim 0.1$ MeV in the first minutes following the Big Bang. Together with the microwave background radiation and the redshift, the light element abundances provide very strong support for the Big Bang hypothesis. The baryon density from the synthesis of the light elements in the first minutes of the universe accounts for only a small part of the total matter density: most of the matter is dark matter.
- The observed strong asymmetry between matter and antimatter has to be ascribed to special baryon number violating and *CP* violating interactions

operating at a very early stage of the Big Bang when the temperature was very high. However, the processes responsible are not presently understood.

Problems

For all constants required refer to Appendix A. The more challenging problems are marked by an asterisk.

(2.1) Assuming that the age of the universe is 14 Gyr and that the total density is equal to the critical density $\rho_c = 9 \times 10^{-27}$ kg m^{-3}, estimate the gravitational binding energy and compare it with the total mass energy of the universe.

(2.2) Calculate the (non-relativistic) escape velocity v of a particle from the surface of a sphere of radius r and uniform mass density ρ. Show that if one assumes Hubble's Law $v = Hr$, the particle will escape provided that $\rho < 3H^2/8\pi G$.

(2.3) Free non-relativistic fermions of rest-mass m in thermal equilibrium at temperature T are described by the FD distribution (2.44). If $kT \ll mc^2$, show that the number density of particles is $g(2\pi mkT/h^2)^{3/2} \exp(-mc^2/kT)$, where g is the number of spin substates.

(2.4) It is estimated that dark vacuum energy today contributes approximately 0.70 to the closure parameter Ω. At what value of the redshift parameter and at what age of the universe would the vacuum energy have been less than 10^{-4} of the energy density of radiation?

(2.5) The Sun has a measured luminosity of 3.9×10^{26} W. It generates its energy from the conversion of hydrogen to helium in thermonuclear fusion reactions, an energy of 26 MeV being liberated for each helium nucleus formed. If the Sun's output has been constant at the above value for 5 Gyr, what is the mass fraction of helium in the Sun?

(2.6) The total amount of energy incident on the Earth's atmosphere from the Sun is 0.135 J m^{-2} s^{-1} (the solar constant). The Earth–Sun distance is $D = 1.5 \times 10^{11}$ m and the solar radius is $R_\odot = 7 \times 10^8$ m. Assuming the Sun is a black body, calculate its surface temperature.

(2.7) It was once proposed that the expansion of the universe could be attributed to an electrostatic repulsion between atoms, on the grounds that the arithmetic values of the electric charges of the electron and the proton might have a very small fractional difference ε. What value of ε would have been necessary? (Note: This hypothesis was quickly disproved by experiment, showing that ε was less than 1% of the value required.)

*(2.8) Find expressions for the dependence of the time t on the density ρ for an expanding 'flat' universe ($K = 0$) dominated (a) by radiation, and (b) by non-relativistic conserved matter. Show that, in either case, t is of the same order of magnitude as the time for the gravitational free-fall collapse of a body of density ρ from rest.

*(2.9) Calculate the expected mass ratio of primordial helium to hydrogen, as in (2.65), but for the case of different numbers of neutrino flavours $N_\nu = 3, 4, 5, 6 \ldots$ Show that each additional flavour will increase the expected ratio by about 5%. Calculate also the expected mass ratio for $N_\nu = 3$ if the neutron–proton mass difference were 1.40 MeV/c^2 instead of 1.29 MeV/c^2, but the free neutron lifetime was unaffected.

*(2.10) In a flat matter-dominated universe ($K = 0$) of age t_0, light from a certain galaxy exhibits a redshift $z = 0.95$. How long has it taken the light signal to reach us from this galaxy? (For a hint, consult equation (5.1).)

*(2.11) In discussing the neutron/proton equilibrium ratio (2.62) it was stated that the rate or width W for the reaction $\nu_e + n \rightarrow e^- + p$ varied as T^5. Verify this directly, referring back to Section (1.8) to compute the cross section for the above reaction as a function of T and using the relevant flux density to compute W from (1.14). Assume that all particles have kinetic energies kT, with $m_e c^2 \ll kT \ll M_p c^2$, that is, treat the nucleons as non-relativistic and essentially stationary, and the leptons as extreme-relativistic. Comparing with the expansion rate (2.47), estimate the temperature at which the neutrons and protons 'freeze out' of equilibrium.

(2.12) What is the minimum value of Ω_v that will result in an accelerated expansion for the case of a flat universe? Neglect the contribution to the energy density from radiation.

(2.13) Prove the statement above equation (2.9), that the gravitational field anywhere inside a spherical shell of uniform density is zero; and that the field outside a spherical distribution of total mass M is equal to that of a point mass M placed at the centre of the sphere. (This is called Newton's Law of Spheres in classical mechanics. In general relativity it is known as Birkhoff's Theorem.)

*(2.14) Derive a formula for the age of an open universe ($\Omega < 1$) with zero cosmological constant and negligible radiation density. Use equation (2.33) and Example 2.3 as guides. Calculate the age for $\Omega = 0.35$. (Hint: Make the substitution $\tan^2 \theta = (1 + z)\Omega/(1 - \Omega)$.)

Conservation rules and symmetries

3.1 Preamble

One of the most important concepts in physics is that of the symmetry or invariance of a system under a particular operation. For example, a snowflake is invariant under a 60° rotation in the plane of the flake, and this tells us something about the physics of the molecular bonding in water. In fact, conservation rules and symmetries have been called the backbone of particle physics. On a broader scale, the conservation of linear momentum in mechanics follows if the energy of the system is invariant under translations in space. For, if there is no change in energy under such a translation, there can be no external forces on the system and the rate of change of momentum must be zero. This last example can be formalized by the Euler–Lagrange equation of classical mechanics:

$$\frac{\mathrm{d}}{\mathrm{d}t}\left(\frac{\partial L}{\partial \dot{q}_i}\right) - \frac{\partial L}{\partial q_i} = 0 \qquad (3.1)$$

where $L = \sum_i (T - V)$ is the difference of kinetic and potential energies of all i particles, q_i is a generalised coordinate, for example that of position, so that $\dot{q}_i = \mathrm{d}q_i/\mathrm{d}t$ is the velocity. Then if L is independent of q_i, $\partial L/\partial q_i = 0$ and the momentum $p_i = \partial L/\partial \dot{q}_i$ is constant. Here, a global symmetry—invariance of L under a space translation—has led to a conservation law.

In relativistic quantum mechanics, the Lagrangian function L is a field **energy density** rather than a sum over the energies of discrete particles, and is furthermore a function of both space and time. The global invariance of L under space–time translations leads to a **conserved current** of four-momentum, which is an example of a more general theorem called Noether's theorem, discussed in Section 3.8. The conservation of the fourth (time) component of this current corresponds to conservation of energy, and of the three space components to conservation of momentum.

The invariance (or non-invariance) of a physical system may occur for **continuous** transformations; for example, a rotation in a phase angle or a translation in space; or it can be a **discrete** transformation, such as the inversion of a spatial or time coordinate, or charge conjugation. For continuous transformations, the associated conservation laws and quantum numbers are additive (the total conserved energy of a system is equal to the sum of the energies of its parts), while for discrete transformations they are multiplicative (e.g. the symmetry under spatial reflection, called the parity, is equal to the product of the parities of the parts of the system).

3.2   Rotations

As an example of a continuous transformation let us consider a spatial rotation through some angle, say ϕ, about the z-axis. The operator of the z component of angular momentum in Cartesian coordinates is defined as

$$J_z = -i\hbar \left(x \frac{\partial}{\partial y} - y \frac{\partial}{\partial x} \right)$$

This operation can also be described by a rotation. Suppose a vector of length r lying in the xy plane makes an angle ϕ with the x-axis. Then, in a rotation through $\delta\phi$, the increments of the Cartesian components are

$$\delta y = r \cos \phi \, \delta\phi = x \, \delta\phi$$

$$\delta x = -r \sin \phi \, \delta\phi = -y \, \delta\phi$$

The effect of this rotation on a function $\psi(x, y, z)$ will be

$$R(\phi, \delta\phi)\psi(x, y, z) = \psi(x + \delta x, y + \delta y, z) = \psi(x, y, z) + \delta x \left(\frac{\partial \psi}{\partial x} \right) + \delta y \left(\frac{\partial \psi}{\partial x} \right)$$

$$= \psi \left[1 + \left(x \frac{\partial}{\partial y} - y \frac{\partial}{\partial x} \right) \delta\phi \right] = \psi \left[1 + \delta\phi \frac{\partial}{\partial \phi} \right]$$

Hence, the J_z operator is given by

$$J_z = -i\hbar \left[x \frac{\partial}{\partial y} - y \frac{\partial}{\partial x} \right] = -i\hbar \frac{\partial}{\partial \phi}$$

A finite rotation is achieved in n infinitesimal steps, that is, $\Delta\phi = n\delta\phi$ where $n \to \infty$, so that

$$R = \lim_{n \to \infty} \left(1 + iJ_z \frac{\delta\phi}{\hbar} \right)^n = \exp \left(iJ_z \frac{\Delta\phi}{\hbar} \right) \tag{3.2}$$

Here J_z is said to be the **generator** of the rotation $\Delta\phi$.

3.3   The parity operation

The inversion of spatial coordinates $(x, y, z) \to (-x, -y, -z)$ is a discrete transformation on the wave amplitude ψ brought about by the parity operator P: $P\psi(\mathbf{r}) = \psi(-\mathbf{r})$. Since in repeating the operation one reverts to the original system, $P^2 = 1$ and the eigenvalues of P must be ± 1. These eigenvalues are referred to as the **parity** of the system ψ. For example, the function $\psi = \cos x$ has $P = +1$ or positive parity since $P\psi = P(\cos x) = \cos(-x) = +\psi$, while if $\psi = \sin x$, $P\psi = -\psi$ so that ψ has negative parity. On the other hand, in the case of a function such as $\psi = (\sin x + \cos x)$, $P\psi = (\cos x - \sin x) \neq \pm\psi$ so this function is not an eigenstate of parity. Parity is a useful concept when dealing with elementary particles since the interactions often have very well-defined properties under the parity operation. This may be contrasted with biological systems, for example, a runner bean or the DNA molecule, which are not eigenstates of parity.

A spherically symmetric potential has the property that $V(-\mathbf{r}) = V(\mathbf{r})$, so that states bound by such a potential—as is usually the case in atoms—can be parity eigenstates. For the hydrogen atom, the wavefunction in terms of the radial coordinate r and the polar and azimuthal angular coordinates θ and ϕ of the electron with respect to the proton is

$$\chi(r,\theta,\phi) = \eta(r)Y_l^m(\theta,\phi)$$

where Y is the spherical harmonic function, with l the orbital angular momentum quantum number and m its z component. Under inversion, $\mathbf{r} \to -\mathbf{r}$, $\theta \to (\pi - \theta)$, while $\phi \to (\pi + \phi)$ with the result that

$$Y_l^m(\pi - \theta, \pi + \phi) = (-1)^l Y_l^m(\theta,\phi)$$

Hence, in this case,

$$P\chi(r,\theta,\phi) = (-1)^l \chi(r,\theta,\phi) \tag{3.3}$$

3.4 Parity conservation and intrinsic parity

In strong and electromagnetic interactions, parity is found to be conserved: the parity in the final state of a reaction is equal to that in the initial state. For example, for an electric dipole (E1) transition in an atom, the change in l is governed by the selection rule $\Delta l = \pm 1$. Thus, from (3.3) the parity of the atomic state must change in such transitions, which are accompanied by the emission of photons of negative parity, so that the parity of the whole system (atom + photon) is conserved. For a (less probable) magnetic dipole (M1) transition, or for an electric quadrupole (E2) transition, the selection rules are $\Delta l = 0$ and 2, respectively, and in either case the radiation is emitted in a positive parity state. In high energy physics, one is generally dealing with pointlike or nearly pointlike interactions and electromagnetic transitions involving small changes in angular momentum ($\Delta J = \pm 1$), in which case photons are emitted with negative parity.

The symmetry of a pair of identical particles under interchange, which was described in Section 1.3, can be extended to include both spatial and spin functions of the particles. If the particles are non-relativistic, the overall wavefunction can be written as a simple product of space and spin functions:

$$\psi = \chi(\text{space})\alpha(\text{spin})$$

Consider two identical fermions, each of spin $s = \frac{1}{2}$, described by a spin function $\alpha(S, S_z)$ where S is the total spin and $S_z = 0$ or ± 1 is its component along the z-(quantization) axis. Using up and down arrows to denote the z components of $s_z = +\frac{1}{2}$ and $-\frac{1}{2}$, we can write down the $(2s + 1)^2 = 4$ possible states as follows:

$$\left.\begin{array}{ll} \alpha(1,+1) = & \uparrow\uparrow \\ \alpha(1,-1) = & \downarrow\downarrow \\ \alpha(1,0) \;\;= (\uparrow\downarrow + \downarrow\uparrow)/\sqrt{2} \end{array}\right\} \quad S = 1, \text{ symmetric}$$

$$\alpha(0,0) = (\uparrow\downarrow - \downarrow\uparrow)/\sqrt{2} \qquad S = 0, \text{ antisymmetric} \tag{3.4}$$

The first three functions are seen to be symmetric under interchange, that is, α does not change sign, while for the fourth one it does. It is seen that the sign of the spin function under interchange is $(-1)^{S+1}$, while that for the space wave-function from (3.3) is $(-1)^L$ where L is the total orbital angular momentum. Hence, the overall sign change of the wavefunction under interchange of both space and spin coordinates of the two particles is

$$\psi \rightarrow (-1)^{L+S+1}\psi \tag{3.5}$$

As an example of the application of this rule, let us consider the determination of the intrinsic parity of the pion. This follows from the existence of the S-state capture of a negative pion in deuterium, with the emission of two neutrons:

$$\pi^- + d \rightarrow n + n \tag{3.6}$$

The deuteron has spin 1, the pion spin 0, so that in the initial state and, therefore, in the final state also, the total angular momentum must be $J = 1$. If the total spin of the neutrons is S and their orbital angular momentum is L, then $\mathbf{J} = \mathbf{L} + \mathbf{S}$. If $J = 1$ this allows $L = 0$, $S = 1$; or $L = 1$, $S = 0$ or 1; or $L = 2$, $S = 1$. Since the neutrons are identical particles it follows that their wavefunction ψ is antisymmetric, so that from (3.5) $L + S$ must be even and $L = S = 1$ is the only possibility. Thus, the neutrons are in a $^3P_1$ state with parity $(-1)^L = -1$. The nucleon parities cancel on the two sides of (3.6), so that the pion must be assigned an **intrinsic parity** $P_\pi = -1$, in order that parity be conserved in this strong interaction.

The assignation of an intrinsic parity to a particle follows if the particle can be created or destroyed **singly** in a parity-conserving interaction, in just the same way that electric charge has been assigned in the same interaction in order to obey charge conservation. Clearly, in the above reaction, the number of nucleons is conserved and so the nucleon parity itself is conventional. It is assigned a parity $P_n = +1$. However, in an interaction it is possible, if the energy is sufficient, to create a nucleon–antinucleon pair, and hence determine its parity by experiment. So while the parity of a nucleon is fixed by convention, the **relative** parity of a nucleon and an antinucleon—or any other fermion–antifermion pair—is not.

In the Dirac theory of fermions, **particles and antiparticles have oppos-ite intrinsic parity.** This prediction was checked in an experiment by Wu and Shaknov, shown in Fig. 3.1, using a ^{64}Cu positron source. Positrons from this source came to rest in the surrounding absorber and formed **positronium**, an 'atomic' bound state of electron and positron, which has energy levels akin to those of the hydrogen atom, but with half the spacing because of the factor 2 in the reduced mass. The ground level of positronium occurs in two closely spaced substates with different mean lifetimes: the spin-triplet ($^3S_1$) decaying to three photons (lifetime 1.4×10^{-7} s), and the spin-singlet state ($^1S_0$) decaying to two photons (lifetime 1.25×10^{-10} s). We consider here the singlet decay:

$$e^+e^- \rightarrow 2\gamma \tag{3.7}$$

The simplest wavefunctions describing the two-photon system, linear in the momentum vector $\mathbf{k}$ and in the polarization vectors (**E**-vectors) $\boldsymbol{\varepsilon}_1$ and $\boldsymbol{\varepsilon}_2$ of the

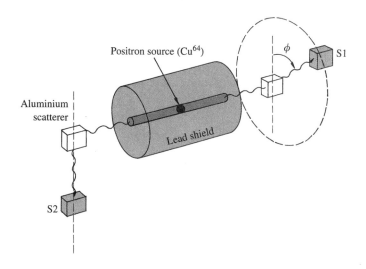

Fig. 3.1 Sketch of the method used by Wu and Shaknov (1950) to measure the relative orientation of the polarization vectors of the two photons emitted in the decay of $^1S_0$ positronium. S1 and S2 are two anthracene counters recording the γ-rays after scattering by aluminium cubes. Their results proved that fermion and antifermion have opposite parity, as predicted by the Dirac theory of the electron.

photons will be

$$\psi_1(2\gamma) = A(\boldsymbol{\varepsilon}_1 \cdot \boldsymbol{\varepsilon}_2) \propto \cos\phi \qquad (3.8a)$$

$$\psi_2(2\gamma) = B(\boldsymbol{\varepsilon}_1 \times \boldsymbol{\varepsilon}_2) \cdot \mathbf{k} \propto \sin\phi \qquad (3.8b)$$

where A and B are constants and ϕ is the angle between the planes of polarization. The first quantity ψ_1 is a scalar and, therefore, even under space inversion ($\phi \to -\phi$), and thus requires positive parity for the positronium system. The quantity ψ_2 is the product of an axial vector with a polar vector, that is, a pseudoscalar quantity, which is odd under inversion. It corresponds to a positronium system of negative parity with a $\sin^2\phi$ distribution of the angle between the polarization vectors. In the experiment, the decays of singlet positronium were selected by observing the two photons emerging in opposite directions from a lead block. The photon polarization was determined indirectly by observing the Compton scattering off aluminium cubes, recorded in anthracene counters as shown in Fig. 3.1. The ratio of the scattering rates for $\phi = 90°$ and $\phi = 0°$ was 2.04 ± 0.08, consistent with the ratio of 2.00 expected for positronium of negative parity. Since the ground states of positronium are S states, the parity measured is the same as that of the electron–positron pair. This experiment therefore confirms that fermions and antifermions have opposite intrinsic parity, as predicted by the Dirac theory.

3.5 Parity violation in weak interactions

While parity is conserved in the strong and electromagnetic interactions, it is violated—what is more, maximally violated—in the weak interactions. This is manifested in the observation that fermions participating in the weak interactions are **longitudinally polarized**. Let σ represent the spin vector of a particle of energy E, momentum $\mathbf{p}$ and velocity v travelling along the z-axis, with $\sigma^2 = 1$. The longitudinal polarization P is the difference in numbers of particles N^+ and N^-, with σ parallel and antiparallel to $\mathbf{p}$ (that is, with spin component

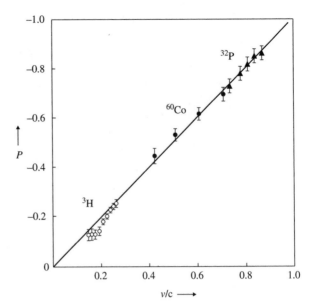

Fig. 3.2 The longitudinal polarization of electrons emitted in nuclear beta decay, plotted as a function of electron velocity v (after Koks and Van Klinken 1976).

$\sigma_z = +1$ or -1), divided by their sum, and is given by

$$P = \frac{N^+ - N^-}{N^+ + N^-} = \alpha \left(\sigma \cdot \mathbf{p} \frac{c}{E} \right) = \alpha \frac{v}{c} \tag{3.9}$$

$$\alpha = -1 \text{ (fermions)} \qquad \alpha = +1 \text{ (antifermions)}$$

This expression for the polarization of fermions in weak interactions was predicted in 1957 by the so-called V–A theory, applying to 'charged current' weak interactions, namely those mediated by $W^{\pm}$ exchange. Figure 3.2 shows the experimental results on the measurement of polarization of electrons emitted in nuclear beta decay, indicating $P = -v/c$ in support of the V–A theory.

Example 3.1 *Prove that a scalar meson ($J^P = 0^+$) cannot decay to three pseudoscalar mesons ($J^P = 0^-$) in a strong or electromagnetic interaction. Can it do so in a weak interaction?*

Let $\mathbf{k_1}, \mathbf{k_2}$, and $\mathbf{k_3}$ be the momenta of the three pseudoscalar mesons in the overall centre-of-momentum frame. Since all the particles are spinless, the decay amplitude can only be a function of their intrinic parities and their three momenta. The two possible linear combinations of the momentum vectors give the following expressions:

$$\mathbf{k_1} \cdot (\mathbf{k_2} \times \mathbf{k_3}) \quad \text{pseudoscalar product}$$

$$\mathbf{k_1} \cdot (\mathbf{k_2} - \mathbf{k_3}) \quad \text{scalar product}$$

Since the parent meson is scalar and the product particles have intrinsic parity $(-1)^3 = -1$, we need to take the pseudoscalar product. Since $\mathbf{k_1} + \mathbf{k_2} + \mathbf{k_3} = 0$, then it follows that $\mathbf{k_1} \cdot (\mathbf{k_2} \times \mathbf{k_3}) = -\mathbf{k_1} \cdot \mathbf{k_2} \times (\mathbf{k_1} + \mathbf{k_2}) = -\mathbf{k_1} \cdot \mathbf{k_2} \times \mathbf{k_1} = 0$, since the three momentum vectors must be coplanar and the amplitude vanishes. If the decay is a weak process, parity is not conserved, so that the scalar product above can be involved and the amplitude can be finite.

3.6   Helicity and helicity conservation

For ultra-relativistic particles with $v = c$, $|pc| = E$, the polarization (3.9) has the simple form

$$H = \boldsymbol{\sigma} \cdot \mathbf{p}/|p| = +1 \quad \text{or} \quad -1 \tag{3.10}$$

where the quantity H is called the **helicity** or handedness. Neutrinos have extremely small masses, that is, velocities $v \approx c$. The momentum and spin vectors define a screw sense, with neutrinos being left-handed (LH) and anti-neutrinos right-handed (RH)—see Fig. 3.3(a). Neutrinos are eigenstates of helicity, with $H = -1$, while antineutrinos have $H = +1$. This is a relativistically invariant description: in transforming from the laboratory frame to another reference frame, necessarily with velocity $v < c$, it is impossible to change the sign of the helicity.

On the other hand, particles with finite mass such as electrons cannot exist in pure helicity eigenstates: they are mixtures of positive and negative helicity states. For example, electrons emitted in weak interactions such as nuclear beta decay are longitudinally polarized, consisting of a combination of LH states with intensity $\frac{1}{2}(1 + v/c)$ and RH states with intensity $\frac{1}{2}(1 - v/c)$, so that the net polarization $P = -v/c$ as in (3.9).

In the interactions of high energy particles, there is a simple rule about helicity. For interactions involving **vector or axial vector fields, helicity is conserved in the relativistic limit**. Note that the strong, electromagnetic and weak interactions are all mediated by vector or axial vector bosons (G, γ, W, or Z

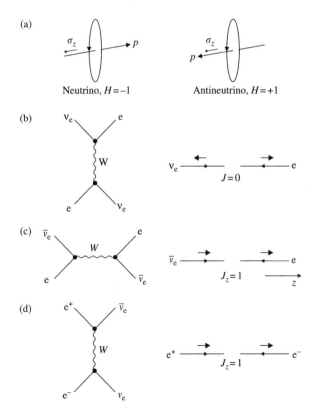

(a)

Neutrino, $H = -1$

Antineutrino, $H = +1$

(b)

$J = 0$

(c)

$J_z = 1$

(d)

$J_z = 1$

Fig. 3.3 (a) A neutrino has helicity $H = -1$, an antineutrino $H = +1$. (b) The reaction $\nu_e + e \rightarrow \nu_e + e$ at high energy mediated by $W^{\pm}$ exchange: viewed in the CMS, both incident particles have $H = -1$ and the total angular momentum $J = 0$. Consequently, the angular distribution of the product particles is isotropic. The total cross section, as given in (1.27), is $\sigma = G_F^2 s/\pi$, where s is the square of the CMS energy. (c) The diagram for anti-neutrino scattering, $\bar{\nu}_e + e^- \rightarrow \bar{\nu}_e + e^-$. As indicated in Example 3.2, the cross section is one-third of that in (b). (d) The diagram for $e^+ + e^- \rightarrow \nu_e + \bar{\nu}_e$. Relative to (c), the cross section is reduced by a further factor of 2.

exchanges). This means that, in any such interaction, and provided the particle involved is relativistic, its helicity is preserved. Thus, a high energy electron of $v \approx c$ in a LH state, for example, will remain in a LH state as it emerges from the interaction. This helicity rule determines the angular distribution in many high energy interactions, and is well illustrated in the interactions of neutrinos and electrons shown in Fig. 3.3(b)–(d).

The cross sections given in the caption to Fig. 3.3 and in the Example 3.2 are for high energy scattering processes mediated by $W^{\pm}$ exchange—the so-called 'charged current' processes. In all these processes, 'neutral current' interactions mediated by Z^0 exchange will also contribute, and indeed this is the only possibility for $e^+ + e^- \rightarrow \nu_\mu + \bar{\nu}_\mu$ or $\nu_\tau + \bar{\nu}_\tau$. Such reactions will involve a weak mixing angle θ_w as described in Section 3.8. Summed over all flavours of neutrinos and for both neutral and charged current reactions, the cross section for electron–positron annihilation to neutrino–antineutrino pairs in the high energy limit ($s \gg m_e^2$) is $\sigma \approx 1.3 G_F^2 s / 6\pi$. This annihilation process is of great astrophysical importance, both in the evolution of the early universe as described in Section 4.6 and in the later, supernova stages of giant stars, discussed in Section 7.8.

Example 3.2   *Calculate the cross section for the process* $e^+ + e^- \rightarrow \nu_e + \bar{\nu}_e$ *via W-exchange, given that the cross section for the process* $\nu_e + e^- \rightarrow \nu_e + e^-$ *is* $G_F^2 s / \pi$. *Assume the electron mass can be neglected at the energies involved.*

Let us do the calculation in two stages. First, we evaluate the cross section for the scattering of antineutrinos by electrons. In the reaction $\bar{\nu}_e + e \rightarrow \bar{\nu}_e + e$, the incident electron and antineutrino have opposite helicities as in (3.9)—see Fig. 3.3(c). Hence $J = 1$ and $J_z = +1$. In this case, back-scattering of the antineutrino in the centre-of-momentum system CMS is forbidden by angular momentum conservation. Of the $2J + 1 = 3$ possible final states, only $J_z = +1$ is allowed by angular momentum conservation. Hence, the cross section, relative to that for neutrino scattering, is reduced by a factor 3 and $\sigma(\bar{\nu}_e + e \rightarrow \bar{\nu}_e + e) = G_F^2 s / 3\pi$.

For the reaction $e^+ + e^- \rightarrow \nu_e + \bar{\nu}_e$, the incident leptons again have opposite helicities as in antineutrino–electron scattering. Just as in that reaction, only the LH helicity state of the electron can be coupled. The difference is that, while the antineutrino exists **only** in the RH state, the positron can have either LH or RH helicity (since it would have originated in an electromagnetic interaction). However, only the RH state of the positron can interact weakly with the (LH) electron, so that the cross section is halved, and $\sigma(e^+ + e^- \rightarrow \nu_e + \bar{\nu}_e) = G_F^2 s / 6\pi$.

3.7   Charge conjugation invariance

The operation of charge conjugation C reverses the sign of the electric charge and magnetic moment of a particle, leaving all other coordinates unchanged. Both strong and electromagnetic interactions are invariant under the C operation. For example, Maxwell's equations are invariant under change of sign of the charge or current and thence of the fields $\mathbf{E}$ and $\mathbf{H}$. In relativistic quantum mechanics, charge conjugation also implies particle–antiparticle conjugation, for example $e^- \leftrightarrow e^+$. As an example of particle–antiparticle symmetry in electromagnetic interactions, a cyclic accelerator can accelerate electrons in a toroidal vacuum tube by means of radio-frequency cavities, and constrains them in, say, a clockwise circular path by means of a magnet ring. The **same** machine will equally accelerate positrons in an anticlockwise direction, and this principle is used in electron–positron colliders, where the accelerated beams with equal energies are arranged, by means of bending and focussing magnets, to meet head-on one or more times per revolution.

On the contrary, weak interactions are not invariant under C. As shown in Fig. 3.3, a neutrino has $H = -1$, and the C operation would transform it into an antineutrino of $H = -1$, which state does not exist. However, the combined operation CP—charge conjugation followed by space inversion—would transform a LH neutrino into a RH antineutrino, which **does** exist (see Section 3.13).

Of course, since both lepton number and baryon number are conserved, there can be no physical process turning an electron into a positron or a neutrino into an antineutrino. However, neutral bosons which are their own antiparticles could be eigenstates of the C operator. For example under the C operation the wavefunction of a neutral pion transforms into itself: $C|\pi^0\rangle \to \eta|\pi^0\rangle$ where, since repeating the process gets us back to the original state, $\eta^2 = 1$ and $C|\pi^0\rangle = \pm|\pi^0\rangle$. The neutral pion decays through an electromagnetic interaction, $\pi^0 \to 2\gamma$. The photon must have $C = -1$ since it is generated by charges and currents which reverse sign under the C operation, and so for a system of n photons, $C = (-1)^n$. Thus the neutral pion must have $C = +1$, and the decay $\pi^0 \to 3\gamma$ is forbidden by C invariance in electromagnetic interactions.

3.8 Gauge transformations and gauge invariance

In Section 3.1 we described examples of translations and rotations in physical space and time. What is just as significant for particle physics are the results of 'internal' symmetry transformations. For example, the plane wavefunction ψ representing a particle with four-momentum $p\,(=p_\mu$ where $\mu = 0, 1, 2, 3)$ can be modified by inserting an arbitrary phase factor α. If $x\,(=x_\mu)$ is the space/time coordinate then the transformation is (in units $\hbar = c = 1$)

$$\psi = \exp(ipx) \to \psi = \exp i(px + \alpha) \tag{3.11}$$

From (3.2) this operation is equivalent to a rotation in some internal 'charge space' of the particle. Clearly, if this phase transformation is **global** (that is, the same over all space), it cannot affect any physical observable. For example, differentiating (3.11), one finds for the expectation value of the momentum of an electron

$$-\psi^* i\frac{\partial \psi}{\partial x} = p \tag{3.12}$$

where $-i\partial/\partial x$ is the momentum operator and the asterisk indicates complex conjugation. The result is independent of the choice of α, since the phase factors cancel. As indicated in Section 3.1, the invariance of the Lagrangian density under such a global phase transformation actually leads to a conserved current, via Noether's theorem. We can illustrate this by writing the above transformation as a small increment ($\alpha \ll 1$):

$$\psi \rightarrow \psi(1 + i\alpha) \tag{3.13}$$

The Lagrangian energy density L of the field ψ appears in the Euler–Lagrange equation analogous to the classical equation (3.1):

$$\frac{\partial}{\partial x}\left(\frac{\partial L}{\partial \psi'}\right) - \frac{\partial L}{\partial \psi} = 0 \tag{3.14}$$

where $\psi' = \partial\psi/\partial x$. If L is invariant under the transformation (3.13), then

$$\delta L = 0 = i\alpha\psi\left(\frac{\partial L}{\partial \psi}\right) + i\alpha\psi'\left(\frac{\partial L}{\partial \psi'}\right)$$

and since

$$i\alpha\frac{\partial}{\partial x}\left(\psi\frac{\partial L}{\partial \psi'}\right) = i\alpha\frac{\partial \psi}{\partial x} \cdot \frac{\partial L}{\partial \psi'} + i\alpha\psi\frac{\partial}{\partial x}\left(\frac{\partial L}{\partial \psi'}\right)$$

then

$$\delta L = 0 = i\alpha\psi\left(\frac{\partial L}{\partial \psi} - \frac{\partial}{\partial x}\left(\frac{\partial L}{\partial \psi'}\right)\right) + i\alpha\frac{\partial}{\partial x}\left(\psi\frac{\partial L}{\partial \psi'}\right)$$

The first term on the right-hand side vanishes, from (3.14), so that the second term must also be zero; that is, if we denote the four-current by

$$J\,(=J_\mu) = \psi\left(\frac{\partial L}{\partial \psi'}\right)$$

this is conserved:

$$\frac{\partial J}{\partial x} = 0 \tag{3.15}$$

From (3.12) we can see that this four-current has the dimensions of a four-momentum. If, for example, we had included the electric charge $|e|$ as a factor in the phase transformation, then the above equation would represent conservation of electric current. Notice that in classical mechanics, invariance of the Lagrangian under some operation (e.g. translation in space) leads to a constant of the motion (in this case, the conserved three-momentum), while in quantum mechanics, invariance of the Lagrangian density—a function of both space and time—under a global phase transformation leads to a conserved current.

So much for **global** phase transformations. However, it is also possible to make a **local** phase transformation, that is one for which $\alpha = \alpha(x)$ is a function of the space/time coordinate. Then, including a factor $|e|$ in the phase to emphasize that we are dealing with electric currents

$$\frac{\partial \psi}{\partial x} = i\left(p + e\frac{\partial \alpha}{\partial x}\right)\psi$$

In this case, physically observable quantities like momentum will be affected by the choice of α and its x-dependence, so local phase invariance does not appear to be a useful concept. However, the electron is a charged particle and will, therefore, be subject to any electromagnetic potential, which will comprise a vector potential $\mathbf{A}$ and a scalar potential Φ. We know that the effect of Φ is to change the energy of the particle from E to $E - e\Phi$, and correspondingly the four-vector potential $A = (\mathbf{A}, \Phi)$ will change the four-momentum from p to $(p - eA)$. Hence, if one includes the effects of an electromagnetic potential of arbitrary magnitude, the above derivative becomes

$$\frac{\partial \psi}{\partial x} = i \left(p - eA + e\frac{\partial \alpha}{\partial x} \right) \psi \tag{3.16}$$

The scale or gauge of the potential A is also arbitrary: one can add to it the gradient of any scalar function, without affecting the values of any physically measurable quantities, namely the associated electric and magnetic fields. This change of the scale or gauge of the potential is called a **gauge transformation**. Choosing α as this arbitrary scalar function, the transformation $A \rightarrow A + \partial\alpha/\partial x$ gives for the derivative

$$\frac{\partial \psi}{\partial x} \rightarrow i \left(p - eA - e\frac{\partial \alpha}{\partial x} + e\frac{\partial \alpha}{\partial x} \right) \psi = i(p - eA)\psi \tag{3.17}$$

so that an observable quantity such as $\psi^* \partial\psi/\partial x$ no longer contains α or $\partial\alpha/\partial x$. The quantity $\partial/\partial x$ on the left-hand side has thus been replaced by $i(p - eA)$ or, in operator notation,

$$\frac{\partial}{\partial x} \rightarrow D = \frac{\partial}{\partial x} - ieA \tag{3.18}$$

called the **covariant derivative**. Note that x and A here are four-vector quantities, that is, the space-time coordinate $x = x_\mu$ ($\mu = 0, 1, 2, 3$ with $x_0 = ct$, $x_1 = x$, $x_2 = y$, $x_3 = z$), and similarly for the four-vector potential $A = A_\mu$ so that written with the indices (3.18) becomes

$$D = \frac{\partial}{\partial x_\mu} - ieA_\mu \tag{3.19}$$

To summarize: by judicious choice of the scalar function the effects of the original local phase transformation on the electron wavefunction and the gauge transformation on the potential cancel exactly. The fact that it is possible to formulate the theoretical description to have this property of local gauge invariance turns out to be vital for the quantum field theory of electromagnetism, called quantum electrodynamics (QED).

Intuitively, one can see on a qualitative basis that these global and local invariances must be consistent with **charge conservation** and the **masslessness of the photon**, respectively. Charge conservation on a global basis comes in, because if the electron were suddenly to lose and then later regain its charge, the above cancellation would not be perfect, since at some value of x the potential A would have no charge to operate on. So, charge must be conserved globally. Second, since the electron can be located anywhere with respect to the source of the potential, the electromagnetic field involved in the local gauge transformation must have an indefinitely long range. From the discussion in Section 1.4 connecting the range of the interaction with the mass of the mediating boson, it follows that if the electromagnetic field has infinite range, the photon must be massless. A corollary is that mass terms cannot occur in the Lagrangian if there is gauge symmetry. As we shall see below, the electroweak theory **does** include massive bosons, and a special mechanism is required to overcome this problem.

We may note here also that a **truly** massless photon is an idealized concept. Real photons have to originate somewhere and end up somewhere else, but the distance they can travel cannot exceed the optical horizon, which is the nominal radius of the observed universe, of order 10^{26} m. If we set this equal to the Compton wavelength λ of the photon, the limit on the mass would be $m_\gamma c^2 < hc/\lambda \sim 10^{-32}$ eV. The best **experimental** limit on the photon mass is based on assuming equilibrium between the magnetic and gravitational fields in the Small Magellanic Cloud, giving a range exceeding 3 kpc (10^{20} m) and hence $m_\gamma c^2 < 10^{-27}$ eV.

Why do we stress the concept of gauge invariance? The point of a gauge invariant theory is that it introduces a symmetry in the calculations, which makes the theory **renormalizable**. This means that it is possible, at least in principle, to make calculations in the form of a perturbation series to all orders in the coupling constant, that is, for a sum over all possible Feynman diagrams, including those involving an arbitrary number of exchanged photons, and not just the one-photon exchange shown in Fig. 1.3.

Figure 3.4 gives examples of how, in (a) an electron can be temporarily dissociated into a 'bare' electron of mass and charge m_0 and e_0, plus a virtual photon, and in (b), into an electron and a photon which converts to an e^+e^- pair. The first diagram involves the coupling α, the second α^2. Because in the second diagram the electric charges of the pair can affect and indeed reduce the field of the parent charge, this last process is referred to as 'vacuum polarization'. Classically, of course, the vacuum contains nothing, by definition. In quantum mechanics, the definition is different: the vacuum is defined as the state of lowest energy of the system. The Uncertainty Principle allows 'vacuum fluctuations', with an energy ΔE (in this example appearing in the form of a pair) provided these are limited to a time $\Delta t \sim \hbar/\Delta E$. Such fluctuations are further discussed in Chapter 4 in connection with dark energy in the universe; in Chapter 5 in the context of the inflationary model of the early universe, where they are postulated to account for the tiny anisotropies in the cosmic microwave background radiation; and in Chapter 7, in connection with Hawking radiation from black holes.

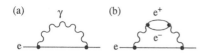

Fig. 3.4 (a) An electron is temporarily dissociated into an electron plus a virtual photon; and (b) into an electron and a virtual electron–positron pair.

In calculating the effects of these so-called 'radiative corrections', a problem arises in that, in principle, the momentum k of the virtual particles involved can go to infinity, and their contribution to the energy of the system, which turns out to be of order $\int dk/k$, is therefore divergent. If the mass of the 'bare' electron is denoted by m_0, this means that when the virtual particles are included, the value of m_0 will become infinite. In fact, this idea of a bare mass is meaningless, since what the experimentalist actually measures is the electron **plus** all the associated virtual processes that can conceivably occur. In fact the same divergences are present in all the processes which the theorist calculates, and can be avoided in all the diagrams and to all orders in the coupling, by re-calibrating or **renormalizing** the (unobservable) bare charges and masses e_0 and m_0, to be their physically measured values, e and m, to be determined of course by experiment.

The accuracy of the answers supplied by QED is illustrated in Table 3.1. The Dirac theory of a point lepton of mass m and charge e predicts a magnetic moment of one Bohr magneton, $\mu_B = e\hbar/2mc$. The actual moment is given

Table 3.1 Anomalous magnetic moments of the electron and muon $(g-2)/2 \times 10^{10}$

	Predicted	Observed
Electron	$11\,596\,524 \pm 4$	$11\,596\,521.9 \pm 0.1$
Muon	$11\,659\,180 \pm 100$	$11\,659\,230 \pm 80$

by $\mu = \mu_\mathrm{B} g s$, where $s = \frac{1}{2}$ is the lepton spin in units of $\hbar$, and $g \approx 2$. The so-called anomaly (the departure from the Dirac value) in QED is

$$\frac{(g-2)}{2} = 0.5 \left(\frac{\alpha}{2\pi}\right) + \text{terms in } \alpha^2, \alpha^3, \ldots \qquad (3.20)$$

where $\alpha = 1/137.06\ldots$ and the terms have been calculated up to those in α^4. The observed and predicted values are seen to agree precisely within errors of less than one part in a million. It may be remarked that the errors shown for the theoretically predicted numbers are larger than those in the observed values, because they depend on the experimental uncertainty in the determination of α.

The success of gauge-invariant theories (QED in this case) can be compared with theories of the past not possessing gauge symmetry, which failed because they contained incurable divergence problems when calculations were made to high orders in the couplings involved. These divergent terms could in principle be cancelled, but only by introducing an indefinitely large number of arbitrary constants, thus losing any predictive power.

So, it is believed that all theories of the fundamental interactions should be renormalizable gauge theories. Unfortunately, no one has yet found a convincing way of extending these ideas to gravity. A quantum theory of gravity does exhibit severe divergences, which can be reduced but not eliminated in a super-symmetric version of the theory (see Section 3.12 below). In **superstring** theory, the pointlike particles responsible for these divergences are replaced by short $(10^{-33}$ cm$)$ strings. Gravity is then renormalizable, but only in 10-dimensional space-time. Incorporating gravity with the other interactions is still an unsolved problem.

3.9   Gauge invariance in the electroweak theory

In QED, we saw that gauge invariance is associated with an infinite set of phase transformations of the wavefunction of the form

$$\psi \to \psi \exp[ie\alpha(x)] \qquad (3.21)$$

These transformations are actually elements of a group called U(1), the 'U' standing for unitary, implying that the norm of the wavefunction is preserved in the transformation, and the '1' refers to a rotation in one dimension as in (3.2). In the electroweak theory, more complicated transformations, belonging to the SU(2) group are also involved. They are of the form

$$\psi \to \psi \exp[ig\boldsymbol{\tau} \cdot \boldsymbol{\Lambda}] \qquad (3.22)$$

Here, the transformation involves the Pauli 2×2 matrices $\boldsymbol{\tau} = (\tau_1, \tau_2, \tau_3)$ and describes rotations about the arbitrary vector $\boldsymbol{\Lambda}$. The Pauli matrices were originally invented to describe spin $\frac{1}{2}$ particles, the '2' in the nomenclature SU(2) referring to the dimension of the matrices, the 'U' indicating that the transformation is again unitary. The 'S' stands for 'special', SU(2) being a subgroup of U(2) in which the matrices are traceless. A fundamental difference between the transformations (3.21) and (3.22) is that U(1) is an Abelian group since $\alpha(x)$ is a scalar quantity. Thus, the effect of two rotations in succession is independent of the order and $\alpha_1 \alpha_2 - \alpha_2 \alpha_1 = 0$, that is, the two operations **commute**. On the other hand, the group SU(2) is non-Abelian, involving the **non-commuting** Pauli operators, for example $\tau_1 \tau_2 - \tau_2 \tau_1 = i \tau_3$.

The electroweak model postulates four massless vector bosons; a triplet w^+, w^-, and w^0 belonging to the SU(2) group and b^0 belonging to the U(1) group, that is, a system with an assumed SU(2) × U(1) symmetry. The neutral component w^0 mixes with the b^0, to form the photon γ and a neutral boson z^0, involving an arbitrary mixing angle θ_W. Finally, scalar bosons called **Higgs scalars** (after their inventor, Higgs (1964)) are postulated, to generate mass by self-interaction, as described below. Three of the four Higgs components are absorbed by the states w^+, w^-, and z^0, to form the massive vector bosons W^+, W^-, and Z^0 introduced in Chapter 1, while the photon γ remains massless. Furthermore, although massive bosons are involved, the theory does remain renormalizable. The **weak and electromagnetic interactions are unified**, and the coupling of the W to leptons, specified by the coupling constant g in (3.22), is given by the relation $e = g \sin \theta_W$. (There are several numerical factors entering into the definition of g, which have arisen historically. The quantity g_w, which we introduced in (1.9) as $g_w^2 = G_F M_W^2$ is related to g by $g_w^2 = \sqrt{2} g^2 / 8$.) The two unknown parameters in the model are the photon mass (zero), which has to be put in 'by hand', and the above mixing angle, which has been measured as $\sin^2 \theta_W = 0.231 \pm 0.001$. The boson masses are then predicted in terms of the Fermi weak interaction constant G_F, e, and the mixing angle:

$$M_W = \left[\frac{g^2 \sqrt{2}}{8 G_F} \right]^{1/2} = \left[\frac{e^2 \sqrt{2}}{8 G_F \sin^2 \theta_W} \right]^{1/2} = \frac{37.4}{\sin \theta_W} \text{ GeV}$$

$$M_Z = \frac{M_W}{\cos \theta_W}$$

(3.23)

The electroweak theory was vindicated by the discovery in 1973 of neutral weak currents, that is the existence of Z^0 exchange as in Fig. 1.3, and by the observation of the W and Z bosons in 1983 (see Figs 1.6 and 1.12). Note that, because the W and Z bosons are massive, compared with the zero mass of the photon, the SU(2) × U(1) symmetry of the model is broken by the Higgs mechanism of mass generation, but because the theory remains renormalizable, cross sections and decay rates mediated by the bosons W and Z can be calculated exactly. All that is missing is the fourth Higgs component, which should exist as a physical particle. A lower limit on the mass is $M_H > 100$ GeV. Finding the elusive Higgs is one of the prime objectives of experimental high energy physics at the present time.

3.10   The Higgs mechanism of spontaneous symmetry breaking

We now discuss briefly the Higgs mechanism for spontaneous symmetry breaking in the electroweak theory. It is relevant to introduce it here, not only because it is an intrinsic part of the very successful electroweak theory, but also because a somewhat similar mechanism has been postulated in connection with the inflationary model of the early universe, which is discussed in Chapter 5.

As stated in Section 3.1, the equation for the Lagrangian energy density L—dimensions (energy)^4—of a field Φ in a quantum-mechanical system is

$$\frac{\partial}{\partial x_\mu}\left(\frac{\partial L}{\partial \Phi'}\right) - \frac{\partial L}{\partial \Phi} = 0 \tag{3.24}$$

where $\Phi' = \partial\Phi/\partial x_\mu$, Φ is the amplitude (with dimensions of energy) of the field particles and x_μ (with $\mu = 0, 1, 2, 3$) is the space-time coordinate (so in units $\hbar = c = 1, x_0 = t, x_1 = x, x_2 = y$, and $x_3 = z$). For free scalar particles of mass μ the Lagrangian function has the form

$$L = T - V = \frac{1}{2}\left(\frac{\partial\Phi}{\partial x_\mu}\right)^2 - \mu^2\frac{\Phi^2}{2} \tag{3.25}$$

which gives for the equation of motion in (3.24) the expression (known as the Klein–Gordon equation—see Appendix D)

$$\left(\frac{\partial^2}{\partial \mathbf{r}^2} - \frac{\partial^2}{\partial t^2} - \mu^2\right)\Phi = 0$$

With the substitution of the operators $E = -i\partial/\partial t$, $\mathbf{p} = -i\partial/\partial \mathbf{r}$, this becomes the usual relativistic relation between total energy, three-momentum and mass:

$$-|\mathbf{p}|^2 + E^2 - \mu^2 = 0$$

Suppose, now, that we are dealing with scalar particles that interact with each other. This means adding an extra term to (3.25), which is of the form Φ^4 (odd powers are excluded because of the symmetry required in the transformation $\Phi \rightarrow -\Phi$, and powers higher than the fourth by the requirement of renormalizability). So the modified Lagrangian is written as

$$L = \frac{1}{2}\left(\frac{\partial\Phi}{\partial x_\mu}\right)^2 - \frac{1}{2}\mu^2\Phi^2 - \frac{1}{4}\lambda\Phi^4 \tag{3.26}$$

where λ is a dimensionless constant representing the coupling of the four-boson vertex. The minimum of the potential V occurs when $\partial V/\partial \Phi = 0$, that is, when

$$\Phi(\mu^2 + \lambda\Phi^2) = 0 \tag{3.27}$$

If $\mu^2 > 0$, the situation for a massive scalar field particle, then $\Phi = \Phi(\text{min})$ when $\Phi = 0$, as is the usual case with the vacuum state having $V = 0$. However,

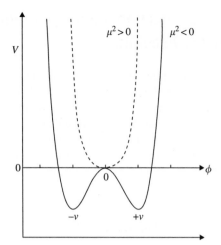

Fig. 3.5 The potential in (3.26) as a function of Φ, the value of a one-dimensional scalar field, for the cases $\mu^2 > 0$ and $\mu^2 < 0$.

it is also possible to consider the case $\mu^2 < 0$, where $\Phi = \Phi(\text{min})$ when

$$\Phi = \pm v = \pm \left(-\frac{\mu^2}{\lambda} \right)^{1/2} \tag{3.28}$$

In this case the lowest energy state has Φ finite, with $V = -\mu^4/4\lambda$ so that instead of being zero, V is everywhere a non-zero constant. The quantity v is called the vacuum expectation value of the field Φ. The situation is illustrated in Fig. 3.5. The minimum at $\Phi = 0$ is referred to as the **false vacuum**, and that at $\Phi = \pm v$ as the **true vacuum**, being the lowest energy state.

In the context of electroweak interactions, one is concerned with **small** perturbations about the energy minimum, so the field variable Φ should be expanded, not about zero but about the chosen vacuum minimum ($+v$ or $-v$ in the above example). If one writes

$$\Phi = v + \sigma(x) \tag{3.29}$$

where σ is the value of the extra field over and above the constant and uniform value v, then, substituting into (3.25) one gets

$$L = \frac{1}{2} \left(\frac{\partial \sigma}{\partial x_\mu} \right)^2 - \lambda v^2 \sigma^2 - \left(\lambda v \sigma^3 + \frac{\lambda \sigma^4}{4} \right) + \text{constant} \tag{3.30}$$

where the constant terms involve powers of v alone. The third term represents the interaction of the σ field with itself. The second term, when compared with the potential in (3.26), is clearly a mass term, with a value for the mass of

$$m = \sqrt{2\lambda v^2} = \sqrt{-2\mu^2} \tag{3.31}$$

So, by making a perturbation expansion about either of the minima $+v$ or $-v$, a **positive real mass** has appeared. Note that the expansion has to be made about **one** of the two minima. Of course, once this is done, the symmetry of Fig. 3.5 is broken. Such a behaviour is called **spontaneous symmetry breaking**. Many examples exist in physics. A bar magnet heated above the Curie point has its elementary magnetic domains pointed in random directions, with zero net moment, and the Lagrangian is invariant under rotations of the magnet in space. On cooling, the domains will set in one particular direction, that of the resultant moment, and the rotational symmetry is spontaneously broken.

The treatment above was of a one-component scalar field. For the more general case of a complex scalar field, $\Phi_1 + i\Phi_2$, the two points $\pm v$ in Fig. 3.5 are replaced by all the points on a circle of radius v obtained by rotating the diagram about a vertical axis. However, the principle of obtaining a real mass associated with the lowest energy 'true' vacuum state by spontaneously breaking the symmetry of the potential remains as before.

The next step is to replace the derivative $\partial/\partial x_\mu$ in (3.30) by the covariant derivative analogous to that in (3.19) but extended to include both the U(1) and SU(2) transformations in (3.21) and (3.22). When this is done one obtains relations for the squares of the masses of the W and Z bosons as in (3.23), and also in terms of the Higgs vacuum term v. The measured values of the boson masses give $v = 246$ GeV, which is thus the scale of the electroweak symmetry breaking. However, the Higgs mass is not directly predicted by the theory, but it should have a mass of the order of the electroweak scale and in any case less than 1 TeV.

The Higgs mechanism has been discussed in a little detail, not just because of its intrinsic importance for the electroweak model of elementary particles, but also because we shall meet a somewhat analogous situation in the inflationary model of the early universe described in Chapter 5.

3.11 Running couplings; the comparison of electroweak theory with experiment

In Section 3.8 it was noted that in gauge theories, perturbation calculations giving finite answers can be carried out to any order in the coupling constant. However, there is a practical limit. For example, in calculating the $(g - 2)$ correction to the electron magnetic moment, there are already 72 Feynman diagrams to be summed over for the term in α^3. The situation is only saved by the smallness of $\alpha \sim 1/137$ and the uncertainty in its experimental value, which together make higher order terms, in α^5 or higher, unimportant or irrelevant in comparing the predicted $(g-2)$ with experiment. Since, for the strong interquark interactions, the coupling α_s is much greater than α, the complications in QCD calculations would be much worse.

Fortunately, to a good level of approximation (called the leading log approximation) it is possible to replace the perturbation series by a single term, an **effective coupling**, which is not constant but depends on the four-momentum transfer q in the process considered. For the electromagnetic interaction, the formula is

$$\alpha(q^2) = \frac{\alpha(\mu^2)}{1 - (1/\pi)\alpha(\mu^2)\ln(q^2/\mu^2)} \tag{3.32}$$

The formula relates the coupling at one momentum transfer q to that at another momentum, μ (incidentally avoiding any problem of the coupling at infinite momentum). The effective coupling is **increasing** with the energy scale. Why is that? Consider a test charge immersed in a dielectric (see Fig. 3.6). The atoms

Fig. 3.6

of the dielectric become polarized, and this produces a **shielding** effect, so that the potential due to the test charge at distances large compared with atomic dimensions is less than it would be without the dielectric. So the effective value of the test charge is reduced at large distances but increases as one probes into smaller distances or equivalently to larger momentum transfers. A similar effect is possible even in a vacuum, since the test charge is continually emitting and re-absorbing virtual pairs—the process called **vacuum polarization** described above—and equation (3.32) gives the quantitative evaluation of this shielding effect, or running of the coupling.

Example 3.3 *The electromagnetic coupling parameter $\alpha \approx 1/137$ at low momentum transfers, $\mu \sim 1$ MeV. Calculate the value of α at the electroweak scale ($q \sim 100$ GeV) and at the GUT scale ($q \sim 3 \times 10^{14}$ GeV).*
From equation (3.32) we have

$$\frac{1}{\alpha(q^2)} = \frac{1}{\alpha(\mu^2)} - \left(\frac{1}{\pi}\right) \ln\left(\frac{q^2}{\mu^2}\right)$$

and substituting for the values of q^2, we find $1/\alpha = 137 - 7.3 \sim 129$ at the electroweak scale, and $1/\alpha = 137 - 25.6 \sim 111$ at the GUT scale. In the latter case, the change is so large that next to leading order terms (in α^2) probably need to be included in (3.32), which is the so-called 'leading log' approximation, applying for small changes to the coupling.

For strong interactions (quantum chromodynamics (QCD)) it turns out that, in addition to the shielding effect of fermion (quark) loops there is also an **anti-shielding** effect, because of the loops containing gluons and the gluon–gluon coupling, as shown in Fig. 3.7. This coupling increasingly 'spreads' the strong colour charge at the larger values of q^2. In this case, the dependence of the strong coupling α_s is found to be

$$\alpha_s(q^2) = \frac{\alpha_s(\mu^2)}{1 + B\alpha_s(\mu^2)\ln(q^2/\mu^2)}$$

$$= \frac{1}{B\ln(q^2/\Lambda^2)} \tag{3.33}$$

where $B = \frac{7}{4}\pi$ and $\Lambda^2 = \mu^2 \exp[-1/B\alpha_s(\mu^2)]$, so that α_s **decreases** with increasing q^2. In the limit of very high q^2, this means $\alpha_s \to 0$, a phenomenon known as **asymptotic freedom**. The experimental evidence for this q^2 dependence is shown in Fig. 3.8.

The running of the couplings is important in performing precision fits of data on electroweak interactions to the Standard Model. The data come from measurements at giant e^+e^- colliders of the W- and Z-boson masses and widths, the forward–backward asymmetry in the decays of these bosons to leptons and

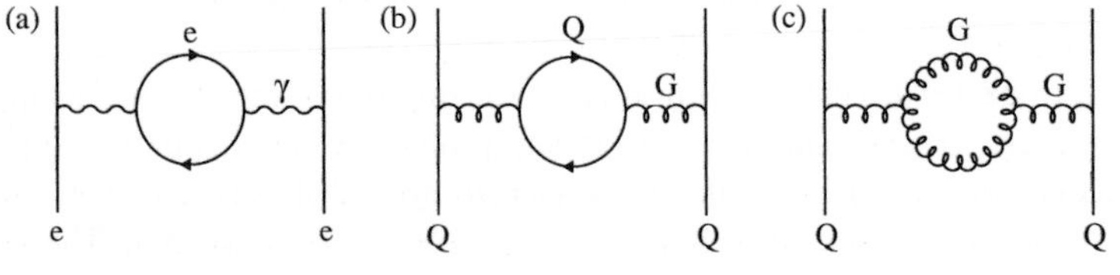

Fig. 3.7 Diagrams involving vacuum polarization effects (a) in QED where loops contain fermions only, and (b) and (c) in QCD, where loops contain both fermions and gluons, and the gluon–gluon coupling produces an anti-shielding effect.

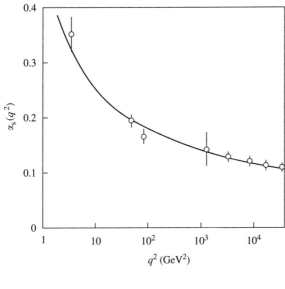

Fig. 3.8 Variation of the QCD 'running coupling' with q^2, the data coming from a variety of sources, including the τ lepton width, inelastic lepton–nucleon scattering, upsilon (=b$\bar{\text{b}}$) decays, Z^0 width, and event shapes and widths in the process $e^+e^- \to$ hadrons. The curve is the prediction for $\Lambda = 200$ MeV in (3.33).

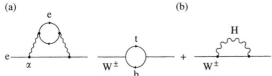

Fig. 3.9 Loop diagrams indicating radiative corrections (a) to α from a virtual fermion loop, and (b) to the mass of the W- or Z-bosons from loops containing a virtual top quark or a Higgs scalar.

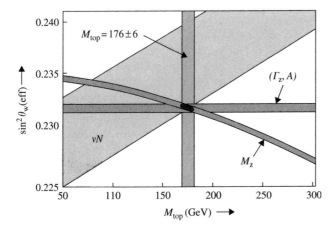

Fig. 3.10 Values of the electroweak mixing angle versus top quark mass, computed from the radiative corrections (as in Fig. 3.9) to various quantities, for example, the Z-boson mass, neutrino–nucleon scattering cross sections (for neutral versus charged currents), and the widths and asymmetries in Z decays. The best fit, where the various curves intersect, is in excellent agreement with the top quark mass measured directly.

to hadrons (via quark pairs), and the cross sections for neutrino and antineutrino scattering on electrons and on nucleons. The different quantities or processes, when evaluated theoretically, will contain different contributions from radiative corrections. Figure 3.9 shows examples of how such corrections can affect α (as described earlier in this section) or the W-boson mass.

Figure 3.10 shows how the Standard Model is tested. Some quantities, such as $M_Z = 91.189 \pm 0.001$ GeV have been measured very accurately. Theoretically, the radiative corrections to the Z-boson mass, and to the quantity $\sin^2 \theta_w$, both depend, for example, on the mass of the top quark. The figure shows the expected variation in $\sin^2 \theta_w$ with M_{top} for the observed value of M_Z. One can also determine $\sin^2 \theta_w$ in other processes with different radiative corrections. The mass of the top quark has also been determined by direct experiment, rather than from a radiative correction. The question then is whether all the data put together can give a unique fit to the model, with a set of best-fit parameters? Clearly this is so; the best fit is indicated by the dark area at the centre of the plot. Although in this plot, the Higgs mass was assumed to be 300 GeV, this quantity can also be determined in the fit, although not very precisely because the radiative corrections depend only logarithmically on the Higgs mass. When this is done, a rather light Higgs mass, $M_H < 160$ GeV is indicated.

3.12   Grand unified theories (GUTs) and supersymmetry

The success of the electroweak theory, unifying the electromagnetic and weak interactions, in describing an enormous range of experimental data, opened up the possibility that unification of the fundamental interactions might be taken one stage further, by incorporating the strong interactions with the electroweak, in what are called **grand unified theories**—GUTs for short. The basic idea is that the SU(2) × U(1) electroweak symmetry (a broken symmetry) plus the (exact) SU(3) colour symmetry of the strong interactions might be encompassed by a more global symmetry, manifested at some high unification energy. Since the effective couplings for the different interactions 'run' in different ways, the possibility arose that they could extrapolate to a universal value, the grand unified coupling α_u.

The first and simplest GUT model was the SU(5) model of Georgi and Glashow in 1974. This incorporated the fermions, both leptons and quarks, into multiplets, inside which, with a common coupling, leptons and quarks could transform into one another via the exchange of massive 'leptoquark' bosons X and Y, with electric charges of $\frac{4}{3}$ and $\frac{1}{3}$ of the elementary charge.

The three couplings involved are denoted by $\alpha_i = g_i^2/4\pi$. Here $g_1 = (\frac{5}{3})^{1/2} e / \cos \theta_w$ and $g_2 = e / \sin \theta_w$, where e is the electron charge and θ_w is the weak mixing angle, and the strong coupling is $g_3 = g_s$ as in (1.7). It may be shown that, in this model, the electroweak parameter has the value $\sin^2 \theta_w = \frac{3}{8}$ at the unification scale, where the three couplings $\alpha_1, \alpha_2, \alpha_3$ all have the same value $\alpha_u = (\frac{8}{3}) \alpha_{em}(M_X)$. So $\alpha_u = \frac{1}{42}$, using the value of $\alpha_{em}(M_X) = \frac{1}{112}$ from Exercise 3.3, for $M_X = 10^{14}$ GeV.

The dashed lines in Fig. 3.11 show how the reciprocal quantities $1/\alpha_{1,2,3}$ vary linearly with the logarithm of the energy scale, as expected from the forms (3.32) and (3.33).

Although the expected value of the unification energy is beyond reach in the laboratory, even at normal energies **virtual** X and Y exchange can take place, and this would lead to the dramatic prediction of **proton decay**, for example, in the mode $p \rightarrow e^+ + \pi^0$, as indicated by the diagram in Fig. 3.12. Because of the strong suppression factor due to the X, Y propagators, the predicted lifetime is very long, of the order of 10^{30} yr. This is in definite contradiction with the experimental lower limit to the lifetime, which exceeds 10^{33} yr. The other problem with this model is that the three couplings (Fig. 3.11) do not exactly meet at a point.

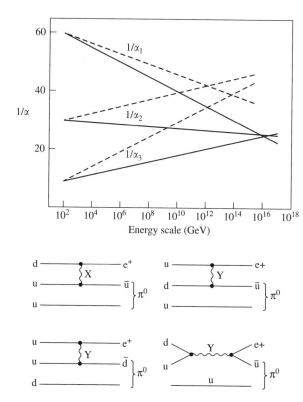

Fig. 3.11 The reciprocal 'running' couplings of the strong, electromagnetic and weak interactions extrapolated to high energies. The dashed lines are the predictions of non-supersymmetric SU(5), and the solid lines, those predicted from supersymmetric grand unified theory.

Fig. 3.12 Feynman diagrams illustrating proton decay in the SU(5) grand unification scheme. The expected lifetime is discussed in Example 3.4 and is of order $\tau \approx 10^{30}$ yr, that is, about one proton decay per kiloton of material per day. This should be easily measurable using underground multikiloton detectors operated over the years. However, proton decay has not yet been observed and the lower limit from five experiments in three continents is $\tau > 10^{33}$ yr.

Example 3.4   *If proton decay is mediated by a boson of mass* $M_X = 3 \times 10^{14}$ *GeV with conventional weak coupling, estimate the proton lifetime (a) given that the muon mass and lifetime are* $m_\mu = 106$ *MeV and* $\tau_\mu = 2.2$ *μs, and (b) using the value of the grand unified coupling from Exercise 3.3.*

(a) For proton decay (Fig. 3.12), the momentum transfers involved are small compared with the W mass, so that the decay amplitude for conventional weak coupling would be $g_w^2/M_W^2 (= G_F)$ as in (1.9), and for a very massive boson it would be g_w^2/M_X^2. Hence, the decay rate would vary as $1/M_X^4$. As compared with muon decay (1.29), the rate would plausibly contain a factor m_p^5 instead of m_μ^5. Thus, a very crude estimate for the proton lifetime would be

$$\frac{\tau_p}{\tau_\mu} \sim \left(\frac{M_X}{M_W}\right)^4 \times \left(\frac{m_\mu}{m_p}\right)^5$$

and inserting the values of proton mass, muon mass, and lifetime, W mass and $M_X = 3 \times 10^{14}$ GeV, gives the value $\tau_p = 9.7 \times 10^{28}$ yr. This estimate is crude in the sense that the various numerical factors entering through parity non-conservation in muon decay are hardly relevant for proton decay, and on the other hand proton decay will contain factors from the quark amplitudes that are not present in muon decay.

(b) A better estimate can be obtained from the formula

$$\tau_p = \frac{M_X^4}{A\alpha_u^2 M_p^5}$$

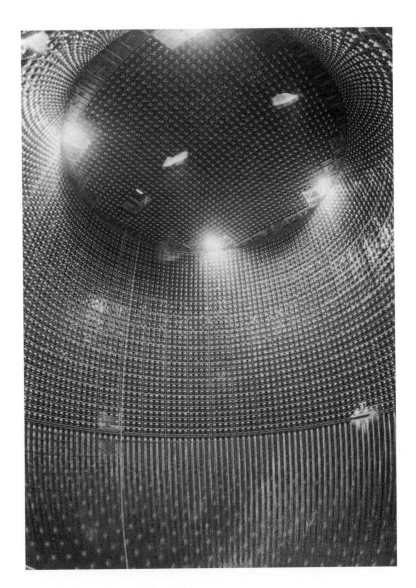

Fig. 3.13 Photograph of the Superkamio-kande water Cerenkov detector used to search for proton decay. For a discussion of the Cerenkov effect, see Section 6.5. The detector consists of a cylinder of 40 m diameter and 40 m depth filled with water, with the surface covered by 11 000 photomultipliers, which record the Cerenkov light produced by relativistic charged particles as they traverse the water. This picture was taken as the tank was being filled with the full volume (50 000 tons) of water. The detector location is the Kamioka mine, Japan, at a depth of 1100 m.

where $A \sim 1$ is a parameter giving the probability of quarks in the proton being close enough for the X-boson to act in the transition $ud \to e^+ + \bar{u}$, for example, see Fig. 3.12. The X-boson mass and proton mass enter in the powers indicated from dimensional arguments as in the previous calculation, and the grand unified coupling $\alpha_u = (\frac{8}{3})\,\alpha_{em}(M_X) = \frac{1}{42}$ from Example 3.3. Inserting these numbers (and recalling that in natural units, $1\,\mathrm{GeV}^{-1} = 9.6 \times 10^{-25}$ s), results in $\tau_p = 4.3 \times 10^{29}/A$ yr, where $A < 1$. The accepted value of the lifetime prediction from minimal SU(5) is $10^{(30\pm0.5)}$ yr.

A modified unification scheme involves the idea of **supersymmetry**, already mentioned in Section 1.3, wherein every fermion has a boson partner, and conversely, for each fundamental boson there is a supersymmetric fermion partner. Supersymmetry was postulated as a way of avoiding the so-called **hierarchy**

problem. In the previous section we noted that the very successful calculation of radiative corrections to the Standard Model involved loops containing virtual fermions and bosons. However, if there exist very massive particles associated with grand unified symmetry schemes, they will be present as virtual states in such loops and lead to divergences in calculating the Standard Model parameters, unless one can arrange cancellation terms. Supersymmetry does just that, since it turns out that the amplitudes for fermion and boson loops have opposite signs. So for every fermion (boson) in a loop there will be another loop with a boson (fermion) partner, and, provided the mass scale of the superpartners is less than about 1 TeV, it can be shown that the divergences are avoided. A bonus of this scheme is that above the SUSY (=supersymmetry) scale, the evolution of the three running couplings is modified and they come closer to meeting at a single point in Fig. 3.11, with a higher unification energy of around 10^{17} GeV.

At the present time, there is no experimental support for supersymmetry or for grand unification. Lower limits on the masses of SUSY particles from accelerators are $\sim$100 GeV. This is of course larger than the masses of most of the known fundamental fermions and bosons. Clearly, supersymmetry is a broken symmetry, and it could be that all the superpartners have masses in the range 100–1000 GeV. A list of some SUSY particles is given in Table 3.2.

Most supersymmetric models postulate an R-symmetry, that is the SUSY particles are produced in pairs with conserved quantum numbers $R = \pm 1$, in much the same way that strange particles are pair-produced with $S = \pm 1$ in strong interactions. Thus, a quark and antiquark with sufficient energy could annihilate to a squark–antisquark pair. A massive SUSY particle would decay, in an R-conserving cascade process, to lighter SUSY particles, and eventually to the lightest superparticle, which would be stable. If this were a photino, for example, its production from squark decay, $\tilde{Q} \rightarrow Q + \tilde{\gamma}$ would be manifest by acoplanarity of the decay and momentum imbalance from the missing photino.

As will be discussed in Chapter 4, one of the major problems in our understanding of the universe is to account for the nature of the 'dark matter', allegedly accounting for over 80% of the total mass. If the dark matter is in the form of elementary particles, then supersymmetric particles, created in the primordial universe, are possible candidates.

Table 3.2 Examples of supersymmetric particles (with spin in units $\hbar$)

Particle		Spin	SUSY Partner		Spin
Quark	Q	$\frac{1}{2}$	squark	$\tilde{Q}$	0
Lepton	l	$\frac{1}{2}$	slepton	$\tilde{l}$	0
Gluon	G	1	gluino	$\tilde{G}$	$\frac{1}{2}$
Photon	γ	1	photino	$\tilde{\gamma}$	$\frac{1}{2}$
Z-boson	Z	1	zino	$\tilde{Z}$	$\frac{1}{2}$
Higgs	H	0	higgsino	$\tilde{H}$	$\frac{1}{2}$
Graviton	g	2	gravitino	$\tilde{g}$	$\frac{3}{2}$

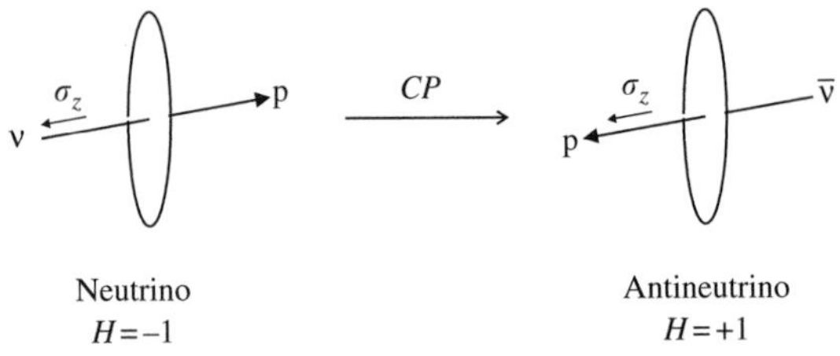

Fig. 3.14 The operation P on a LH neutrino transforms into a RH neutrino state, which is not observed. The C operation on a LH neutrino state transforms into a LH antineutrino state, which is also not observed. The combined CP operation, however, transforms a LH neutrino into a RH antineutrino, which is observed.

Neutrino
$H = -1$

Antineutrino
$H = +1$

3.13   *CPT* theorem and *CP* and *T* symmetry

The operations of charge conjugation C, spatial inversion P, and time reversal T are connected through the very important **CPT theorem**. This states that **all** interactions are invariant under the three operations C, P, and T taken in any order. The theorem predicts that the masses, magnetic moments, lifetimes etc. of particles and antiparticles should be identical, a prediction which is verified to very high accuracy. For example, the difference in masses of the neutral kaon K^0 and its antiparticle $\bar{K}^0$ is less than 1 in 10^{19}, while the difference in absolute values of the magnetic moments of the positron and electron is less than one part in 10^{12}. The *CPT* theorem also predicts the spin-statistics relation, that integral and half-integral spin particles obey Bose–Einstein and Fermi–Dirac statistics, respectively.

While *CPT* invariance is, as far as we know, universal, *CP* and *T* symmetries are not. Let us recall from (3.9) and Fig. 3.3(a) that while the weak interactions are not invariant under C or under P, the operation CP does transform a LH neutrino state into the RH state of its charge conjugate, the antineutrino—see Fig. 3.14. In fact, for a time it was thought that the *CP* symmetry might be universal, but then the evidence for *CP* violation was observed in the decay of neutral kaons, as we now discuss.

3.14   *CP* violation in neutral kaon decay

The kaons are the lightest mesons formed from the combination of a strange quark or antiquark with a non-strange antiquark or quark. They are produced in strong interactions of hadrons and occur in four states, all of spin-parity $J^P = 0^-$ and with masses of 0.494 GeV/c^2 for $K^+(= u\bar{s})$ and $K^-(= \bar{u}s)$, and 0.498 GeV/c^2 for $K^0(= d\bar{s})$ and $\bar{K}^0(= \bar{d}s)$. The states with a strange quark have $S = -1$, while those with a strange antiquark have $S = +1$. All the kaon states are unstable. The charged kaons, being particle and antiparticle, have the same mean lifetime of 12.4 ns. For the neutral kaons, however, two different lifetimes are observed. The state called K_S has $\tau = 0.089$ ns and that called K_L has $\tau = 51.7$ ns (the subscripts standing for 'short' and 'long'). The existence of two lifetimes arises because the decaying states the experimentalist detects are superpositions of K^0 and $\bar{K}^0$ amplitudes. This mixing occurs through virtual 2π and 3π intermediate states and involves a **second-order weak interaction of** $\Delta S = 2$:

$$K^0 \underset{3\pi}{\overset{2\pi}{\rightleftarrows}} \bar{K}^0$$

First, we can form *CP* eigenstates from the neutral kaon states as follows:

$$K_S = \sqrt{\tfrac{1}{2}}(K^0 + \bar{K}^0) \quad CP = +1$$
$$K_L = \sqrt{\tfrac{1}{2}}(K^0 - \bar{K}^0) \quad CP = -1$$

$$(3.34)$$

where, since the kaons have spin zero, the operation *CP* on the wavefunction has the same effect as that of charge conjugation, *C*. Taking into account the negative intrinsic parity of the pion mentioned in Section 3.4, the decay modes will be $K_S \rightarrow 2\pi$ where the final state consists of two pions in an S-state with $CP = +1$, and $K_L \rightarrow 3\pi$ with $CP = -1$. Thus, while the neutral kaons are **produced** as eigenstates of strangeness, K^0 and $\bar{K}^0$, they **decay** as superpositions of these states, which are actually eigenstates of *CP*.

In 1964, it was found by Christenson *et al.* that the above states were in fact **not** pure *CP* eigenstates. If we denote a pure $CP = +1$ state by K_1, and a pure $CP = -1$ state by K_2, the K_L and K_S amplitudes are written as

$$K_S = N(K_1 - \varepsilon K_2)$$
$$K_L = N(K_1 + \varepsilon K_2)$$

$$(3.35)$$

where the normalizing factor $N = (1 + |\varepsilon|^2)^{-1/2}$ and $\varepsilon \approx 2.3 \times 10^{-3}$ is a small parameter quantifying the level of *CP* violation. The experiment commenced with a beam of K^0 generated in a strong interaction. After coasting for many K_S mean lives, the experimenters were left with a pure K_L beam. It was observed that a small proportion of the K_L decays were to a two-pion state, with $CP = +1$ (see Fig. 3.15). *CP* violation is also demonstrated in the leptonic decay modes of K_L. If we denote the rate for $K_L \rightarrow e^+ + \nu_e + \pi^-$ by R^+, and for $K_L \rightarrow$

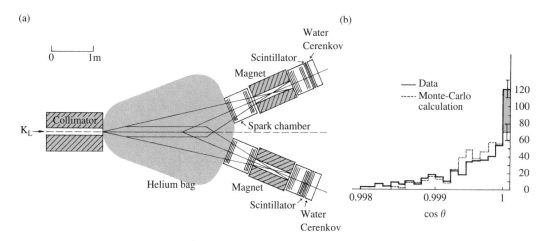

Fig. 3.15 Arrangement of the experiment by Christenson *et al.* (1964) demonstrating the *CP* violating $K_L \rightarrow \pi^+\pi^-$ decay. The charged products of the decays were analysed by two magnet spectrometers instrumented with spark chambers and scintillators. The rare two-pion decays are distinguished from the common three-pion decays by requiring that the two-pion invariant mass should be consistent with the kaon mass, and that the resultant vector momentum of the two pions should be in the beam direction. The distribution in $\cos\theta$ is that expected for three-pion decay, plus some fifty events collinear with the beam and attributed to the rare two-pion mode.

$e^- + \bar{\nu}_e + \pi^+$ by R^-, then it is observed that

$$\Delta = \frac{R^+ - R^-}{R^+ + R^-} = (3.3 \pm 0.1) \times 10^{-3} \qquad (3.36)$$

One of the most striking features of the universe is the very large asymmetry between matter and antimatter, as already discussed in Section 2.11. There, we already noted that *CP* violating interactions are necessary in order to generate a baryon–antibaryon asymmetry. The result (3.36) emphasizes that *CP* violation is actually required in order to differentiate unambiguously between matter and antimatter on a cosmic scale. Here, on Earth, we define the positron of antimatter as having a positive charge and the electron as negative. But these are just names and what we define as positive or negative charge is quite arbitrary. All physical results would have been the same if we had defined the electron as positive and the positron as negative. So we need an unambiguous way of defining what we call matter and antimatter to an intelligent being in a far corner of the universe. *CP* violation in neutral kaon decay now provides the answer. The positron is defined as that charged lepton which is more prolific (by 0.3%) in the long-lived K_L decay.

Example 3.5 *If the annihilation of proton and antiproton proceeds through an S-state, show that* $p\bar{p} \to K_1 + K_2$ *can occur, but not* $p\bar{p} \to K_1 K_1$ *or* $K_2 K_2$, *where* K_1 *and* K_2 *are eigenstates of* $CP = +1$ *and* -1, *respectively.*

A proton–antiproton system with total angular momentum L and total spin S has symmetry $(-1)^{L+S}$ under interchange of space and spin coordinates. But this is equivalent to charge conjugation or particle–antiparticle conjugation, leaving space and spin alone. Hence, the system has $C = (-1)^{L+S}$ and parity $P = (-1)^{L+1}$, taking account of the opposite parities of particle and antiparticle. Hence the initial state of proton and antiproton has

$$CP = (-1)^{2L+S+1} = (-1)^{S+1} \quad \text{for all } L \text{ values}$$

Let J be the total angular momentum of the two kaons, where $|L + S| \geq J \geq |L - S|$. Measured in their rest-frames, the K_1 has $CP = +1$ and the K_2 has $CP = -1$. If the orbital angular momentum of the pair is J, this introduces a factor $(-1)^J$ for the parity. Hence, in the final state,

$$\text{for } 2K_1 \qquad CP = (+1)(+1)(-1)^J = (-1)^J$$

$$\text{for } 2K_2 \qquad CP = (-1)(-1)(-1)^J = (-1)^J$$

$$\text{for } K_1 + K_2 \quad CP = (-1)(+1)(-1)^J = (-1)^{J+1}$$

For annihilation from an **S-state**, $L = 0$ and $J = S$, so the initial state has $CP = (-1)^{J+1}$ where $J = 0$ or 1. Thus, annihilation to $K_1 + K_2$ is allowed and $2K_1$ or $2K_2$ is forbidden.

For annihilation from a **P-state**, $L = 1$ and if $S = 1, J = 0, 1$ or 2. In this case, $CP = +1$ in the initial state so that $J = 0$ or 2 allows $2K_1$ or $2K_2$ in the final state, while if $J = 1$ only $K_1 + K_2$ is allowed. If $S = 0, J = 1$, the initial value of $CP = -1$ and only the states $2K_1$ or $2K_2$ are allowed.

Experimentally, it is observed that for annihilation at rest only $K_1 K_2$ is observed, as expected if an $L = 0$ state is involved.

3.15 *CP* **violation in the Standard Model**

There are in fact **two** sources of *CP* violation in neutral kaon decay. First, the states (3.35) with definite lifetimes are not pure *CP* eigenstates. This is known as **indirect** *CP* violation, occurring in the mass eigenstates themselves through a second-order transition of $\Delta S = 2$. But also, *CP* violation occurs in the actual decay process, which of course involves a first-order $\Delta S = 1$ transition. This is known as **direct** *CP* violation. It happens that in neutral kaon decay, the direct *CP*-violating amplitude ε' is very small compared with the indirect amplitude ε. Indeed, it was to take over thirty years from the first observation of *CP* violation to establish the existence of the direct process and measure it reliably. The ratio $\varepsilon'/\varepsilon = (16.6 \pm 1.6) \times 10^{-4}$.

The Standard Model of particle physics makes some predictions about the level of direct *CP* violation. To introduce this, let us go back to the Fermi coupling in the weak interactions. The **leptons** are coupled to the $W^{\pm}$ mediating boson via a universal coupling specified by the Fermi constant G_F—see (1.9). However, for the **quarks**, the coupling to the $W^{\pm}$ is to weak interaction eigenstates that are **admixtures of flavour eigenstates**. The quark doublets analogous to the lepton doublets

$$\begin{pmatrix} \nu_e \\ e^- \end{pmatrix} \quad \text{and} \quad \begin{pmatrix} \nu_\mu \\ \mu^- \end{pmatrix}$$

are written as

$$\begin{pmatrix} u \\ d' \end{pmatrix} \quad \text{and} \quad \begin{pmatrix} c \\ s' \end{pmatrix}$$

where

$$\begin{matrix} d' = d\cos\theta_c + s\sin\theta_c \\ s' = -d\sin\theta_c + s\cos\theta_c \end{matrix} \quad \text{or} \quad \begin{pmatrix} d' \\ s' \end{pmatrix} = \begin{pmatrix} \cos\theta_c & \sin\theta_c \\ -\sin\theta_c & \cos\theta_c \end{pmatrix} \begin{pmatrix} d \\ s \end{pmatrix} \quad (3.37)$$

The mixing angle $\theta_c = 12.7°$ is called the Cabibbo angle. Thus, for neutron decay, which in quark language is written as $d \rightarrow u + e^- + \bar{\nu}_e$, the coupling is $G_F \cos\theta_c = 0.975G_F$, while for the decay of a strange Λ-hyperon, $\Lambda \rightarrow p + e^- + \bar{\nu}_e$, or in quark symbols $s \rightarrow u + e^- + \bar{\nu}_e$, the coupling is $G_F \sin\theta_c = 0.22G_F$. Here, we have deliberately included only two of the three lepton and quark doublets in the above equation. When one includes all three families, that is, the doublets (ν_τ, τ^-) and (t, b), the transformation replacing (3.37) will be a 3×3 matrix called the CKM matrix after its proponents Cabibbo, Kobayashi, and Maskawa. This is written as

$$\begin{pmatrix} d' \\ s' \\ b' \end{pmatrix} = V_{\text{CKM}} \begin{pmatrix} d \\ s \\ b \end{pmatrix}$$

where the absolute magnitudes of the (generally complex) elements of the matrix are

$$V_{\text{CKM}} = \begin{vmatrix} V_{\text{ud}} & V_{\text{us}} & V_{\text{ub}} \\ V_{\text{cd}} & V_{\text{cs}} & V_{\text{cb}} \\ V_{\text{td}} & V_{\text{ts}} & V_{\text{tb}} \end{vmatrix} \approx \begin{vmatrix} 0.975 & 0.221 & 0.004 \\ 0.221 & 0.975 & 0.039 \\ 0.008 & 0.038 & 0.999 \end{vmatrix} \tag{3.38}$$

The extreme off-diagonal elements V_{td} and V_{ub} are very small and not well determined.

The important point is that an $N \times N$ matrix contains $N(N-1)/2$ Euler (mixing) angles, which is 3 for $N = 3$ (compared with 1 for $N = 2$), and $(N-1)(N-2)/2$ arbitrary phases (i.e. 1 for a 3×3 matrix and none for a 2×2). This phase, say δ in the CKM matrix enters the wavefunction as $\exp[i(\omega t + \delta)]$, which is not invariant under time reversal $t \to -t$. So this is a possible T-violating, or equivalently CP-violating, amplitude in the Standard Model. The existence of this phase implies that some of the elements of the CKM matrix must be complex. Since on the hypothesis of universal Fermi coupling, the matrix V is unitary, the off-diagonal elements of the product V^*V must be zero. So, for example, multiplying the top row of the complex transpose matrix V^* by the last column of V, we get for the right-hand top corner element of the product matrix

$$V_{\text{ud}}^* V_{\text{ub}} + V_{\text{cd}}^* V_{\text{cb}} + V_{\text{td}}^* V_{\text{tb}} = 0 \tag{3.39}$$

and this can be plotted as a 'unitarity triangle' in the complex plane as shown in Fig. 3.16. The three angles α, β, and γ can be found by measurements on the decays of neutral B-mesons, that is the quark–antiquark combinations $\text{b}\bar{\text{d}}$ and $\text{d}\bar{\text{b}}$, known as B_{d}^0 and $\bar{\text{B}}_{\text{d}}^0$, and the combinations $\text{b}\bar{\text{s}}$ and $\text{s}\bar{\text{b}}$, called B_{s}^0 and $\bar{\text{B}}_{\text{s}}^0$. For B-meson decays, the direct CP violation process is dominant over the indirect one. Pair production of B-mesons in enormous numbers has been achieved at electron–positron colliders especially built for the purpose and called 'B-factories' at Stanford, USA and KEK, Japan. Unlike the neutral kaons, the more massive B-mesons have very many decay modes. The level of CP violation is obtained by measuring the difference in decay rates of B^0 and $\bar{\text{B}}^0$ to the small fraction of the decay modes that are common to both. For example, the decays B_{d}^0 to ψK_{s}, where ψ is the ground state $\text{c}\bar{\text{c}}$ meson resonance, measure $\sin 2\beta$, while decays to $\pi^+\pi^-$ measure $\sin 2\alpha$, and the decay B_{s}^0 to ρK_{s} measures $\sin 2\gamma$. The object of these measurements is to determine whether the observed rates of CP violation are consistent with the constraints of the

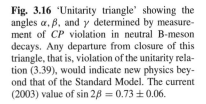

Fig. 3.16 'Unitarity triangle' showing the angles α, β, and γ determined by measurement of CP violation in neutral B-meson decays. Any departure from closure of this triangle, that is, violation of the unitarity relation (3.39), would indicate new physics beyond that of the Standard Model. The current (2003) value of $\sin 2\beta = 0.73 \pm 0.06$.

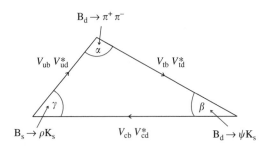

Standard Model. Although none has so far been observed, any departure from the unitarity relation (3.39)—non-closure of the triangle in Fig. 3.16—would be a signal of new physics beyond the Standard Model.

The important question for astrophysics is whether the level of *CP* violation included in the Standard Model is sufficient to account for, or is relevant to, the observed *CP* violation and matter–antimatter asymmetry on a universal scale. Current thinking is that the matter asymmetry must have been generated in the very early universe, at energies far in excess of those associated with the Standard Model and laboratory experiments.

3.16 Summary

- Symmetries and invariance principles give rise to conservation rules. Invariance of the Lagrangian function under a global phase transformation leads to a conserved current (Noether's Theorem).
- Transformations of the wave function under inversion of the space coordinates defines the parity of the system. Parity is conserved in electromagnetic and strong interactions, but not in weak interactions. As a result, fermions involved in charged-current weak interactions possess longitudinal polarization $P = \alpha(\boldsymbol{\sigma} \cdot \mathbf{p})/E = \alpha\,(v/c)$, where $\boldsymbol{\sigma}$ is the spin vector (with $\sigma^2 = 1$), $\mathbf{p}$ and E are the three-momentum and total energy of the particle, and $\alpha = -1$ for fermions and $+1$ for antifermions.
- Particles created singly in parity-conserving interactions have to be assigned an intrinsic parity. Fermions and antifermions have opposite intrinsic parities.
- The helicity of a particle is a well-defined quantum number for ultra-relativistic particles; $H = \boldsymbol{\sigma} \cdot \mathbf{p}/|\mathbf{p}| = +1$ or -1 (i.e. RH or LH). In vector or axial-vector interactions, the helicity of a relativistic particle is preserved, that is, it has the same value before and after an interaction. Neutrinos have helicity -1 and antineutrinos, helicity $+1$.
- It is believed that all successful field theories must have the property of invariance under local gauge (or phase) transformations. This leads to renormalizability of the theory, giving finite predictions to all orders in the coupling constant. In QED, local gauge invariance leads to the masslessness of the photon.
- In QED, the gauge transformation belongs to the group U(1). In the electroweak theory, the gauge transformations belong to the group SU(2) involving non-commuting operators.
- Although the mediating bosons W and Z in the electroweak theory are massive, the theory remains renormalizable as a result of the Higgs mechanism.
- In the simplest electroweak model, one physical scalar Higgs boson is predicted. The present lower limit on the mass is 100 GeV.
- The summation over Feynman graphs of higher orders can be approximated by single boson exchange with an effective coupling which 'runs' with the momentum transfer involved.
- Taking account of the virtual processes involved (radiative corrections), the electroweak theory predicts relations between the various parameters

(particle masses, mixing angles, decay asymmetries etc.) and these have been tested experimentally to high accuracy.

- GUTs have been postulated, unifying electroweak and strong interactions at high energy ($\sim 10^{14}$ GeV or more).
- The so-called hierarchy problem has led to the postulate of supersymmetry, that all fermions (bosons) will have boson (fermion) partners. The experimental mass limits on all SUSY particles are above 100 GeV. At present there is no direct evidence, either for GUTs or for supersymmetry, but SUSY particles have been postulated as candidates for dark matter.
- All interactions are invariant under the C, P, and T operations taken in any order. This *CPT* invariance results in the same mass and lifetime for particles and antiparticles and for the spin-statistics relation.
- *CP* invariance holds good in strong and electromagnetic interactions, but is violated in weak interactions. *CP* violation is allowed in the Standard Model of particle physics, with weak transformations between three families of quarks and leptons. *CP* violation is required to account for the matter–antimatter asymmetry of the universe.

Problems

(3.1) Show that, if the pions are in a state of zero relative orbital angular momentum (S-state), then $\pi^+\pi^-$ is an eigenstate of $CP = +1$ and $\pi^+\pi^-\pi^0$ is an eigenstate of $CP = -1$.

(3.2) Explain why the π^+ and π^- mesons are of equal mass, while the baryons Σ^+ and Σ^-, both of strangeness $S = -1$, have masses of 1189.4 MeV/c^2 and 1197.4 MeV/c^2, respectively.

(3.3) The neutral non-strange mesons ρ^0 (spin $J = 1$, mass 770 MeV) and f^0 ($J = 2$, mass 1275 MeV) can both decay to $\pi^+\pi^-$. What are their C and P parities? State which of the decays $\rho^0 \rightarrow \pi^0\gamma$ and $f^0 \rightarrow \pi^0\gamma$ are allowed, and estimate the branching ratio.

(3.4) Show that the reaction $\pi^- + d \rightarrow n + n + \pi^0 + Q$ (where $Q = 1.1$ MeV) cannot proceed for pions at rest.

(3.5) The flux of relativistic cosmic-ray muons at sea-level is approximately 250 m^{-2} s^{-1}. Their rate of ionization energy loss as they traverse matter is about 2.5 MeV gm^{-1} cm^2. Estimate the annual human body dose due to cosmic ray muons, in grays or rads (1 gray = 100 rad = 6.2×10^{12} MeV kg^{-1}).

If protons underwent decay and their total mass energy (0.94 GeV) then appeared in the form of ionizing radiation, calculate the value of the mean proton lifetime that would result in a radiation dose equal to that from cosmic-ray muons.

*(3.6) At energies of a few GeV, the cross section for the electromagnetic process $e^- + p \rightarrow e^- +$ hadrons is much larger than that for the weak process $e^- + p \rightarrow \nu_e +$ hadrons. However, at high energies and at high enough values of the momentum transfer, the two processes may have comparable cross sections. These conditions would obtain, for example, at the HERA collider at DESY, Hamburg, where 30 GeV electrons collide head-on with 820 GeV protons.

(a) Calculate the total collision energy at HERA in the overall centre-of-momentum frame of the electron and proton.

(b) If the primary collision is treated as between the electron and a quasi-free u-quark carrying 25% of the proton momentum, what is the CMS energy of the electron–quark collision?

(c) At approximately what value of the four-momentum transfer squared (q^2) between electron and quark will the electromagnetic and weak cross sections become equal? Refer to Section 1.9 for cross section formulae.

(d) Write down any other process of electron–proton scattering that will be important at high q^2.

*(3.7) On coming to rest in matter, a positron forms an 'atomic' S-state e^+e^- with an electron, called positronium, which is observed to decay to two or three photons, with two distinct lifetimes.

(a) What are the quantum numbers (total angular momentum J, parity P, and charge conjugation parity C) of these states?

(b) The energy levels of the hydrogen atom are given by the formula $E_n = -\alpha^2 \mu c^2 / 2n^2$, where n is the principal quantum number and $\mu = mM/(m+M)$ is the reduced mass of the proton, mass M and the electron, mass m. Calculate the $n=2 \rightarrow n=1$ level spacing in eV in positronium ($M = 938$ MeV/c^2, $m = 0.511$ MeV/c^2, $\alpha = \frac{1}{137}$).

(c) Try to estimate the lifetimes of the two decay modes, based on the fact that electron and positron wavefunctions have to overlap to annihilate, and that the Bohr radius in hydrogen is $a = \hbar/(\mu c \alpha)$.

(3.8) Electron–positron annihilation at the appropriately high 'resonant' energy can result in the formation of the Υ-meson (the upsilon meson) of mass 9460 MeV/c^2, which is a bound state of a 'bottom' quark and anti-quark: $e^+ e^- \rightarrow b\bar{b} \rightarrow$ hadrons (see Fig. 1.9). Assuming the quark pair is in a state of orbital angular momentum $L = 0$, what are the quantum numbers J^{PC} of the Υ-meson?

Energy levels due to radial excitations of the $b\bar{b}$-system are observed above the ground state, the first such level being one of mass 10 023 MeV/c^2. The corresponding 2^3S–1^3S level separation in positronium is 5.1 eV (see the previous question). Estimate the value of the strong coupling α_s binding the quark and antiquark, assuming for simplicity a $1/r$ (Coulombic) interquark potential (i.e. the first term in equation (1.7)).

*(3.9) Write down how the following quantities will transform under the P (space inversion) and T (time reversal) operations:

Position coordinate	$\mathbf{r}$
Momentum vector	$\mathbf{p}$
Spin/angular momentum vector	$\boldsymbol{\sigma} = \mathbf{r} \times \mathbf{p}$
Electric field	$\mathbf{E} = -\nabla V$
Magnetic field	$\mathbf{B} = \mathbf{i} \times \mathbf{r}$
Electric dipole moment	$\boldsymbol{\sigma} \cdot \mathbf{E}$
Magnetic dipole moment	$\boldsymbol{\sigma} \cdot \mathbf{B}$
Longitudinal polarization	$\boldsymbol{\sigma} \cdot \mathbf{p}$

Show that an electric dipole moment for the neutron would violate T-invariance. Try to estimate an upper limit to such a dipole moment, assuming the appropriate level of CP invariance is that observed in neutral kaon decay.

Estimate the expected level of asymmetry in the scattering of polarized protons by polarized protons (the polarization being longitudinal).

(3.10) All of the following decays are allowed by energy conservation. Which of them is allowed by other conservation laws? (Note: The ρ-meson has $J^P = 1^-$. The π- and η-mesons have $J^P = 0^-$, and their principal decay modes are to two photons):

$$\rho^0 \rightarrow \pi^+ + \pi^- \qquad \rho^0 \rightarrow \eta + \gamma$$

$$\rho^0 \rightarrow \pi^0 + \pi^0 \qquad \rho^0 \rightarrow \pi^0 + \eta$$

$$\pi^0 \rightarrow \gamma + e^+ + e^- \qquad \eta \rightarrow e^+ + e^-$$

<table>
<tr><td>

4   Dark matter and dark energy in the universe

</td></tr>
</table>

In Chapter 2 we already noted that it appears that a large fraction of the matter in the universe is dark (i.e. non-luminous) matter. The need to postulate such dark matter was noted as early as the 1930s by Zwicky, who observed that galaxies in the Coma cluster seemed to be moving too rapidly to be held together by the gravitational attraction of the visible matter. Obviously, we can hardly be satisfied with our picture of the universe until the nature and distribution of such vast quantities of matter has been settled. For example, an important question is whether this dark matter is in the form of new types of elementary particles, which have been roaming around since the earliest stages of the Big Bang: and if so, what are such particles, and why have we not met with them in accelerator experiments? Or, could it be that some of the dark matter is agglomerated in the form of non-luminous stellar objects made out of the same matter as ordinary stars, or as mini black holes or some other objects.

According to present ideas, the quark and lepton constituents of matter with which we are familiar in experiments at accelerators, produced in the numbers foreseen by the model of nucleosynthesis in the early universe described in Chapter 2, can account for only about 5% of the present energy density of the universe. Dark matter is estimated to account for some 25% of the total, but the bulk of the energy density—that is, some 70%—has to be assigned to 'dark energy', which in Chapter 2 was identified with vacuum energy. However, the source of the dark energy—like that of the dark matter—is unknown at present. In particular, the equation of state of the dark energy may be different from that in Table 2.2, and the ratio of pressure to density could even be time-dependent. However, to be specific we shall fix our ideas on energy associated with the vacuum state, as this has already been discussed, in a different context, for the Higgs mechanism in Chapter 3, and will be called upon again in the context of inflation in Chapter 5.

In this chapter, we shall first present the indirect evidence for the existence of dark matter, then describe briefly some of the possible dark matter candidates and the attempts to detect them directly, and finally discuss the evidence for the dark energy component.

4.1   Dark matter in galaxies and clusters

The measurement of the rotation curves of velocity versus radial distance for stars and gas in spiral galaxies gives strong, if indirect, indications for the existence of 'missing' mass, in the form of non-luminous matter. Consider for example a star of mass m at a distance r from the galactic centre, moving with tangential velocity v as shown in Fig. 4.1. Then, equating gravitational and centrifugal forces we obtain

$$\frac{mv^2}{r} = \frac{mM(<r)G}{r^2} \tag{4.1}$$

where $M(<r)$ is the mass inside radius r. A spiral galaxy such as our own has most of the luminous material concentrated in a central hub, plus a thin disc. For a star inside the hub, we expect $M(<r) \propto r^3$ and, therefore, $v \propto r$, while for one located outside the hub, $M \sim$ constant and, therefore, we expect $v \propto r^{-1/2}$. Hence, the velocity should increase at small r and decrease at large r. On the contrary, for many spiral galaxies, the rotation curves are quite flat at large r values. An example is shown in Fig. 4.2. This has led to the suggestion that the bulk of the galactic mass—typically 80–90%—is in the form of dark matter in a halo, as in Fig. 4.1.

Surveys of galaxy clusters show that much of the visible mass is in the form of very hot, X-ray emitting gas. The gas temperature (typically 10^7–10^8 K), estimated from the X-rays measured with the ROSAT satellite, implies velocities of gas particles far in excess of the escape velocities as deduced from the visible mass. If the gas is bound by gravitational forces, suggested by the fact

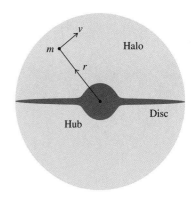

Fig. 4.1 An end-on view of a spiral galaxy, consisting of a central hub, a disc, and a possible halo of dark matter.

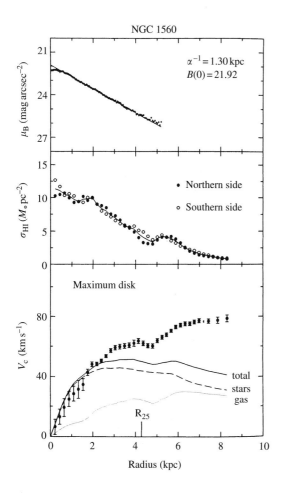

Fig. 4.2 Example of rotation curves for the spiral galaxy NGC 1560. In the top panel the luminosity is plotted against radial distance, showing an exponential fall-off. The middle panel shows the luminosity in the Hα line. In the bottom panel, the points show the observed tangential velocities of stars in this galaxy as a function of radial distance. The curves show the expected values obtained by numerical integration of the mass inside a particular radius as in (4.1), with the contributions from stars and gas shown separately. They are clearly unable to account for the observed velocities at large radii (from Broeils 1992).

that it appears concentrated towards the cluster centres, the greater part (at least 80%) of the total mass must be dark matter.

The major surveys of galaxies and galaxy clusters, such as the infrared IRAS satellite survey, comparing the motional energy with the gravitational energy, also provide indisputable evidence for dark matter. The masses of galaxy clusters can also be estimated directly by their effects on the images of more distant quasars, due to the process of gravitational lensing, which is discussed in the following section.

Dark matter also seems to be required from quite independent considerations of the level of fluctuations in the cosmic microwave background, as discussed in the next chapter. These density fluctuations are observed to be of order $\Delta\rho/\rho \sim 10^{-5}$, and fluctuations 2–3 orders of magnitude larger would have been necessary if formation of galaxy and galaxy clusters was to be achieved by gravitational collapse of baryonic matter alone, once it had decoupled from radiation at $z \sim 1000$. On the other hand, as shown in Example 2.6, the existence of (cold) dark matter with $\Omega(0) \sim 0.25$ would have led to dominance of matter over radiation at a higher redshift ($z \sim 5000$) and more effective gravitational collapse of matter (with dark matter dragging baryonic matter along with it) from a much earlier epoch.

4.2   Gravitational lensing

Important information regarding the amount and location of dark matter has come from gravitational lensing. The gravitational deflection of photons passing by a point mass M at a distance of closest approach b is given by the formula

$$\alpha = \frac{4GM}{c^2 b} \tag{4.2}$$

This is the deflection predicted by Einstein's general theory of relativity, being exactly a factor 2 larger than the deflection one obtains according to Newtonian mechanics (see Problem 4.1). In fact, using heuristic arguments, it is also possible to reproduce the result (4.2) just using the special theory of relativity and the equivalence principle (see, e.g. the very clear derivation in Adler, Balzin and Schiffer 1959). The accuracy of the Einstein prediction was first demonstrated by the 1919 solar eclipse expedition, which measured the deflection of light from stars close to the Sun's limb.

The gravitational deflection of light implies that massive objects may act as **gravitational lenses**, as foreseen by Einstein even before the relation (4.2) had been tested experimentally. Suppose in Fig. 4.3 that S is a source of light (star) and that the rays to the observer O pass close to a massive point lensing object L of mass M. The diagram represents the situation in the plane defined by O, S and L, and is the gravitational analogue of a thin lens system in optics. In the general case, the source and lens will not be collinear with the observer and there will then be two images of the source, S1 and S2. Then, if α denotes the gravitational deflection and b the closest distance of approach, we have from (4.2)

$$\alpha D_{LS} = D_S(\theta_1 - \theta_S)$$

$$\theta_S = \theta_1 - \frac{4GM}{bc^2}\frac{D_{LS}}{D_S} = \theta_1 - \frac{4GM}{c^2}\frac{D_{LS}}{D_S D_L}\frac{1}{\theta_1} \tag{4.3}$$

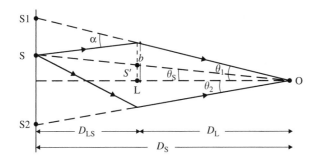

Fig. 4.3 The two images S1 and S2 of a source S formed from gravitational lensing by the point mass L.

In the collinear case, $\theta_S = 0$. Then, we can write

$$\theta_1 = \theta_E = \left(\frac{4GM}{c^2} \frac{D_{LS}}{D_S D_L} \right)^{1/2} \tag{4.4}$$

where θ_E is the angle of the so-called **Einstein ring**. In this collinear case, the image of S is a ring of light centred on the line of sight. For finite θ_S and a point lensing mass, however, one obtains just two images lying in the plane defined by the source, lens, and observer, with angles that are solutions of the quadratic (4.3):

$$\theta_{1,2} = \left[\theta_S \pm \sqrt{\theta_S^2 + 4\theta_E^2} \right] / 2 \tag{4.5}$$

The above analysis assumes a point lensing mass. Often, the lensing object or objects will be extended in space, and more complex, multiple images are then formed. Examples of lensing were first observed for very intense and very distant sources called quasi-stellar objects or **quasars**, which are in fact the most powerful radio and optical sources known. Quasars are examples of galaxies with very active nuclei (AGNs), and are probably powered by the gravitational energy from massive black holes (see Section 7.12). In the case of quasars the lensing mass is a 'foreground' galaxy or galaxy cluster. An example of a doubly-imaged quasar is shown in Fig. 4.4.

Since in multiply-imaged events, the different images involve different light paths, time delays will be involved. The path lengths are proportional to the distance scale, that is, to the inverse $1/H_0$ of the Hubble parameter, so that study of multiply-imaged quasars offers a method for the determination of H_0. However, the important thing is that by measurement of the multiple images of such distant quasars, the total mass of the foreground galaxy or cluster can be estimated. The total mass density of the universe found in this way indicates a value for the closure parameter associated with the matter content of $\Omega_m \approx 0.3$, as quoted in (2.29).

Example 4.1 *Estimate the value of the Einstein radius for lensing by a stellar-type mass in the local galaxy, by considering the specific case of a pointlike lensing object of* $10M_\odot$, *situated midway between the observer and a source at a distance of* 2 pc (6×10^{16} m)—*a typical interstellar distance.*

Inserting the above numbers in (4.4), the value of the Einstein radius is found to be $\theta_E = 0.32$ μrad $= 0.065$ arcsec. One has to compare this angle with the resolving limit of a telescope. An earth-bound optical telescope has a resolution of about 1 arcsec, and even that of the Hubble Space Telescope

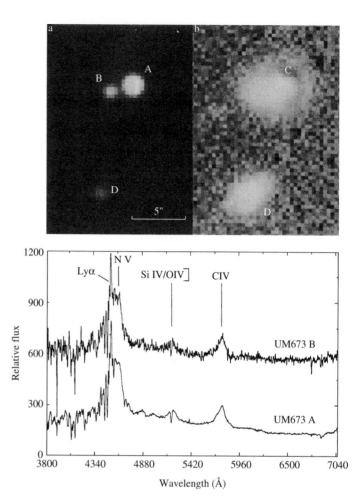

Fig. 4.4 An example of the double image of a quasar, observed by the European Southern Observatory, as it is lensed by a foreground galaxy. The top picture shows (at left) the quasar CCD image split into two parts, A and B. Subtracting these images from the frame reveals (at right) the lensing galaxy marked C. Object D is a background galaxy.

The plot of the wavelength response at bottom shows that the two images A and B, separated by 2.2 seconds of arc, have identical spectra (from Surdej *et al.* 1987).

is good to only 0.1 arcsec. So, resolving separate optical images of sources lensed by objects of stellar mass is not feasible.

While the above example shows that images due to gravitational lensing by objects of typical stellar masses are too close to be resolved, distinguishing separate images is possible for massive lensing objects, that is, for galaxies or galaxy clusters. For example, for a cluster of mass $10^{14} M_\odot$, with $D_{LS} = D_L = D_S/2 = 100$ Mpc, $\theta_E \sim 65$ arcsec (assuming one can treat the cluster as a point mass), which is quite measurable.

4.3   Amplification by gravitational lenses: microlensing and MACHOs

Even when the images of a source produced by gravitational lensing cannot be resolved, an amplification of the intensity may occur, in what is called a **microlensing event**. Suppose that a pointlike lensing mass is moving at velocity v normal to the line of sight, and that the source subtends an angle θ_S at the observer. All quantities are measured in the plane defined by the observer, lens, and source, with the notation of Fig. 4.3. In this case, the angle θ_S will be a function of time, with a minimum value when the lens is at the closest distance of approach to the line of sight to the source. From Fig. 4.5, the right-angled triangle AS′L gives us $LS'^2 = AS'^2 + AL^2$, where $LS' = D_L\theta_s$, $AS' = D_L\theta_s(\text{min})$ and $AL = vt$, where time t is measured from the moment of closest approach of the lensing object to the line of sight to the source. Dividing through by $(D_L\theta_E)^2$ and with $x = \theta_S/\theta_E$, $x(\text{min}) = \theta_S(\text{min})/\theta_E$, we obtain

$$x^2 = x^2(\text{min}) + \left(\frac{vt}{D_L\theta_E}\right)^2 = x^2(\text{min}) + \frac{t^2}{T^2} \tag{4.6}$$

where we have defined $T = D_L\theta_E/v$. When the two images are not separated, there results in an amplification of the (single) signal. From Liouville's theorem, the phase-space density, that is, the number of photons per unit solid angle, is unaffected by the imaging, so that if θ is the angle of the image, the amplification will be $A = d\Omega/d\Omega_S = \theta\, d\theta/(\theta_S\, d\theta_S)$. Since from (4.5) the amplification from each image is found to be

$$A_{1,2} = \left[1 + (x^2/2) \pm x\sqrt{1 + x^2/4}\right] \Big/ \left[2x\sqrt{1 + x^2/4}\right] \tag{4.7}$$

it follows that, adding the amplitudes from the two (unresolved) images, the net amplification becomes

$$A = \left(1 + \frac{x^2}{2}\right) \Big/ \left[x\sqrt{1 + \frac{x^2}{4}}\right] \tag{4.8}$$

with x^2 defined in (4.6). Figure 4.6 shows how the signal depends on time, for a few cases of the ratio $x(\text{min})$. For $x(\text{min}) \ll 1$, the peak value of A is approximately equal to $1/x(\text{min})$. Figure 4.7 shows an example of a microlensing event, in which a massive dark object amplifies the light signal from a star in the Large Magellanic Cloud (a nearby mini-galaxy).

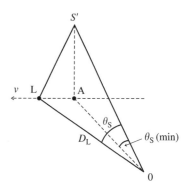

Fig. 4.5 A point lensing mass L moving with velocity v perpendicular to the line of sight: O is the observer and S′ is the projected position of the source in the plane of the lens.

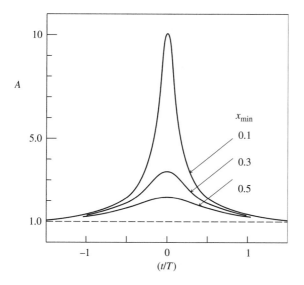

Fig. 4.6 Examples of the dependence of the amplification for microlensing events calculated from (4.8) for different values of $x(\text{min})$ as a function of t/T.

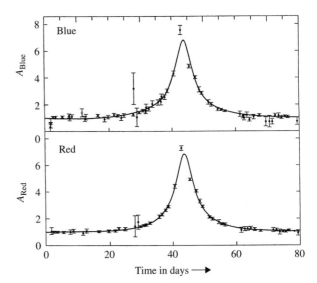

Fig. 4.7 Example of a microlensing event, the source being a star in the Large Magellanic Cloud, at a distance of some 50 kpc. Note the same signal is observed in blue and in red light (after Alcock *et al.* 1993).

Example 4.2   *Calculate the typical time T for lensing by a pointlike object of mass $0.1M_\odot$ moving at a velocity of $v = 200$ km s^{-1} normal to the line of sight, and situated half way to a source star at a distance of 50 kpc.*

Inserting these numbers in the above equations gives $\theta_E = [(4GM/c^2) (D_{LS}/D_S D_L)]^{1/2} = 6.2 \times 10^{-10}$ rad and $T = D_L \theta_E / v = 2.39 \times 10^6$ s ~ 28 days.

MACHOs (or massive astrophysical compact halo objects) is the name given to dark matter in the form of microlensing objects with masses of the order of stellar masses in our galaxy. Typically their masses lie in the range 0.001– $0.1M_\odot$. Several hundred MACHOs have been observed, for example by their microlensing of light from stars in the Large Magellanic Cloud, as in Fig. 4.7. A characteristic of these events is that the same amplification is observed in blue and red light, a fact which distinguishes them from variable stars. The reason for the achromaticity is clear. If the photon momentum is p, its effective gravitational mass is p/c, so that it will receive a transverse momentum $\Delta p \propto p$ from a gravitational field. Hence, the deflection $\Delta p/p$ will be independent of wavelength h/p.

4.4   The lensing probability: optical depth

The probability that a particular source will undergo gravitational lensing as a measurable effect is called the **optical depth**. This is defined as the probability that at some instant of time, the line of sight to an individual star will be within the Einstein radius of a lens in the intervening distance. If N_L is the density of lenses per unit volume, and they are distributed uniformly, then, since an Einstein ring extends over an area of $\pi(D_L \theta_E)^2$ it follows that the optical depth will be

$$\tau = \int \pi D_L^2 N_L dD_L \theta_E^2$$

where the integral extends from $D_L = 0$ to $D_L = D_S$. Substituting for θ_E from (4.4) and with $y = D_L/D_S$ and $\rho = N_L M$ for the mass density of lenses, the integral becomes

$$\tau = 4\pi G \left(\frac{D_S}{c}\right)^2 \int \rho(y)y(1-y)\,\mathrm{d}y$$

with y running from 0 to 1. If ρ is constant, this expression simplifies to

$$\tau = 2\pi G \left(\frac{D_S}{c}\right)^2 \frac{\rho}{3} \tag{4.9}$$

which depends only on the distance to the source and the average mass density of lensing objects between the observer and the source. Inserting typical values of density for our galaxy, and considering sources near the periphery of the central bulge at ∼5 kpc, yields a value of $\tau \sim 10^{-7}$. Thus, lensing will be a comparatively rare occurrence, and to detect dark matter in the form of 'dark stars'—that is, MACHOs as described above—one needs to examine the light curves of many millions of stars over months and years. This has required computerized search techniques, which were first used in the old automated analysis systems used in scanning bubble chamber film in particle physics experiments.

Finally, we remark that the effect of a MACHO on the light from a more distant star is not only angular deflection, but also a **time delay** (the effective velocity of light is reduced by the gravitational field) usually referred to as the 'Shapiro delay'. Such an effect was first observed in the 1960s when return radar pulses bouncing off planets were found to be delayed if they passed close to the Sun. So, in principle, a MACHO could introduce a measurable phase delay in the very regular radio or X-ray signals from a pulsar (see Section 7.10).

The magnitude of the magnification involved in microlensing varies inversely as the impact parameter between the lensed star and the MACHO, and so detection of microlensing typically involves examination of millions of stars, as discussed above. On the other hand, the Shapiro delay falls off only logarithmically with impact parameter, so even with only a few hundred known pulsars, an effect may be detectable, although so far none has been claimed.

We now discuss the various possible candidates that have been proposed to constitute dark matter, as well as the experimental methods employed to search for them.

4.5   Baryonic dark matter

Some of the dark matter **must** be baryonic, since the value $\Omega_{\text{baryon}} \approx 0.05$ deduced from nucleosynthesis in the Big Bang is an order of magnitude larger than the closure parameter associated with visible stars, gas, and dust, of $\Omega_{\text{lum}} \approx 0.01$—see (2.27) and (2.28). Some at least of this baryonic dark matter has been accounted for in the form of compact halo objects (MACHOs) described above. It is not yet known if such objects can account for all or even the bulk of the baryonic dark matter. Leaving this aside, baryonic matter makes only a small contribution to the overall density of the universe and a less than 25% contribution to the estimated total density of dark matter.

4.6   Neutrinos

The favoured hypothesis is that non-baryonic dark matter is made up of elementary particles, created at an early hot stage of the universe, and stable enough to have survived to the present day. As indicated in Section 2.8, the neutrinos ν_e, ν_μ, and ν_τ and their antiparticles, together with electrons, positrons, and photons would have been produced prolifically in the early universe, and present in comparable numbers, according to the equilibrium reactions

$$\gamma \leftrightarrow e^+ + e^- \leftrightarrow \nu_i + \bar{\nu}_i \tag{4.10}$$

where $i = e, \mu, \tau$. As indicated in Section 3.6, the cross section for electron–positron annihilation to a neutrino–antineutrino pair is a weak process with a cross section of order $\sigma \sim G_F^2 s/6\pi$, where s is the square of the CMS energy (see Example 3.2 in Section 3.6). The collision rate for this reaction is $W = \langle \rho \sigma v \rangle$ where ρ is the number density of electrons or positrons and v is their relative velocity. Since $\rho \sim T^3$ is the number density of relativistic particles at temperature T, the rate $W \sim T^5$, while the universal expansion rate $\dot{R}/R = H$, the Hubble constant at that era. In a radiation dominated universe, $H \sim T^2$ (see (2.47)). Hence, as T falls during the expansion, neutrinos must decouple as soon as $W < H$. Inserting numerical values (see Problem 4.2) one finds for the critical temperature, $kT \sim 3$ MeV. So for $t > 0.1$ s, neutrinos are decoupled and the neutrino fireball expands and cools independently of the other particles or radiation (apart, of course, from the universal redshift).

The number density of the neutrinos will be comparable with that of the photons. However, when $kT < 1$ MeV, the photons will be boosted by the annihilation process $e^+ + e^- \rightarrow \gamma + \gamma$, which converts the energy content of electrons and positrons into photons. The entropy per unit volume of the particle gas will be $S = \int dQ/T$, where Q is the energy content per unit volume of photons, electrons, and positrons at temperature T, and from (2.37) and the integrals (2.40)

$$S = \left(\int \frac{4aT^3\, dT}{T} \right) \times \left(1 + \frac{7}{8} + \frac{7}{8} \right) = \frac{4aT^3}{3} \times \frac{11}{4} \tag{4.11}$$

where $a = 4\sigma_{st}/c$ is the radiation constant and σ_{st} is Stefan's constant. After annihilation the photons have attained a temperature T_1 with entropy

$$S_1 = \left(\frac{4a}{3} \right) T_1^3$$

but since the expansion is adiabatic (isentropic) $S_1 = S$ so that

$$T_1 = \left(\tfrac{11}{4} \right)^{1/3} T$$

So if the temperature today of the microwave photons is T_γ, that of the relic neutrinos, which received no boost will be

$$T_\nu = \left(\tfrac{4}{11} \right)^{1/3} T_\gamma = \left(\tfrac{4}{11} \right)^{1/3} \times 2.73 = 1.95 \text{ K} \tag{4.12}$$

The corresponding number density of neutrinos plus antineutrinos is easily shown to be

$$N_\nu = \left(\tfrac{3}{11}\right) N_\gamma = 113 \text{ cm}^{-3} \tag{4.13}$$

per neutrino flavour, to be compared with a number density of 411 cm^{-3} for the microwave photons. We note from this number density that the total energy density of neutrinos would be equal to the critical density (2.23) if the sum of the masses of the three flavours has the value

$$\sum_{e,\mu,\tau} m_\nu c^2 = 47 \text{ eV} \tag{4.14}$$

So relic neutrinos with masses in the few eV mass range could make significant contributions to dark matter. Of course, unlike essentially all other dark matter candidates, they have the great advantage that they are known to exist. However, evidence from neutrino oscillations (Sections 6.9–6.11) suggests very much smaller neutrino masses than indicated by (4.14). Another problem with neutrinos as dark matter candidates is that they would constitute 'hot' dark matter. With the critical temperature of $kT \sim 3$ MeV, neutrinos were relativistic when they decoupled from other matter and also when the structures in the universe were forming. Consequently, they would stream rapidly under gravity and, just like the photons, tend to iron out any primordial density fluctuations. So if large-scale structures are to form, computer simulations indicate that the fraction of dark matter that is 'hot' can only be of the order of 30% or so.

Aside from the question of relic neutrinos forming dark matter, the existence of some 340 neutrinos, inferred above, in each cubic centimetre throughout space, with energies in the meV range, poses a truly formidable challenge to the experimentalist to detect them.

4.7 Axions

The axion is a very light pseudoscalar particle (spin-parity 0^-) postulated in connection with the absence of *CP* violation in strong interactions quantum chromodynamics (QCD). In principle, complex phases can occur in the quark wavefunctions in QCD, and these would be *T*-violating or *CP*-violating (as indeed they are in the weak interaction sector). The fact is however that the upper limit to the electric dipole moment of the neutron is nine orders of magnitude less than strong *CP* violation predicts. To cancel this undesirable feature and to account for the smallness of any posssible violation, Peccei and Quinn (1977) proposed a new spontaneously broken global U(1) symmetry, with an associated massless boson (a so-called Goldstone boson), which receives a small mass by mixing with other neutral pseudoscalar mesons (e.g. π^0, η^0). There is no experimental laboratory evidence for such a particle, but it is a possible candidate for dark matter, since axions, if they exist, would have formed as a condensate in the early universe. The characteristics of axions—the mass m_a and the vanishingly small coupling to other particles—depend on just one parameter, which is the (unknown) scale of the symmetry-breaking interaction.

Like the neutral pion, the axion is expected to decay to two photons (however, the lifetime is proportional to $1/m_a^5$ and exceeds the age of the universe for any mass below 10 eV). This implies that in a suitable magnetic field (supplying an incoming photon), one should be able to observe the conversion of an axion to a photon. Attempts to observe this conversion in a microwave cavity have so far failed to detect any signal. Equally, if axions constitute the bulk of dark matter in our local galaxy, their decay to two photons each of energy $m_a/2$—despite the long lifetime—should be revealed as spectral lines in optical telescopes if the mass is in the few eV region, and absence of such effects excludes such a mass range. Axion masses above 1 eV are also excluded because they would result in a drastic reduction in the number of visible red giant, helium-burning stars. These would lose energy by axion emission, to be compensated by increased nuclear fusion and a shortening of the lifetime of the star, and hence a reduction in the numbers visible at any one time.

In order to account for dark matter, that is, to reach an energy density of the order of the critical density, one requires axion masses in the region of 10^{-6}–10^{-3} eV/c^2. Axions have the advantage (if they exist) that they would contribute to the 'cold' dark matter, so the hypothesis has to be taken seriously.

4.8   WIMPs

The most popular hypothesis, currently, for dark matter particles is that they are weakly interacting massive particles (WIMPs), moving with non-relativistic velocities at the time of freeze-out and thus constituting cold dark matter.

First however, we should ask if these particles could be massive neutrinos. If they were, and had masses such that they were still relativistic at freeze-out, the closure parameter would increase with neutrino mass m_ν as in (4.14), and for $m_\nu \sim 1$ MeV would reach the unphysical value of $\Omega \sim 10^4$. For higher masses, the heavy neutrinos would become non-relativistic at decoupling, and as shown below, this then results in $\Omega \propto 1/m_\nu^2$ instead of $\Omega \propto m_\nu$. The closure parameter falls back below $\Omega = 1$ for a mass above 3 GeV. So, the fact that the measured value of Ω is not large compared with unity, excludes the mass range 50 eV to 3 GeV, while experiments at the LEP electron–positron collider at CERN prove conclusively that there are no 'extra' conventional neutrinos of mass below $\frac{1}{2}M_Z = 45$ GeV. Conventional neutrinos of $m_\nu > 45$ GeV, if they existed, would however make no significant contribution to Ω (< 0.01, from the $1/m_\nu^2$ dependence). The dependence of the closure parameter on mass is illustrated in Fig. 4.8, drawn for WIMPs generally, that is, for massive non-relativistic particles with conventional weak couplings.

For several reasons, it is considered that **supersymmetric particles** could be the most likely WIMP candidates. As described in Section 3.11, such SUSY particles are expected to be created in pairs, with opposite values $R = \pm 1$ of a conserved quantum number called R-parity. Heavier SUSY particles would decay to lighter ones in R-conserving processes, ultimately ending up with the lightest SUSY particle (LSP), which we denote by the symbol χ. This particle is assumed to be stable and, therefore, to have survived from the primordial era of the universe. The LSP is usually identified with the **neutralino**, a neutral fermion, which is the lightest of the states arising from a linear combination of the photino, zino and two higgsinos (see Table 3.2). Since such anomalously heavy particles are not constituents of atoms or nuclei, they cannot have either electromagnetic or strong coupling and are assumed to interact only weakly.

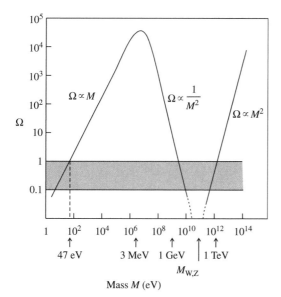

Fig. 4.8 Variation of the closure parameter with WIMP mass, assuming conventional weak coupling. The shaded region, corresponding to $\Omega = 0.1$–1, is that in which the contribution to the closure parameter from massive neutrinos or WIMPs must lie, thus excluding the range of masses 100 eV–3 GeV. Accelerator experiments suggest that WIMPs must have masses exceeding $M_Z/2 = 45$ GeV, otherwise Z-bosons could decay into WIMP–antiWIMP pairs. However, for masses which are large compared with the Z-boson mass, the weak cross section falls rapidly because of propagator effects, so that WIMPs in the TeV mass range are possible dark matter candidates, depending on the precise values of the WIMP coupling.

Although neutralinos are stable, they can, of course, disappear by annihilation with their antiparticles, which would have been generated with the same abundance as the particles. There are many free parameters in SUSY models, which means that the LSP mass as well as the annihilation cross section and cosmological abundance can vary over quite wide ranges, and it is probably this flexibility that is part of the attraction of such models.

Let us now look more closely at the constraints that WIMP models have to fulfil. First, we are searching for **cold** dark matter, since this must form the bulk of all dark matter if one is to successfully account for the development of structures in the universe, as explained in Chapter 5. So the WIMPs must be non-relativistic when they 'freeze out'. This freeze-out occurs when the rate of $\chi\bar{\chi}$ annihilation falls below the expansion rate, that is, when

$$N\langle\sigma v\rangle \leq H \tag{4.15}$$

where N is the WIMP number density, v is the relative velocity of particle and antiparticle, σ is the WIMP–antiWIMP annihilation cross section and H is the Hubble parameter at the time of freeze-out. It will be seen that the WIMP abundance varies inversely as the annihilation cross section, so that weaker interactions lead to earlier freeze-out and, consequently, higher densities and larger contributions to the closure parameter. Since the WIMPs are massive and non-relativistic, $M \gg T$ where M is the WIMP mass and T is the temperature at freeze-out in energy units. Then, the density will be given by the Boltzmann relation (see Problem 2.3):

$$N(T) = \left(\frac{MT}{2\pi}\right)^{3/2} \exp\left(\frac{-M}{T}\right) \tag{4.16}$$

The exact value of the $\chi\bar{\chi}$ annihilation cross section is of course unknown, but if it is of the same order as the weak cross section then on dimensional grounds we could set $\langle\sigma v\rangle \sim G_F^2 M^2$—see (1.27). Inserting (4.16) in (4.15) and with

$H = 1.66g^{*1/2}T^2/M_{PL}$ as given by (2.47) in the radiation-dominated universe, the freeze-out condition becomes

$$(MT)^{3/2} \exp\left(\frac{-M}{T}\right) G_F^2 M^2 \leq \frac{fT^2}{M_{PL}} \tag{4.17}$$

where f includes the numerical constants involved and is of the order of 100. The Fermi constant squared is $G_F^2 \approx 10^{-10}$ GeV^{-4} while the Planck mass $M_{PL} \approx 10^{19}$ GeV. Inserting these numbers, one can easily solve numerically for the value of $P = M/T$ at freeze-out. It varies slowly and logarithmically, from around $P = 20$ for $M = 1$ GeV to around $P = 30$ for $M = 100$ GeV. In the following we take this ratio as a constant, $P = 25$. Now going back to (4.15) and (2.47), and recalling that the expansion parameter $R \propto 1/T$, the WIMP density $N(0)$ today, when the temperature is $T_0 = 2.73$ K, will be

$$N(0) \sim \frac{(T_0/T)^3 \times (T^2/M_{PL})}{\langle \sigma v \rangle} \tag{4.18}$$

The corresponding WIMP energy density will be

$$\rho_{wimp} = MN(0) \sim \frac{PT_0^3}{(M_{PL}\langle \sigma v \rangle)} \sim \frac{6 \times 10^{-31}}{\langle \sigma v \rangle} \text{ GeV s}^{-1}$$

with σv in cm^3 s^{-1}. Dividing by the critical energy density $\rho_c = 3H_0^2 c^2/8\pi G \approx 5 \times 10^{-6}$ GeV cm^{-3} from (2.23) we get for the closure parameter

$$\Omega_{wimp} = \frac{\rho_{wimp}}{\rho_c} \sim \frac{10^{-25}}{\langle \sigma v \rangle} \text{ cm}^3 \text{ s}^{-1} \tag{4.19}$$

Note that for a given cross section, the closure parameter is independent of WIMP mass, apart from small corrections due to the weak dependence of $P = M/T$ on the mass. At freeze-out the WIMP velocity will be given by $\frac{1}{2}Mv^2 = 3T/2$, or $v/c \sim (3/P)^{1/2} \sim 0.3$ so that (4.19) indicates that an annihilation cross section of the order of 10^{-35} cm^2 would lead to a closure parameter of order unity. It is perhaps quite remarkable that this cross section is of the order of magnitude expected for the weak interactions, since there is no *a priori* connection between the closure density of the universe and the Fermi constant. In any case, if this is a mere coincidence it is a bonus, in the sense that one does not have to invent new couplings as well as new particles in trying to account for dark matter.

However, as indicated previously for neutrinos, for the conventional weak coupling, the annihilation cross section will rise as M^2 and, therefore, the closure parameter falls as $1/M^2$, and at high values of M the WIMPs could make no substantial contribution to the energy density. This state of affairs holds until the WIMP mass becomes comparable with or larger than the mass of the mediating weak bosons, W and Z. Then, as indicated in (1.9) the boson mass in the propagator term becomes less important. For $M \sim M_Z$, annihilation to WW or ZZ pairs (as well as to more exotic SUSY particles) at first increases the cross-section (as in Fig. 1.9 for the corresponding electromagnetic process), but eventually at large enough $s = 4M^2$ it falls off as $1/s$, so that

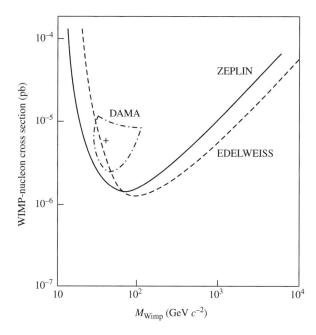

Fig. 4.9 Upper limits on the WIMP–nucleon scattering cross section as a function of WIMP mass from the EDELWEISS (Benoit *et al.* 2002) and ZEPLIN (Smith *et al.* 2002) experiments. The cross section inferred by the DAMA group (Bernabei *et al.* 2002) from annual modulation is shown by the closed contour.

$\Omega_{\text{wimp}} \sim M^2$—see Fig. 4.8. However, depending on the exact details of the SUSY model, any WIMP masses below about 100 TeV could contribute values of $\Omega_{\text{wimp}} < 1$. Of course, the hierarchy problem (Section 3.12) requires that SUSY particles cannot have masses much above one TeV.

4.9   Expected WIMP cross sections and event rates

There are two distinct possibilities for detection of WIMPs. Direct detection of dark matter relies on observation of the scattering of the WIMPs inside the detector, while indirect detection relies on the observation of the annihilation products of WIMPs in, for example, the halo of the galaxy, or as a result of their gravitational trapping in the core of the Sun or the Earth. In the latter cases of course, the only secondary products that could be detected would be neutrinos. In fact, no evidence for an extra flux of high energy neutrinos from the direction of the Sun or from the Earth's core has been found.

In the case of direct detection, the WIMP rate may be expected to exhibit some angular and time dependence. For example, if WIMPs predominantly populate the galactic halo, there might be a daily modulation because of the shadowing effects of the Earth when turned away from the galactic centre. An annual modulation in the event rate would also be expected as the Earth's orbital velocity around the Sun adds to or subtracts from the velocity of the Solar System with respect to the galactic centre, so that both the velocity distribution of WIMPs and the cross section for detection change with time.

We now discuss the expectations for elastic scattering of WIMPs by nuclei in the detector, signalled by the nuclear recoil. The WIMP velocities are expected to be of the order of galactic escape velocities, that is $v \sim 10^{-3}c$, so that we can use non-relativistic kinematics. Then, if $E = Mv^2/2$ is the kinetic energy of a WIMP of mass M, colliding with a nucleus of mass $M_N = mA$ where A is the mass number and m the nucleon mass, it is straightforward to show that the total CMS energy is (refer to Appendix B for relativistic transformations)

$$\varepsilon = [(M + M_N)^2 + 2M_N E]^{1/2}$$

$$\approx [M + M_N]\left[1 + \frac{M_N E}{(M_N + M)^2}\right] \tag{4.20}$$

where in the second line we use the fact that $E \ll M_N$ or M. If p^* denotes the (equal and opposite) momentum of each particle in the CMS, then, in the non-relativistic approximation

$$\varepsilon = \left(M_N + \frac{p^{*2}}{2M_N}\right) + \left(M + \frac{p^{*2}}{2M}\right) \tag{4.21}$$

so these two equations give

$$p^{*2} = \frac{2\mu^2 E}{M} = \mu^2 v^2 \tag{4.22}$$

where $\mu = M_N M/(M_N + M)$ is the reduced mass. The laboratory kinetic energy E_r of the recoiling nucleus is maximum when its CMS vector momentum is reversed in the collision, so that it is scattered in the forward direction with laboratory momentum $2p^*$ and $E_r(\text{max}) = 2p^{*2}/M_N = 2\mu^2 v^2/M_N$. This has a value varying from $v^2 M_N/2$ when $M_N = M$ to $2v^2 M_N$ when $M \gg M_N$. Since the CMS angular distribution at these low velocities will be isotropic, the recoil energy distribution will vary uniformly between zero and $E_r(\text{max})$. So with $v \sim 10^{-3} c$ and $M_N \sim A$ GeV, we obtain recoil energies $E_r \sim A$ keV or less. Hence, a sensitive detector is needed to observe such small recoil energies.

The scattering cross section of the target nucleus depends on details of the SUSY model parameterization. For guidance, we again assume a conventional weak cross section. From (1.18) with $|T_{if}| = G_F$ the cross section per target nucleus will be

$$\sigma \approx \frac{G_F^2 p^{*2} K}{\pi v_r^2} = \frac{G_F^2 \mu^2 K}{\pi} \tag{4.23}$$

where the relative velocity of incident particle and target nucleus in the CMS is $v_r = v = p^*/\mu$. The quantity K is a numerical model-dependent factor. For spin-independent coupling, the scattering amplitudes from the different nucleons in the target nucleus should add coherently, so that K will contain a factor A^2. However, the momentum transfer is of order $p^* = \mu v \sim 10^{-3} A$ GeV, while the nuclear radius $R = 1.4 A^{1/3}$ fm $\sim 7 A^{1/3}$ GeV^{-1}. The nucleus can only recoil coherently if $p^* R \ll 1$, or $A \ll 50$, otherwise K will contain a suppression factor (the square of the so-called form-factor).

The other possibility is spin-dependent (axial-vector) coupling, for which the amplitudes from different nucleons do not add since most of the nucleon spins cancel out, and the cross section is smaller by a factor of order A^2 than that for coherent scattering. As examples, for WIMPs identified with sneutrinos (see Table 3.3) the interaction is scalar and coherent, while if the WIMP is the LSP (neutralino) with spin $\frac{1}{2}$, the interaction will be mostly incoherent.

The event rate to be expected depends on the WIMP number density and the scattering cross section. Because of their gravitational concentration in the galaxy and particularly the disc and halo, the WIMP energy density in the solar system is estimated to be some 10^5 times that in the universe at large, at $\rho_{wimp} \sim 0.3$ GeV cm^{-3}, yielding a flux of $\varphi_{wimp} \sim 0.3 \, v/M$ cm^{-2} s^{-1}, where the WIMP mass M is in GeV. The reaction rate per target nucleus will be $W = \sigma \varphi_{wimp}$, as in (1.14), and the event rate per unit target mass from (4.23) will be

$$R = \frac{W}{M_N} \sim \frac{10K}{AM} \quad \text{events kg}^{-1} \text{ day}^{-1} \tag{4.24}$$

Typical values of $M = 100$ GeV and $A = 20$ predict $R \sim 0.01$ events per kg per day for incoherent scattering and $R \sim 1$ per kg per day for coherent. The cross sections and rates depend on the many free parameters in SUSY models, and so the above numbers are only indicative: but they suffice to emphasize the severe experimental problems of detecting signals from low energy recoils at extremely low rates, against cosmic ray and radioactive background effects.

4.10   Experimental WIMP searches

Direct detection of WIMPs via the recoil of the scattering nucleus has been attempted by a number of different methods. The ionization from the recoil as it traverses the detector material can be recorded as a pulse in a semiconductor counter (Ge or Si), which has excellent sensitivity to recoils in the keV energy range, or in the form of scintillation light from scintillating materials such as NaI or liquid Xe. However, the bulk of the energy lost by the recoil will appear in the form of lattice vibrations (phonons) in the medium. These can be recorded through cryogenic detectors operating at low temperature (<1 K). The phonon pulse results in a local rise in temperature, which will affect the resistance of a thermistor attached to the detector and can be recorded as a voltage pulse. The phonon pulses are very slow in comparison with electrical pulses from ionization, and therefore random background noise can be more of a problem.

As stated earlier, the signals from WIMPs have to be distinguished from those due to background radioactivity and the interactions of cosmic-ray induced neutrons and photons. For this reason, emphasis has to be on very pure materials and on locating the detectors deep underground to reduce the cosmic ray muon flux. The separation of genuine from background events can be achieved in several ways. For example, the energy spectrum and event rate of recoils will be different for detectors with nuclei having different A values and/or different nuclear spins. Some discrimination is also possible on the basis of pulse length in scintillators. Electrons produced from photon or radioactive backgrounds have longer pulse lengths than nuclear recoils of the same energy. Similarly, the ratio of ionization energy loss to lattice (phonon) energy loss is also different for recoil nuclei and for electrons. Finally, WIMP recoils should show a small seasonal dependence of the signal. The latter arises from the fact that the Sun orbits the Galaxy with $v \sim 200$ km s^{-1}, while the Earth orbits the Sun with $v \sim 30$ km s^{-1}. The two velocities add vectorially to give a maximum in summer (on June 3) and minimum in winter. There results a small annual change in WIMP fluxes, detector cross sections and event rates, of the order of 5%.

Figure 4.9 shows current experimental upper limits on WIMP–nucleon scattering cross sections, assuming coherent nuclear scattering. These limits already exclude the cross section ranges expected for LSPs in some versions of the supersymmetric models. However, more refined experiments are continually under development and will eventually have the necessary sensitivity to explore the entire range of expected cross sections. So far, only one experiment has claimed an annual modulation signal. Using a large (100 kg) NaI detector, the DAMA group report a 5% annual modulation for low recoil energies, less than 6 keV, with a significance level of about 2 standard deviations. However, the EDEL-WEISS experiment with a cryogenic Ge detector, and the ZEPLIN experiment using liquid Xe, set limits incompatible with the DAMA result.

4.11   Dark energy: the Hubble plot at large red shifts

As stated in Chapter 2, the total energy density ρ_{tot} appearing in the Friedmann equation (2.8) may have three separate sources—matter, radiation, and vacuum energy density—as listed in Table 2.2. For non-relativistic matter, $\rho \propto R^{-3}$ while for radiation or any ultra-relativistic particles, $\rho \propto R^{-4}$. In either case, as indicated from this table, the density falls off with time as t^{-2}. On the other hand, the vacuum energy density—if indeed that is the source of dark energy—is constant, so that however small it may be relative to other forms of energy density at early times, eventually it must begin to dominate at large enough values of t. From the expression (2.35) for the deceleration parameter q it is apparent that if at some epoch the vacuum energy density $\rho_v > \rho_r + \rho_m/2$, the universe will **accelerate**.

The evidence for a substantial dark energy component comes from several sources: galaxy redshift surveys; the age of the universe (see Example 2.3); but most dramatically from the measurement of the Hubble flow at large redshifts from the analysis of Type Ia supernovae. Large redshifts mean enormous distances and hence, a look backwards in time to the remote past.

First, we should note that there are in fact several different types of supernovae. Type II supernovae are associated with the final stages of very massive stars and their subsequent transformation to neutron stars and black holes, as discussed in Chapter 7. A Type Ia supernova, which is what concerns us here, is distinguished from other types by its very high luminosity and the absence of lines from the hydrogen spectrum. It is believed that it develops from stars which have burned all their hydrogen, and have reached the white dwarf stage with a carbon/oxygen core, as described in Chapter 7. The flash is assumed to be due to the thermonuclear explosion of the white dwarf, which is part of a binary system and has steadily accreted matter from its main sequence companion until it eventually exceeds the critical Chandrasekhar mass (see Section 7.8) and explodes. The result is that in a matter of seconds the stellar material is converted largely to nickel and iron by thermonuclear fusion, with a tremendous release of energy. The dispersed nickel nuclei subsequently decay to cobalt and iron over a period of months, setting the timescale for the (roughly) exponential decay of the light curve.

The light output from a Type Ia supernova typically grows over a period of a few weeks, before reaching a maximum and thereafter falling off exponentially. There are variations in the maximum light output between different supernovae, but there is found to be a dependence of the peak luminosity on the timescale τ to reach the maximum, varying approximately as $\tau^{1.7}$. Presumably, this dependence arises because the brighter supernovae originate from more massive stars. They are more energetic and the ensuing fireball has to expand for longer in order for the opacity to drop enough for the photons to escape. After making this empirical correction based on the 'width' of the light curve, the estimated light output from different supernovae shows remarkably small dispersion, only of the order of 6%. Incidentally, it may be remarked that allowance must also be made at high redshift for the fact that the decay curves will be 'stretched' by the time dilation factor $(1 + z)$. It should also be noted at this point that the procedure for normalizing supernova output is entirely empirical.

Before describing the experimental results, let us first ask how different cosmological parameters can change the slope of the Hubble plot as a function of redshift. The actual plot is made of the distance as estimated from the luminosity or apparent magnitude of the star, which is the so-called luminosity distance D_L defined in equation (2.3), the expected value of which can be calculated as a function of redshift z from the presently measured value of the Hubble parameter H_0, and assumed values of the various contributions to the closure parameter Ω from matter, vacuum energy, and the curvature term, as defined in Section 2.5. We neglect radiation here, since it is important only at very large redshifts $(z > 1000)$. As indicated in Appendix C.4, the results for different scenarios then follow from straightforward integration. Table 4.1 gives the expressions for the dimensionless quantity $D_L(z)[H_0/c]$, and the values of $D_L(z)$ are plotted in Figs 4.10 and 4.11 together with the experimental data.

Figure 2.3 gave the results from the measurements of supernovae at low redshifts $(z < 0.1)$, using Cepheid variables in the same galaxies to calibrate the distance/luminosity scale. The results for this region of z indicated a very linear and uniform Hubble flow with $H_0 = 72$ km s^{-1} Mpc^{-1}. The fact that the data fell on a straight line with little dispersion gives confidence that the normalization methods employed are adequate.

Because of this reproducibility and the brightness of the supernovae, which allows one to probe to large distances and redshifts, Type Ia supernovae have come to be regarded as 'standard candles', so that their brightness or apparent magnitude, coupled with the decay curve, fixes the distance from the Earth. However, such events only occur at the rate of order one per century per galaxy.

Table 4.1 Luminosity distance versus redshift

Dominant component	Ω_m	Ω_v	Ω_k	$D_L H_0/c$
Matter (Einstein–de Sitter universe)	1	0	0	$2(1 + z)[1 - (1 + z)^{-1/2}]$
Empty universe	0	0	1	$z(1 + z/2)$
Vacuum	0	1	0	$z(1 + z)$
Flat	0.35	0.65	0	From numerical integration (see Appendix C.4)

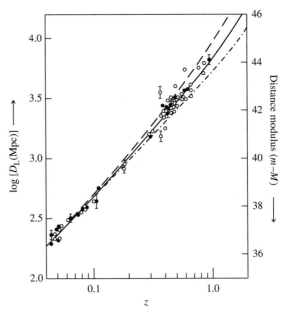

Fig. 4.10 Hubble plot from Type Ia supernovae at low and high redshifts, to be compared with Fig. 2.3. The data are from the High z SN Search (Riess *et al.* 2000), denoted by solid circles, and from the Supernova Cosmology Project (Perlmutter *et al.* 1999)), shown as open circles. Typical errors are shown for a few points. The solid line shows the expected variation for an empty universe ($\Omega_k = 1, \Omega_m = \Omega_v = 0$) that is neither accelerating nor decelerating—see (2.8) and (2.35). The dot-dashed line is for a flat, matter-dominated (so-called Einstein–de Sitter) universe ($\Omega_m = 1, \Omega_v = \Omega_k = 0$) that is decelerating relative to the expansion at early times, while the dashed line is for the vacuum-dominated case ($\Omega_v = 1, \Omega_m = \Omega_k = 0$) of perpetual acceleration. In calculating these curves from Table 4.1, radiation contributions to the energy density were ignored, and H_0 was taken as 65 km s^{-1} Mpc^{-1} (slightly less than the value taken from data at low redshifts and other sources as in Fig. 2.3). The vertical scales are of the logarithm of the luminosity distance (at left) and distance modulus (2.3) on the right.

The method employed was to scan a strip of the sky containing around ten thousand galaxies, then to repeat the survey three weeks later and by taking the difference, to detect the dozen or so supernovae that had developed in the meantime. Once identified, their light curves could be studied in detail.

The data in Fig. 4.10 suggest an acceleration of the expansion, manifested by a small increase in the slope of the Hubble plot, that is an increase, for $z > 0.1$, in the distance for a given redshift, relative to what would be expected for a universe that is neither accelerating nor decelerating (shown by the solid curve). Notice that for a non-accelerating universe, where H is obviously constant, the plot in Fig. 4.10 is **not** a straight line, because of the way in which distance (or magnitude) is measured from the observed brightness of the source in an expanding universe.

In Fig. 4.11 the linear trend has been divided out by plotting the **difference** in magnitude (or equivalently, in the logarithm of distance) as compared with the case of an empty universe with $\Omega_m = \Omega_v = 0$, and hence $\Omega_k = 1$. This shows more clearly the effects itemized in Table 4.1 for different cosmological parameters. The case of an empty universe that is neither accelerating nor decelerating is in this case shown by the horizontal straight line.

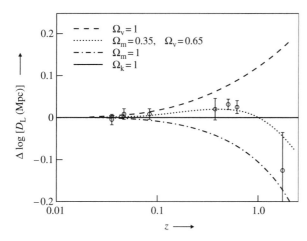

Fig. 4.11 By dividing out the Hubble slope in the previous figure, the trends with z become clearer. This plot is actually of the *difference* in magnitude, or in the logarithm of the distance, from the value for an empty, non-accelerated universe, shown as a horizontal line. Curves with an upward slope correspond to acceleration, and those with a downward slope to deceleration. The long-dashed curve is for a vacuum-dominated universe, the dot-dashed line for a matter-dominated universe, and the dotted line is for a flat universe with $\Omega_m = 0.35$ and $\Omega_v = 0.65$. In this case the vacuum term Ω_v dominates at low $z < 0.4$, corresponding to an acceleration, while at larger z values, the matter term $\Omega_m(1 + z)^3$ is more important and leads to deceleration (see Example 4.3). The experimental points represent the data from Fig. 4.10, combined into a few bins of redshift. The point on the extreme right is from a single supernova (SN 1997ff) at redshift 1.7 (after Straumann 2002).

The results in Fig. 4.11 appear to exclude a flat matter-dominated universe ($\Omega_m = 1$). The best fit to this data is for $\Omega_m = 0.35, \Omega_v = 0.65$, in good agreement with independent estimates from observations of large-scale galaxy surveys combined with analysis of anisotropies in the microwave background radiation, described in Chapter 5, as well as estimates from the Hubble parameter and independent determinations of the age of the universe (see Example 2.3 and Fig. 5.11). The best-fitting curve in this case indicates an upward slope for $z < 0.4$ (i.e. an acceleration) and a downward slope (i.e. a deceleration) for larger z values and earlier times in the universe.

Example 4.3 *In a flat universe, with $\Omega_m(0) = 0.35$ and $\Omega_v(0) = 0.65$, at what value of z will the acceleration/deceleration be zero?*
From (2.35), $q(z) = \frac{1}{2}\Omega_m(0)(1 + z)^3 - \Omega_v(0)$, which is zero when $(1 + z) = [2\Omega_v(0)/\Omega_m(0)]^{1/3}$, or $z = 0.549$, which is the position of the flat maximum in the dotted curve in Fig. 4.11. A universe that is neither accelerating nor decelerating is often said to be 'coasting'.

Note that the analysis above involves a comparison of high redshift with low redshift supernovae, so that the absolute luminosity scale, which is of course necessary in order to measure the Hubble parameter, is not required when comparing slopes.

However, there are some complications in comparing the luminosities of supernovae at different redshifts. First, they occur at different epochs and the metal content in early stars will be less than that in more recent ones, which have formed from the recycled debris of earlier stellar generations, and this can affect the opacity and hence the luminosity. This and other possible differential effects, such as dimming due to absorption or scattering by dust, which will be more important the more distant the star, can either be excluded because the effects go in the wrong direction, or because they can be corrected for. Nevertheless, the recent (2002) results, indicating an expanding universe which today is actually accelerating, need to be treated with some caution.

4.12   Vacuum energy: the Casimir effect

In Chapter 2 we already mentioned that present observations, such as those described in the previous section, appear to require for their interpretation a major contribution from dark vacuum energy to the present energy density of the universe. This vacuum energy is postulated to arise through quantum fluctuations, that is, the spontaneous appearance and disappearance of virtual particle–antiparticle pairs and quanta, as required by the Uncertainty Principle. That this concept is not just a figment of the physicist's imagination was already demonstrated many years ago, when Casimir (1948) predicted that by modifying the boundary conditions on the vacuum state, the change in vacuum energy would lead to a measurable force, subsequently detected and measured by Sparnay (1958) and more recently and comprehensively by Lamoreaux (1997) and Roy *et al.* (1999).

Essentially, the Casimir effect in its original configuration arises when two perfectly conducting, flat parallel plates at zero temperature are placed close together with a very small separation a (see Fig. 4.12). The vacuum energy between the plates is different from that in the same volume with the plates absent, because the plates introduce boundary conditions on the fluctuating field. For example, if the virtual quanta are those of the electromagnetic field, there are boundary conditions on the associated electric and magnetic fields (the component of **E** parallel to the plates and of **B** normal to the plates must vanish at the surface, so that if the x-axis is normal to the plates, wavenumbers $k_x < \pi/a$ are forbidden). This difference in vacuum energy corresponds to a force between the plates which is actually attractive in this particular configuration (the sign of the force, in general, depends on the geometry).

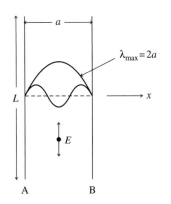

Fig. 4.12 End-on view of parallel plates A and B in an experiment to measure the Casimir effect, demonstrating the existence of vacuum energy density. The electric field E must vanish at the surface of perfectly conducting plates, so that $\lambda(\text{max}) = 2a$ or $k_x(\text{min}) = \pi/a$, and such boundary conditions change the value of the vacuum energy and give rise to an attractive force. Successful attempts to verify the effect have used the experimentally simpler configuration of a plate and a hemisphere rather than two plates.

On the basis of dimensional arguments alone, one can understand that the force per unit area in Fig. 4.12 must be of order $\hbar c/a^4$. Planck's constant times the velocity of light must enter, as it does in all Uncertainty Principle problems, and gives dimensions of energy times length, which has to be divided by the fourth power of a length in order to get a force per unit area. If the plates are of side $L \gg a$, the only length of relevance is the separation a. We simply quote here the result of a full calculation (see e.g. Itzykson and Zuber 1985)

$$ F = -\frac{\pi^2}{240}\frac{\hbar c}{a^4} \sim \frac{13}{a^4}\ \mu\text{g wt cm}^{-2} \tag{4.25} $$

where the plate separation a is in microns (μm). This tiny force, of order micrograms weight per square centimetre, and its dependence on plate separation, has been measured and the above formula verified to within 1% accuracy. Of course, the effect does not measure the absolute value of the vacuum energy density, but only the change when the topology is altered. On the other hand, the gravitational field couples to the **absolute** values of energy and momentum, and the total vacuum energy can only be measured via its gravitational effect.

The Casimir effect has implications outside quantum field theory and cosmology, for example in electromechanical systems on sub-micron scales, where it could lead to malfunctions of the system. There are also classical macroscopic analogues of the Casimir effect. The most famous is known to all mariners. Under certain wave conditions, two ships sailing close together beam-to-beam experience a force of attraction, due to the fact that the wave pattern between the ships is affected by the presence of the hulls and certain wavelengths are again suppressed (see Buks and Roukes 2002 for references).

4.13 Problems with the cosmological constant and dark energy

The cosmological constant $\Lambda = 8\pi G\rho_{\text{vac}}$ presents one of the major—if not the major—conceptual problems in cosmology, and has done so ever since Einstein introduced it.

It has long been argued that the dark energy density associated with the cosmological constant ought to have a 'natural' value determined by the scale of gravity. This natural unit is then the Planck mass energy $M_{\text{PL}}c^2 = (\hbar c^5/G)^{1/2} = 1.2 \times 10^{19}$ GeV, in a cube of side equal to the Planck length $\hbar/M_{\text{PL}}c$ (see (1.12)), that is, an energy density

$$\frac{(M_{\text{PL}}c^2)^4}{(\hbar c)^3} \sim 10^{123} \text{ GeV m}^{-3} \qquad (4.26)$$

a truly gigantic number, which of course is nonsensical since it would imply that the universe could only be a few seconds old at most. So, it is perhaps instructive to see in more detail how this number is arrived at.

In quantum field theory one can describe the vacuum fluctuations as due to an ensemble of simple harmonic oscillators of different frequencies. The energy of one such oscillator is $(n + \frac{1}{2})\hbar\omega$ where ω is an angular frequency and $n = 0, 1, 2, \ldots$ We are concerned here with the ground state, that is, the so-called 'zero-point energy' $E = \frac{1}{2}\hbar\omega$. In a sense, it is a matter of choice whether one takes this zero-point energy seriously or simply ignores it, since measurements are usually about energy differences, and only when we come to gravity do we have to worry about the absolute energy value. However, if we try to identify it with the mysterious, dark vacuum energy, then we have to sum over all oscillators in the volume. From (1.16), the number of possible quantum states in a spatial volume V, with wave numbers $k = p/\hbar$ lying in the element $k \to k + dk$ and integrated over all directions, is $4\pi Vk^2\,dk/(2\pi)^3$. So the total energy per unit volume of all the oscillators will be

$$\varepsilon = E/V = \frac{\hbar}{4\pi^2} \times \int k^2\,dk\,\omega_k \qquad (4.27)$$

The angular frequency is related to the wavenumber by $\omega_k^2 = k^2c^2 + m^2c^4/\hbar^2$ where m is the oscillator mass. Obviously this integral is divergent, but let us cut it off at some value k_{m} or $E_{\text{m}} \gg mc^2$. Then, with $\omega_k \approx kc$ in the relativistic approximation,

$$\varepsilon = \frac{\hbar c}{16\pi^2} k_{\mathrm{m}}^4 = \frac{E_{\mathrm{m}}^4}{16\pi^2(\hbar c)^3} \tag{4.28}$$

Here, the cut-off is arbitrary. For example, we can place it at an energy scale where we expect quantum field theory to start to fail, and that is the Planck scale $E_{\mathrm{m}} = M_{\mathrm{PL}}$ of quantum gravity. Including the numerical constants left out in (4.26), this gives $\varepsilon \sim 10^{121}$ GeV m^{-3} as before, to be compared with the critical energy density in (2.23) of $\rho_{\mathrm{c}} = 5$ GeV m^{-3}, of which only a part can be assigned as vacuum energy. So why is the observed vacuum energy/cosmological constant only about 10^{-121} of the naïve expectation? Of course, one might vary the cut-off in (4.28), but there seems to be nothing to gain thereby, as the following example shows.

Example 4.4 *What would a 'reasonable' value for the vacuum energy density imply for the cut-off in maximum vacuum oscillator energy ?*
  One could reduce the value of E_{m} so that the vacuum energy density is of the same order as the critical density, that is $\varepsilon = \rho_{\mathrm{c}} = 5$ GeV m^{-3}. Then, substituting in (4.28) would give $E_{\mathrm{m}} \sim 0.01$ eV only. This is just as unacceptable as the result (4.26), since it is ridiculously small in comparison with the masses of practically all known elementary particles, or of atomic energy levels.

Twenty years ago, before the importance of dark energy was fully apparent, it was believed that the matter density was such that $\Omega_{\mathrm{m}} = \Omega_{\mathrm{tot}} = 1$ and that the cosmological constant might even be identically zero. In view of the above argument, the difficulty then was to understand why the cosmological constant was so incredibly small, or even zero. Here, at least, one can say that zero is a natural number, for which a reason might be found. For example, the masslessness of the photon is associated with a symmetry principle, namely the local gauge invariance of the electromagnetic interaction as described in Section 3.7. However, no symmetry principle is known that could set $\Lambda = 0$. Indeed, the finiteness of the vacuum energy/cosmological constant seems to follow inexorably from quantum field theory, for the very simple reason that the virtual states of the real particles which contribute to Ω_{m} **must** contribute to Ω_{v}. However, we should emphasize that the above integral over the elementary oscillators will include the summed effects of all types of elementary particles, with somewhat different amplitudes and phases, and that there can also be cancellations. For example, the contributions of fermions and bosons come in with opposite signs. Thus, in a theory of **exact** supersymmetry, where every boson is matched by a fermion of the same mass and vice versa, there could even be exact cancellations, with a vacuum energy of zero. However, we know that in the real world, even if supersymmetry turns out to be valid, it must be a badly broken symmetry. While at very large values of k in (4.27), well above the supersymmetric scale, there could be exact cancellations, this would not be the case at lower k values.

The actual situation is even worse than this, since the supernovae results present us with a finite, non-zero number for the vacuum energy density, inconceivably small in comparison with what might be expected, but one for which the relative contribution to the overall energy density apparently changes with time. For example, the ratio of vacuum energy density to matter energy density today is $\rho_{\mathrm{v}}/\rho_{\mathrm{m}} \sim 2$, but while ρ_{v} is constant, $\rho_{\mathrm{m}} \propto R^{-3} \propto (1+z)^3$. Thus, $\rho_{\mathrm{v}}/\rho_{\mathrm{m}} \sim 2/(1+z)^3$, and at the time of decoupling of matter and radiation, when

$(1 + z) \sim 1100$, the ratio ρ_v/ρ_m would have only been 10^{-9}. Equally, in the future, the ratio will become very large, as the matter density falls off as $1/R^3$ and the scale parameter R will eventually increase exponentially with time.

So, another puzzle is the fact that despite these huge variations in the fractional contribution of the vacuum density with time, at the present epoch it just happens to be within a factor 2 of the matter energy density. One possibility to avoid this strange coincidence is to postulate a new type of fundamental interaction, which does involve a time dependence for the dark energy. As we shall see in Chapter 5, in connection with inflation in the early universe, conventional field theories do predict changes in the vacuum energy with time, but these involve **abrupt** changes connected with phase transitions. Gentle changes would require a new type of force. Since this would postulate a fifth form of fundamental interaction it has been called **quintessence**. This new type of interaction would involve an equation of state (see Table 2.2) such that the ratio $w = P/\rho c^2$ lies in the interval $-1 < w < -1/3$ so as to ensure an accelerated expansion, with the possibility that $w = w(t)$ is time dependent.

Finally, a totally different and perhaps more desperate (or more sensible?) approach to the problem has been to appeal to the anthropic principle, namely that life exists only when the laws of physics allow it. In this case, it is the value for Λ at the present epoch. Had it differed by just an order of magnitude or so, there would have been no human race to ponder on the problem. As the saying goes, we live in the best of all possible worlds. This argument becomes more plausible in the context of inflationary models of the early universe, described in Chapter 5. These suggest that our particular universe is just one of an enormous number of universes, so that the human race evolved in the one where conditions happened to suit it.

In summary, the phenomenon of the cosmological constant or dark energy, accounting at the present time we believe for the bulk of the energy in the universe, is simply not understood, and this, like our incomprehension of the matter–antimatter asymmetry of the universe, could be ranked as a major failure in the subjects of cosmology and particle physics. These failures have not grown up overnight. The problems of dark matter and of the vacuum energy/cosmological constant have been lurking for at least seventy years, but they have become more acute in the last decade because of the vastly improved quality and quantity of the experimental data. Optimistically, however, these problems are to be seen as presenting great challenges for the future.

4.14 Summary

- The rotation curves of stars in spiral galaxies imply that the bulk of the matter (80–90%) is non-luminous, and located in a galactic halo.
- Studies of X-rays from galactic clusters indicates velocities of the gas particles emitting the X-rays that are far in excess of escape velocities based on the visible mass.
- Dark matter is also required in cosmological models of the early universe, if the structure of galaxies and galaxy clusters is to evolve from the very small primordial density fluctuations deduced from inhomogeneities in the microwave radiation (discussed in Chapter 5).
- Independent evidence for dark matter is found from the gravitational lensing of distant galaxies and clusters by foreground galaxies. Gravitational

micro-lensing of individual stars, appearing as a temporary, achromatic enhancement in luminosity, shows that some of the galactic dark matter is baryonic, this matter appearing in the form of so-called MACHOs, which are dark star-like objects with masses of 0.001–0.1 $M_\odot$.

- Baryonic dark matter makes less than 25% contribution to the total dark matter density, and the bulk of dark matter is non-baryonic.
- The most likely candidates for dark matter are WIMPs, that is, very massive, weakly interacting particles, constituting 'cold' dark matter. Until the nature of such particles is established, the most common suggestion is that they are supersymmetric particles such as neutralinos.
- Several experiments are under way to detect WIMPs directly by observing nuclear recoils from elastic scattering of WIMPs.
- Observations of Type Ia supernovae at high redshifts ($z \sim 1$) suggest that in earlier times the expansion rate (i.e. the Hubble parameter H) was less than it is today; or that, relative to earlier times, the universal expansion is now accelerating.
- The present acceleration is interpreted in terms of a finite value for the cosmological constant, or for the existence of dark (vacuum) energy. This dark energy seems to account for some two-thirds of the total energy density of the universe today.
- The reality of vacuum energy is evidenced by the laboratory observation of the Casimir effect, which is a manifestation of a change in the vacuum energy when boundary conditions are imposed upon it.
- There is no satisfactory explanation for the observed magnitude of the dark energy. Other possible sources of dark energy, such as a completely new type of interaction, cannot be excluded.

Problems

More challenging or longer questions are denoted by an asterisk.

(4.1) Estimate the angular deflection of a photon by a point mass M, according to Newtonian mechanics. Express the result in terms of b, the closest distance of approach

*(4.2) Show that, as indicated in Section 4.6, the temperature at which neutrinos decouple from other matter and radiation in the early, radiation-dominated universe, is $kT \sim 3\,\text{MeV}$. Hints for the stages in solving this problem are as follows:

  (a) Start with the cross section for the process $e^+ + e^- \rightarrow \nu_e + \bar{\nu}_e$, via W exchange, which is given in Example 3.2 as $\sigma = G_F^2 s/(6\pi)$, where s is the CMS energy squared and it is assumed that masses can be neglected (i.e. $\sqrt{s} \gg m_e c^2$). Evaluate this cross section in cm^2 when s is in MeV^2.
  (b) Show that, treating the electrons and positrons as a Fermi gas of relativistic particles the mean value

of s at temperature T is given by $\langle s \rangle = 2\langle E \rangle^2$, where $\langle E \rangle$ is the mean energy of the particles in the distribution, which can be found from equation (2.40).

  (c) Calculate the density N_e of electrons and positrons as a function of kT, and hence the rate for the above reaction, $W = \langle \sigma v \rangle N_e$ where v is the relative velocity of the particles, as a function of kT.
  (d) Using (2.38) and (2.45) calculate the time t of the expansion as a function of kT, and setting this equal to $1/W$, deduce the value of kT at neutrino decoupling.

(4.3) Calculate an expression for the tangential velocity v of a star near the edge of the disc of a spiral galaxy of radius R and mass M, and thus find an expression for the optical depth τ for microlensing in terms of v. Give numbers for the Milky Way, with a mass of $1.5 \times 10^{11} M_\odot$ and a disc radius of 15 kpc.

(4.4) Obtain an expression for the kinetic energy E_R of a nucleus of mass M_R recoiling in an elastic collision with a dark matter particle of mass M_D and incident kinetic energy E_D, in terms of the angle of emission relative to the incident direction. Find the limiting values of recoil energy in terms of M_D and M_R. Calculate the maximum recoil energy of a nucleus of 80 proton masses, in collision with a dark matter particle of mass 1000 times the proton mass, travelling with a typical galactic velocity of 200 km s^{-1}.

(4.5) Assume that the universe is flat, with $\Omega_m(0) = 0.35$, $\Omega_v(0) = 0.65$. What is the numerical value of the accel-

eration or deceleration with respect to the Earth, of a galaxy at redshift $z = 0.03$? Compare this with the local acceleration (g) due to the Earth's gravity. Neglect the 'peculiar velocity' of the Earth with respect to the Hubble flow and assume $H_0 = 70$ km s^{-1} Mpc^{-1}.

(4.6) Show that, if the vacuum energy density and matter energy density today are comparable in magnitude, then, when the universe was a fraction f of its present age, the relative contribution of the vacuum energy would have been f^2.

Development of structure in the early universe

The Big Bang model described in Chapter 2 seems to give a rather convincing description of the development of the universe. It is underpinned by three striking successes:

1. The observation of the redshift of distant galaxies.
2. The prediction of the abundances of the light elements via primordial nucleosynthesis.
3. The existence of the all-pervading cosmic microwave background radiation.

This success is all the more remarkable since the principal tenets of the model—isotropy and homogeneity of the 'cosmic fluid'—are to be contrasted with the universe today, characterized by a decidedly non-isotropic, non-homogeneous nature—galaxies, galactic clusters, voids, and so forth. The question arises: How did we get from the uniformity of the Big Bang model to the present universe with its extremely lumpy structure, out to the very largest scales?

As indicated below, it can plausibly be argued that this structure had its origins in quantum fluctuations in energy density that occurred in the very early universe and were then 'frozen out' when the universe underwent an exponential and superluminal expansion stage called inflation. These tiny fluctuations in density and temperature—typically at the 10^{-5} level—then acted as seeds for the development of much greater fluctuations in density via the subsequent process of gravitational collapse during the epoch of matter domination.

We shall begin, therefore, with a brief outline of the inflation scenario, which was postulated two decades ago in response to some difficulties with the Big Bang model, mainly with respect to the initial conditions that are apparently required. We first discuss two of the principal ones, the horizon and the flatness problems.

5.1 Horizon and flatness problems

The **particle horizon** is defined as the distance out to which one can observe a particle, by exchange of a light signal. In other words, the horizon and the observer are causally connected. More distant particles are not observed, they are beyond the horizon. The horizon is finite because of the finiteness of the velocity of light and the finite age of the universe. In a static universe of age t, we expect to be able to observe particles out to a horizon distance $L_H = ct$. As time passes, L_H will increase and more particles will move inside the horizon. At the present time the universe has age $t_0 \sim 1/H_0$—see (2.4). The quantity

$ct_0 \sim c/H_0$ is usually referred to as the **Hubble radius**, that is, the product of the Hubble time and the velocity of light.

In an expanding universe, it is apparent that the horizon distance will be somewhat greater than ct. We are assuming here that on very large scales, light travels in straight lines, that is, we are dealing with a flat universe with zero curvature ($K = 0$). Suppose that a light signal leaves a point A at $t = 0$ (see figure below) and arrives at the point B at $t = t_0$. By the time $t = t_0$, A will have moved, relative to B, to the point C.

```
t=t₀     t=0                               t=t′                        t=t₀
 x -------- x -- ---- →  -------------|←cdt′→|--------------------x
 C      A                                                            B
 ← --------------------Lн(t₀)----------------------------→
```

Consider the time interval dt' at time t', where $0 < t' < t_0$. The light signal will cover a distance $c\,dt'$, but because of the Hubble expansion, by the time $t = t_0$, this will have expanded to $c\,dt'R(0)/R(t')$, where $R(t)$ is the scale parameter in (2.5) and $R(0)$ is its present value. Hence, the horizon distance will be

$$L_H(t_0) = R(0) \int \frac{c\,dt'}{R(t')} \tag{5.1}$$

For most cosmological models, $R(t) \propto t^n$ where $n < 1$, so that integrating from $t = 0$ to $t = t_0$ the above formula gives

$$L_H(t_0) = \frac{ct_0}{(1-n)} \tag{5.2a}$$

One observes that the ratio

$$\frac{L_H(t)}{R(t)} \propto t^{(1-n)} \tag{5.2b}$$

so that the fraction of the universe that is causally connected was once much smaller than it is now.

Example 5.1 *Calculate the particle horizon distance for a flat universe dominated by (a) matter and (b) radiation.*
Refer to Table 2.2; for a matter-dominated universe $n = \frac{2}{3}$ and hence $L_H = 3ct_0 = 2c/H_0$, while for the case of radiation domination, $n = \frac{1}{2}$ and $L_H = 2ct_0 = c/H_0$, where we have used the fact that the ages for the matter- and radiation-dominated cases are $t_0 = 2/(3H_0)$ and $1/(2H_0)$, respectively (see Example 2.3). With $1/H_0 = 14$ Gyr, the corresponding horizon distances are 2.5×10^{26} and 1.25×10^{26} m, respectively.

In particular, the time of decoupling of matter and radiation was $t_{dec} = 3 \times 10^5$ yr (see Section (2.9)), and the horizon size then would have been approximately $2ct_{dec}$. By now, this would have expanded to $2ct_{dec}(1 + z_{dec})$ where from (2.57) $z_{dec} = 1100$. Hence, the angle subtended by that horizon distance at the earth for the case of a flat universe would be, with $t_0 = 1.4 \times 10^{10}$ yr,

$$\theta_{dec} \sim \frac{2ct_{dec}(1+z_{dec})}{c(t_0 - t_{dec})} \sim 2° \tag{5.3}$$

This formula shows that only the microwave radiation observed over small angular scales, of order one degree or so, corresponding to the time of the last interaction of these photons, could ever have been causally connected and in thermal equilibrium with other matter. On the contrary, after allowing for a dipole anisotropy associated with the 'peculiar velocity' of the Earth with respect to the microwave radiation, the temperature of the radiation is found to be uniform to within one part in 10^5, out to the very largest angular scales. This is the horizon problem.

The **flatness problem** arises as follows. From (2.23) and (2.24) the fractional difference between the actual density and the critical density is

$$\frac{\Delta\rho}{\rho} = \frac{\rho - \rho_c}{\rho} = \frac{3Kc^2}{8\pi GR^2\rho} \tag{5.4}$$

If we consider the radiation-dominated era, $\rho \propto R^{-4}$. So from (5.4) it follows that $\Delta\rho/\rho \propto R^2 \propto t$. So at early times, $\Delta\rho/\rho$ must have been much smaller than it is today, when $t \sim 4 \times 10^{17}$ s and it is of order unity. For example, for $kT \sim 10^{14}$ GeV, a typical energy scale of grand unification, $t \sim 10^{-34}$ s, and at that time $\Delta\rho/\rho$ would have been $\sim 10^{-34}/10^{18} \sim 10^{-52}$ (and even smaller than this if we include the period of matter dominance). How then could $\Omega = \rho/\rho_c$ have been so closely tuned so as to be of the order of unity today?

In short, these two problems require a mechanism that allows thermal equilibrium outside conventional particle horizons, and can reduce the curvature K in (5.4) by a huge factor. A possible answer was supplied by Guth in 1981 (Guth 1981). He postulated an extremely rapid exponential expansion by a huge factor as a preliminary stage of the Big Bang, a phenomenon known as **inflation**. Since that time, there have been a number of inflationary models—old inflation, new inflation, chaotic inflation, eternal inflation, etc.—none of them yet fully capable of a completely satisfactory description. However, there seems to be little doubt that some sort of inflationary scenario is an obligatory first stage of the birth pangs of the universe.

5.2   Inflation

In this section we give a brief and qualitative description of the inflation scenario. First we recall the Friedmann equation (2.8):

$$\left(\frac{\dot{R}}{R}\right)^2 = 8\pi G\frac{\rho + \rho_v}{3} - \frac{Kc^2}{R^2}$$

where $\rho = \rho_m + \rho_r$ is the energy density of matter and radiation, and ρ_v is the vacuum energy density, which as explained in Chapter 2 is a space- and time-independent quantity. Suppose a situation arises in which ρ_v dominates the other terms on the right-hand side of the equation. Then, the fractional expansion rate becomes constant and one obtains **exponential growth** over some time interval between t_1 when inflation commences and t_2 when it terminates:

$$\left(\frac{\dot{R}}{R}\right)^2 = 8\pi\frac{G\rho_v}{3}$$

and

$$R_2 = R_1 \exp[H(t_2 - t_1)] \tag{5.5}$$

where

$$H = \left(8\pi \frac{G\rho_v}{3}\right)^{1/2} = H_0 \left(\frac{\rho_v}{\rho_c}\right)^{1/2}$$

Also, since RT is constant, the temperature will fall exponentially during the inflation era, that is, the energy per particle is red-shifted away by the expansion:

$$T_2 = T_1 \exp[-H(t_2 - t_1)] \qquad (5.6)$$

As stated above, the horizon distance at the likely timescale ($t = 10^{-34}$ s) of the GUT era, for example, was $ct \sim 10^{-26}$ m. If we take as the notional size of the present day universe, when $t_0 \sim 4 \times 10^{17}$ s, the value $ct_0 \sim 10^{26}$ m, the radius at $t = 10^{-34}$ s would have been $(10^{-34}/4 \times 10^{17})^{1/2} \times 10^{26} \sim 1$ m, which is enormously larger than the horizon distance at that time. However, in the inflationary scenario, the physical size of the universe before inflation is postulated to be smaller than the horizon distance, so that there was time to achieve thermal equilibrium by causal interactions, which can take place over time intervals entirely dictated by the speed of light. During the inflationary period this tiny region has to expand and encompass the 1 m size of the universe which commences the conventional Big Bang 'slow' expansion, with $R \propto t^{1/2}$. This evolution therefore requires that

$$\exp H(t_2 - t_1) > 10^{26} \quad \text{or} \quad H(t_2 - t_1) > 60 \qquad (5.7)$$

If this condition can be achieved, the horizon problem disappears, since even the most distant parts of the universe would once have been in close thermal contact, and it was only the superluminal expansion of space, far above the speed of light, which necessarily left them disconnected. The flatness problem is also taken care of, since the curvature term in (2.8) is reduced by a factor

$$\left(\frac{R_2}{R_1}\right)^2 = \exp[2H(t_2 - t_1)] \sim 10^{52}$$

so that if $\Omega(t_1)$ is only of the **order** of unity at the beginning of inflation, at the end of inflation it will be incredibly close to unity:

$$\Omega(t_2) = 1 \pm 10^{-52} \qquad (5.8)$$

and on large enough, supergalactic scales the universe should be equally flat and uniform at the present day. An analogy can be made with the inflation of a rubber balloon: as it inflates, the curvature of the surface decreases and in the limit a portion of the surface appears quite flat.

There is one other problem solved by inflation. **Magnetic monopoles** were suggested by Dirac in 1932, and are definitely predicted to exist in grand unified theories (where quantization of electric charge and, therefore, of magnetic charge appears naturally). The monopole masses would be of the order of the GUT mass scale and they should have been created in the early universe with a number density comparable with photons. Had this been really so, the energy density would have been so large that the age of the universe would have been reduced by a factor of the order of one million. Searches for magnetic monopoles have met with no success and the observed upper limit on monopole

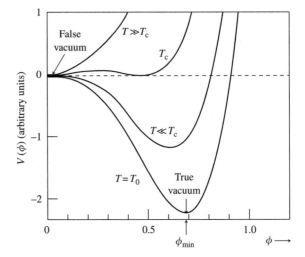

Fig. 5.1 Potential $V(\phi)$ of the 'inflaton' field plotted against the field vacuum expectation value $\langle\phi\rangle$ at different temperatures, in the early model of inflation. The critical temperature is denoted by T_c. For temperatures slightly less than this value, a transition can be made from the 'false' to the 'true' vacuum via quantum-mechanical tunnelling. Inflation takes place while the system is in the 'false' vacuum state and ends when it reaches the 'true' vacuum.

density is many orders of magnitude below the above figure. Provided however that monopoles, because of their large masses, can only be created at very high temperatures, before the inflationary process commences, the monopole problem is also solved, since the monopole number density will fall by an exponential factor through inflation and typically there would be only one monopole left in our entire universe. After inflation, the temperature would be far too low to lead to monopole creation.

In its original form, the inflation mechanism was likened to the Higgs mechanism of the self-interactions of a scalar field and spontaneous symmetry breaking in the very successful theory of the electroweak interaction described in Chapter 3, only at a much higher energy scale, for example the GUT scale, of perhaps 10^{14} GeV. The scalar field involved in the present case is referred to as the 'inflaton' field. Suppose that one were to start off with an intensely hot microscopic universe near the Planck temperature $kT \sim 10^{19}$ GeV, expanding and cooling as in (2.38), and that the initial evolution suddenly became dominated at $t = t_1$ by such an 'inflaton' field ϕ, consisting of scalar particles of mass m. For $kT \gg m$, the field is assumed to be in the ground state with a vacuum expectation value $\langle\phi\rangle = 0$ as in Fig. 5.1. This state is referred to as the 'false vacuum' state. At temperatures below a critical value $kT_c \sim m$, however, through a process of spontaneous symmetry breaking, the vacuum expectation value of the field can become different from zero, with $\langle\phi\rangle = \phi_{min}$ and a lower potential energy. The system will therefore try to make the transition from the metastable state of the 'false' vacuum to the 'true' vacuum . The inflationary phase occurs while the system is in the false vacuum state, during the period $t_1 \rightarrow t_2$ when the energy density ρ_v is approximately constant. The inflationary expansion is of course driven by the vacuum energy.

It will be recalled that in Chapters 2 and 4 we saw that a large fraction of the energy density in the universe today appears to be in the form of vacuum energy, which is independent of the temperature (see Table 2.2). In the distant past this would have been a vanishingly small fraction, since the energy densities of radiation and matter vary as T^4 and T^3, respectively. In the context of inflation discussed here, we are calling upon a **second** and quite separate

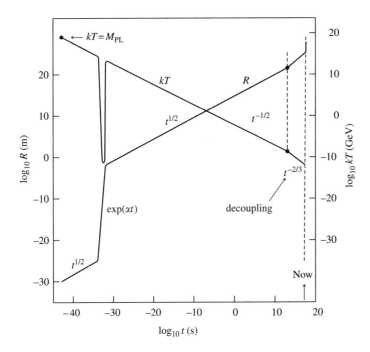

Fig. 5.2 Sketch of the variations of R and T with time in the inflationary scenario.

source of vacuum energy, existing only on an enormously high energy scale and disappearing as soon as the inflationary stage is completed.

The inflationary phase is terminated in this model when the transition to the true vacuum occurs on account of quantum-mechanical tunnelling through the potential barrier between the false and the true vacuum. 'Bubbles' of the true vacuum then develop, and these are supposed to merge into each other and stop the inflation. The energy density $\rho_v \sim m^4$, which is liberated as the inflation ends and the system enters the true vacuum state, is the 'latent heat' which reheats the supercooled inflationary universe, so that it reverts to the conventional Big Bang model with 'slow' expansion and cooling. This reheating is analogous to the heat liberated when supercooled water suddenly undergoes a first-order phase transition to form ice, the supercooled water being the analogue of the false vacuum and the ice that of the true vacuum. The variations of R and T with time, in this model, are sketched in Fig. 5.2.

We already noted at the beginning of Chapter 2 that the gravitational potential of the universe today is almost equal to its mass energy, so the total energy is near zero. It is important to emphasize that, in the inflationary scenario, the universe starts out essentially from nothing, with zero total energy, as in equation (2.10) for a flat universe with $K = 0$. As the inflation proceeds, more and more positive energy appears in the rapidly expanding region occupied by the scalar field ϕ: eventually, after the transition to the true vacuum, the 'reheating' phase will lead to the creation of the enormous numbers of particles ($\sim 10^{88}$!), which eventually form the material universe. As this is happening, more and more negative energy appears in the form of the gravitational potential energy of the expanding region. The total energy remains at a small and possibly zero value, with $K \approx 0$. The enormous energy associated with the expansion and particle creation is simply provided by the gravitational potential energy of the

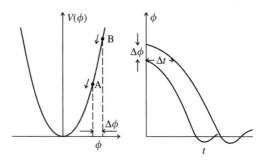

Fig. 5.3 A typical potential curve (on left) for the chaotic inflationary scenario. No phase transition is involved in this case. Quantum fluctuations in ϕ mean that different points in the universe, say at A and B, begin and end inflation at different times (see the right-hand diagram), separated by an interval $\Delta t = \Delta\phi/\dot{\phi}$ as in (5.19).

expanding material. It is a bit like cycling down a steep hill, starting from rest at the top. The large kinetic energy acquired upon reaching the bottom is exactly offset by the loss of potential energy due to the change in height.

The early model of inflation sketched above suffered because it did not seem possible to obtain the necessary inflationary growth as well as to terminate the inflation efficiently so as to end up with a reasonably homogeneous universe. Wherever the transition between 'false' and 'true' vacuum takes place via quantum-mechanical tunnelling, 'bubbles' of true vacuum form and inflation ends. These bubbles will then grow slowly via causal processes, whereas outside them, exponential inflation continues, and one ends up with a very lumpy situation.

5.3 Chaotic inflation

The above problems are avoided in the **chaotic inflation** model, due originally to Linde (1982, 1984). The basic idea is that conditions in different parts of the space-time domain vary in a random and unpredictable fashion, so that some regions attain the condition of inflation before others, and each such 'bubble' becomes a universe on its own. The inflaton potential is assumed to be a smooth function as in Fig. 5.3 (in this case, the quadratic function in (5.14)). No phase transition is involved, and it turns out that the termination of inflation is achieved more easily than in the previous model.

Let us begin by writing down the Lagrangian energy of the inflaton field:

$$L(\phi) = T - V = R^3[\dot{\phi}^2/2 - V(\phi)] \tag{5.9}$$

where ϕ is the amplitude of the field, which in natural units, $\hbar = c = 1$, has the dimensions of mass, as in the case of the Higgs field in Chapter 3 (see (3.25)), and R is the expansion factor. This equation involves the difference of the kinetic and potential energies T and V of the field as in (3.1). The total energy density of the field is then

$$\rho_\phi = \frac{T + V}{R^3} = \frac{\dot{\phi}^2}{2} + V(\phi) \tag{5.10}$$

The Euler–Lagrange equation (3.1) for the system takes the form

$$\frac{\partial}{\partial t}\left(\frac{\partial L}{\partial \dot{\phi}}\right) - \frac{\partial L}{\partial \phi} = 0 \tag{5.11}$$

Applying this to (5.9) and dividing through by R^3 gives

$$\ddot{\phi} + 3H\dot{\phi} + \frac{dV}{d\phi} = 0 \tag{5.12}$$

This equation is similar to that for a ball rolling to and fro in a saucer, or that of a simple pendulum oscillating in a very dense gas, the middle term corresponding to friction losses, that is, to the reheating mechanism at the end of inflation. If, at the beginning of the inflation process, the kinetic energy of the field is small compared with the potential energy, $\ddot{\phi} \approx 0$ and $\dot{\phi}$ is small, so that $\phi \approx \phi_0$, a more or less constant value, and $V = V(\phi_0) \approx \rho_\phi$. Then, the Friedmann equation (2.8) takes the form (using units $\hbar = c = 1$ and the relation $G = 1/M_{\text{PL}}^2$)

$$H^2 = \frac{8\pi G \rho_\phi}{3} = \frac{8\pi V(\phi_0)}{3M_{\text{PL}}^2} \tag{5.13}$$

so that the universe inflates exponentially with an almost constant expansion factor as in (5.7). In this scenario, the potential is usually taken to be of the simple quadratic form

$$V(\phi) = \tfrac{1}{2}m^2\phi^2 \tag{5.14}$$

As the inflation proceeds, ϕ changes slowly as V 'rolls' gently down the curve in Fig. 5.3. With $\ddot{\phi} \approx 0$, (5.12) and (5.14) give

$$\dot{\phi} = -\frac{m^2\phi}{3H} \tag{5.15}$$

and integrating we obtain

$$\phi = \phi_0 \exp\left(-\frac{m^2 \Delta t}{3H}\right) \tag{5.16}$$

where $\Delta t = t_2 - t_1$ is the period of inflation as in (5.7). Clearly, ϕ should not fall too rapidly or the full expansion will not be obtained, and inserting the limit from (5.7) we, therefore, find from (5.16) that

$$60 < H\Delta t < \frac{3H^2}{m^2}$$

and hence from (5.13) and the condition $V(\phi_0) < M_{\text{PL}}^4$, that

$$\frac{m}{M_{\text{PL}}} < \left(\frac{2\pi}{15}\right)^{1/2} \tag{5.17}$$

This is the condition, that a large enough inflation factor is obtained, consistent with the energy density in the inflaton field being less than M_{PL}^4, at which level unknown quantum gravitational effects could become important. Eventually, the system rolls into the potential well of the true vacuum and inflation ceases, and as explained above, the to-and-fro oscillations in the well correspond to the reheating phase.

While none of the models of inflation has yet been totally successful in providing exactly the conditions required, there seems little doubt that an inflationary scenario of some sort—or a mechanism which produces the same effect—is an essential stage in the early development of the universe. There are two strong predictions of such a model. The first is that the closure parameter Ω_{tot} must be extremely close to unity, that is, the curvature parameter K must be near zero. The second is that only one particular patch of the early universe would have been in the 'false' vacuum state at the chosen time when the quantum fluctuation described in the next section took place: there must have been many other universes growing from other such patches. So, enormous as our universe is, inflation suggests that it is but a dot in the ocean, a tiny part of a much larger space domain.

5.4   Quantum fluctuations and inflation

It is believed that quantum fluctuations are at the heart of anisotropies in the early universe. In Chapter 3, we saw that quantum fluctuations in elementary particle physics, in the form, for example, of the creation and annihilation of virtual electron–positron pairs, were able to account for the anomalous magnetic moments of the electron and the muon, and that such fluctuations are a vital part of the very successful electroweak theory. Quantum fluctuations are also involved, for example, in connection with Hawking radiation from black holes (see Section 7.12), and as the source of the vacuum energy described in Chapter 4, and as found in the laboratory with the famous Casimir effect.

Quantum fluctuations arise as a result of the Uncertainty Principle. In a particular time interval Δt the energy of a system cannot be specified to an accuracy better than ΔE, where $\Delta t \Delta E \sim \hbar$. Let us apply this to the case that inflation is connected with a transition in the 'inflaton field' at the energy scale $m \sim M_{GUT}$, for example, and consider the time just before inflation takes place. Referring to (2.38) appropriate to relativistic particles and ignoring numerical factors, we find that the age of the universe when the energy per particle is $kT = m$ is of order $t \sim M_{PL}/m^2$ (here and in what follows we use units $\hbar = c = 1$). The fluctuation in energy per particle is then $\Delta E = 1/\Delta t \sim m^2/M_{PL}$ and the fractional fluctuation is therefore

$$\frac{\Delta E}{E} \sim \frac{m}{M_{PL}} \tag{5.18}$$

Since the energy density $\rho \propto E^4$ for relativistic particles, the density fluctuations $\Delta\rho/\rho$ should be of similar magnitude. This admittedly naïve and simplistic estimate has to be modified in a more serious calculation, but it makes the interesting point that the magnitude of the density fluctuation (observed to be at the 10^{-5} level) depends largely (but, it turns out, not entirely) on the choice of the inflaton mass scale.

More quantitatively, fluctuations in the field amplitude ϕ can be thought of as due to the different times at which different 'bubble' universes complete inflation as in Fig. 5.3, via the relation

$$\Delta t = \frac{\Delta\phi}{\dot\phi} \tag{5.19}$$

When discussing fluctuations in the microwave background radiation in Section 5.9, the amplitude of the fluctuations at the horizon scale are important, and they are determined by the different amounts by which the universes have expanded:

$$\frac{\Delta\rho}{\rho} = \delta_{\text{hor}} = H\Delta t \sim \frac{H^2}{\dot\phi} \tag{5.20}$$

where the Hubble time is $1/H$ and we have used the relation $\Delta\phi \sim H$ from the Uncertainty Principle. Using equations (5.13) and (5.14) for H^2 and (5.15) for $\dot\phi$, we thus obtain for the estimated density fluctuation

$$\frac{\Delta\rho}{\rho} \sim \left(\frac{m}{M_{\text{PL}}}\right) \times \left(\frac{\phi}{M_{\text{PL}}}\right)^2 \tag{5.21}$$

modifying our first estimate (5.18). We repeat that experimentally this quantity is of order 10^{-5}. Ideally, of course, it would be nice to **predict** the magnitude of the fluctuations from the inflation model, but at the present time this does not seem possible, since the number expected depends on the precise form assumed for the inflaton potential $V(\phi)$.

Until a really convincing model of inflation is perfected (perhaps as a result of future detailed studies of the microwave anisotropies described in Section 5.9), it is not possible to state what the true predictions are regarding the level of quantum fluctuations. Even so, the idea that the material universe, extending now to the order of 10^{26} m, had its origins in a quantum fluctuation, which started off space/time as we know it as a microcosm of radius 10^{-27} m is certainly interesting, and at least to the physicist, quite an appealing one.

5.5 The spectrum of primordial fluctuations

The quantum fluctuations referred to above are 'zero-point' oscillations in the cosmic fluid. As soon as inflation commences, however, at superluminal velocity, most of the fluid will move **outside the horizon scale** $1/H$. (We recall here that the horizon distance is of order ct, where t is the time after the beginning of the expansion, and in units $c = 1$ is equal to the reciprocal of the expansion rate $1/H$.) This means that there will no longer be communication between the crests and the troughs of the oscillations: the quantum fluctuations are therefore **'frozen' as classical density fluctuations at the super-horizon scale**. We also note from (5.18) and (5.21) that since no particular distance scale is specified for the fluctuations, the spectrum of fluctuations will follow a power law, which (unlike an exponential, e.g.) does not involve any absolute scale. Although we refer here to the fluctuations as density fluctuations, in cosmology they are also referred to as perturbations in the metric of space-time associated with variations in the curvature parameter—see Appendix C. (We should also mention that there are different possible types of fluctuations: it is usually assumed that the perturbations are **adiabatic**, that is, that the density variations are the same in different components—baryons, photons, etc. Inflation may also be violent enough to produce gravitational waves as well as density fluctuations, and these could induce **polarization** of the microwave photons, which will however be very difficult to detect.)

We can see how the fluctuations depend on the index of the power law determining the balance between small and large scales, using an argument due to Barrow (1988). This is based on the idea that an exponential expansion is invariant under a time translation. No matter at what time one fixes the start of the exponential growth, the universe will look the same at every epoch. Thus, the expansion rate H is constant, the density ρ_v is constant, the horizon distance $1/H$ is constant, and the universe is effectively in a **stationary state**. No time or place can then have significance over any other, with the result that the amplitude of the perturbations in the metric structure must be the same on all length scales as they enter the horizon—otherwise, a change in the magnitude of the perturbation could be used to indicate a time sense. This metric (curvature of space) is determined by the gravitational potential Φ, and in the absence of time dependence, this will obey Poisson's equation

$$\nabla^2 \Phi = 4\pi G \rho \tag{5.22}$$

Since in spherical polar coordinates, $\nabla^2 \Phi = (1/r^2)[\partial(r^2 \partial\Phi/\partial r)/\partial r]$, the solution is

$$\Phi(r) = 2\pi G \rho r^2 / 3$$

On the scale of the horizon distance $r_{\text{Hor}} = 1/H$, which is the only natural length in the problem, we have therefore

$$\Phi = \frac{2\pi G \rho}{3H^2}$$

while on some arbitrary scale, $\lambda < 1/H$, fluctuations in the gravitational potential due to fluctuations $\Delta\rho$ in density will be

$$\Delta\Phi = 2\pi G \Delta\rho \lambda^2 / 3$$

Hence, the fractional perturbation in the gravitational potential on the scale λ has the value

$$\frac{\Delta\Phi}{\Phi} = H^2 \lambda^2 \frac{\Delta\rho}{\rho} \tag{5.23}$$

As explained above, in a stationary state, $\Delta\Phi/\Phi$ must be some constant independent of the arbitrary distance scale λ. Since H is also approximately constant, it follows that the density fluctuation, as it comes inside the horizon (and specifically its root-mean-square value) must have a spectrum with the power law dependence on λ of

$$\langle \delta_\lambda^2 \rangle^{1/2} = \left(\frac{\Delta\rho}{\rho} \right)_{\text{rms}} \sim \frac{1}{\lambda^2} \tag{5.24}$$

known as the PHZ (Peebles–Harrison–Zeldovich) spectrum, typical of the inflationary scenario. In words, this spectrum gives the universe the same degree of 'wrinkliness' and the same amplitude for the perturbations on the horizon, independent of the epoch, as would be expected for a stationary state. For this reason the above spectrum is called **scale-invariant**.

Notice that the fluctuations predicted are actually smaller than the purely statistical fluctuations of the number N of particles contained in the volume λ^3, since according to (5.24), $\Delta N/N \propto N^{-2/3}$, while for a statistical fluctuation $\Delta N/N \propto N^{-1/2}$. This smoothing out of fluctuations on large scales is an example of a general rule that, as we shall see below, in an accelerating universe perturbations decay, while in a decelerating universe perturbations grow with time.

5.6 Large-scale structures: gravitational collapse and the Jeans mass

In Chapter 2 the early universe was described as a homogeneous, isotropic, and perfect primordial fluid (a perfect fluid being one in which frictional effects are negligible), undergoing a universal expansion. In contrast, the universe today is 'grainy' with the matter clumped into billions of individual galaxies, each containing of the order of 10^{11} stars, and separated from their neighbours by enormous voids in space. Starting off from the Big Bang, we have to ask what were the physical processes taking place which led to such structures. The developments on the smallest scales, that of the stars themselves, are dealt with in Chapter 7. Here, we discuss the large-scale structures. Such structures were, we believe, originally seeded by tiny fluctuations in the inflationary phase described above, which are detectable today in observations on the microwave background radiation, as described in Section 5.9. Before discussing those observations and their interpretation, however, we consider the general conditions necessary for gravitational collapse of a classical gas cloud, as originally enunciated by Jeans.

Let us first estimate the time required for a cloud of ordinary gas to collapse under gravity, assuming, to begin with, that gas pressure can be neglected. Suppose that the cloud is spherical, of constant mass M and of initial radius r_0, and that it begins gravitational contraction. When the radius has shrunk to r, a small mass m in the outermost shell will have lost gravitational potential energy $GMm(1/r - 1/r_0)$ and gained kinetic energy $(m/2)(dr/dt)^2$, assuming that it was initially at rest. Equating these two, we get for the time of free fall from $r = r_0$ to $r = 0$

$$t_{FF} = \int \frac{dr}{dr/dt} = \int \left(\frac{2GM}{r} - \frac{2GM}{r_0} \right)^{-1/2} dr \qquad (5.25)$$

Substituting $r = r_0 \sin^2 \theta$ and with the limits $\theta = \pi/2$ and 0, this integral gives

$$t_{FF} = \frac{\pi}{2} \left(\frac{r_0^3}{2GM} \right)^{1/2} = \left(\frac{3\pi}{32G\rho} \right)^{1/2} \qquad (5.26)$$

where ρ is the mean initial density of the cloud. Note that the result is independent of the radius, for a given initial density. This free fall time, it may be observed, is comparable with the circulation time of a satellite in close orbit about the initial cloud, equal to $(3\pi/G\rho)^{1/2}$.

As the cloud of gas condenses, gravitational potential energy will be transformed into kinetic (heat) energy of the gas particles. If these are atoms or molecules, this motional kinetic energy may be absorbed through collisional dissociation of molecules or ionization of atoms, as well as resulting in atomic excitation, which can be radiated away as photons if the cloud is transparent. These processes absorb and then re-emit the gravitational energy liberated and allow the cloud to contract further, but eventually hydrostatic equilibrium will be attained when the pressure of the heated gas balances the inward gravitational pressure. The total kinetic energy of the gas at temperature T will be

$$E_{kin} = \frac{3}{2} \frac{MkT}{m} \tag{5.27}$$

where m is the mass per particle, M/m is the total number of particles and $3kT/2$ is the mean energy per particle at temperature T. The gravitational potential energy of a sphere of mass M and radius r is

$$E_{grav} \approx \frac{GM^2}{r} \tag{5.28}$$

where there is a numerical coefficient of order unity, depending on the variation of density with radius (and equal to $\frac{3}{5}$ if the density is constant). Comparing these two expressions, we find that a cloud will condense if $E_{grav} \gg E_{kin}$, that is, if r and ρ exceed the critical values

$$r_{crit} = \frac{2MGm}{3kT} = \frac{3}{2}\left(\frac{kT}{2\pi\rho Gm}\right)^{1/2}$$

$$\rho_{crit} = \frac{3}{4\pi M^2}\left(\frac{3kT}{2mG}\right)^3 \tag{5.29}$$

Example 5.2 *Calculate the critical density and radius of a cloud of molecular hydrogen with a mass of 10 000 solar masses at a temperature of 20 K.*

Inserting the values in SI units of $M_\odot = 2 \times 10^{30}$, $G = 6.67 \times 10^{-11}$, $k = 1.38 \times 10^{-23}$ and $m = 3 \times 10^{-27}$ into equations (5.29) gives the values

$$r_{crit} = 1.06 \times 10^{19}\ \text{km} = 0.34\ \text{kpc}$$

$$\rho_{crit} = 3.18 \times 10^{-23}\ \text{kg m}^{-3} = 9600\ \text{mol m}^{-3}$$

These are typical temperature and density values for clouds of gas in **globular clusters**, each of which contain of the order of 10^5 stars. Individual stars will form as a result of density fluctuations in the cloud, requiring from (5.29) gas densities about 10^8 times larger.

From the viewpoint of the development of large-scale structure in the universe, we would like to determine which criteria lead to a cloud of gas condensing as a result of an upward fluctuation in density in one part of it. In terms of the density ρ, there is a critical size of the cloud called the **Jeans length** with a value

$$\lambda_J = v_s \left(\frac{\pi}{G\rho}\right)^{1/2} \tag{5.30}$$

obtained essentially by multiplying the sound velocity by the free-fall time. The mass of a cloud of diameter equal to the Jeans length is called the **Jeans mass**:

$$M_J = \frac{\pi\rho\lambda_J^3}{6} \tag{5.31}$$

Here, v_s is the velocity of sound in the gas. The typical time for sound waves (propagated as a result of any density perturbations) to cross a cloud of size

L is L/v_s, and this is less than the gravitational collapse time (5.26) when $L \ll \lambda_J$. So the perturbation just results in sound waves oscillating to and fro, and there is no preferred location towards which matter can gravitate. On the other hand, if $L \gg \lambda_J$, sound waves cannot travel fast enough to respond to density perturbations and the cloud will start to condense around them. For a cloud of non-relativistic matter, the lengths λ_J in (5.30) and r_{crit} in (5.29) are, of course, one and the same (up to numerical factors of order unity). For, then,

$$v_s^2 = \frac{\partial P}{\partial \rho} = \frac{\gamma kT}{m} \tag{5.32}$$

where γ is the ratio of specific heats, equal to $\frac{5}{3}$ in neutral hydrogen. In that case,

$$\lambda_J = \left(\frac{5\pi kT}{3G\rho m} \right)^{1/2} \tag{5.33}$$

Thus, in terms of the temperature of non-relativistic gas particles, both r_{crit} in (5.29) and λ_J in (5.30) are of the order of magnitude $(kT/G\rho m)^{1/2}$.

5.7 The growth of structure in an expanding universe

We now apply the ideas in the preceding section, based on classical density perturbations, to fluctuations in the early universe. Suppose that an upward fluctuation in density occurs at some point in a static (i.e. non-expanding) homogeneous and isotropic fluid, that is the density increases by a small amount $\Delta\rho$ above the unperturbed density ρ, where $\Delta\rho \ll \rho$. The gravitational force which the perturbation exerts, and consequently the inflow of material attracted towards the perturbation per unit time will both be proportional to $\Delta\rho$, so that $d(\Delta\rho)/dt \propto \Delta\rho$. Hence—provided gas pressure does not oppose it—this simple argument suggests that the density perturbation would be expected to grow exponentially with time. However, in the case of a non-static, expanding universe, the gravitational inflow can be counterbalanced by the outward Hubble flow. It then turns out that the time dependence of growth of the density fluctuation is a power law rather than an exponential. Intuitively, one can guess that, if the perturbation is small so that all effects are linear, and it is expressed in terms of the so-called **density contrast** $\delta = \Delta\rho/\rho$, this dimensionless quantity can only be proportional to the other dimensionless number associated with the Hubble flow, namely, the expansion parameter ratio $R(t_2)/R(t_1)$.

Quantitatively, we have to enquire whether the growth of cosmic structures on the largest scales can be understood in terms of the tiny anisotropies (temperature and density fluctuations at the 10^{-5} level) observed in the cosmic microwave radiation, already mentioned in Section 2.7 and discussed in more detail in Section 5.9. The fluctuations in the temperature of the microwave radiation observed today must be essentially the same as those obtaining at the time when the microwave photons had their last interaction, namely the time of decoupling of baryonic matter and radiation, about 300 000 years after the Big Bang, when $z_{dec} \sim 1000$. Thereafter, the plasma of protons and electrons was replaced largely by neutral atoms, practically transparent to radiation—hence astronomy!

The standard treatment of the growth of small fluctuations in density by means of perturbation theory is rather lengthy and is given in Appendix E. Here, we derive the principal result by means of a short cut, treating the initial upward density fluctuation as a matter-dominated, closed 'micro-universe' of mass M and positive curvature ($K = +1$), as described by the lower curve in Fig. 2.4 and by Example 2.2. Then, from equation (2.14) the values of R and t in parametric form are

$$R = a(1 - \cos\theta) = \frac{a\theta^2}{2}[1 - \theta^2/12 + \cdots]$$

$$t = b(\theta - \sin\theta) = \frac{b\theta^3}{6}\left[1 - \frac{\theta^2}{20} + \cdots\right]$$

(5.34)

where $a = GM/c^2$ and $b = GM/c^3$, and the expansion on the right is for very early times, that is, for $\theta \ll 1$. Taking the $\frac{2}{3}$ power of the second equation to find θ^2 as a function of $t^{2/3}$, and inserting in the first equation one obtains

$$R(t) = \frac{a}{2}\left(\frac{6t}{b}\right)^{2/3}\left[1 - \frac{(6t/b)^{2/3}}{20} + \cdots\right]$$

(5.35)

We see that when $t \ll b/6$, $R(t) \propto t^{2/3}$, that is, the increase in radius with time is the same as that in a flat, matter-dominated universe of $\Omega = 1$ (see Table 2.2)—a result that is true for any value of K. For larger, but still small values of t, the density enhancement, compared with the flat case, grows linearly with the expansion factor $R(t)$:

$$\delta = \frac{\Delta\rho}{\rho} = -3\frac{\Delta R}{R} = +\frac{3}{20}\left[\frac{6t}{b}\right]^{2/3} \propto R(t) \propto 1/(1+z)$$

(5.36)

just as we anticipated. Incidentally, we may note here that, had we done the same exercise for an open universe as in equation (2.15), the value of δ would have come in with the opposite sign, with the density perturbation decreasing with time.

According to the simple linear dependence in (5.36), the primordial (10^{-5}) fluctuations in the microwave radiation at the time of decoupling ($z_{dec} \sim 1000$), would by now have grown by some three orders of magnitude in a matter-dominated universe. This, however, is not enough to account for the much larger density fluctuations in the material of the present universe. Of course, the model we have used to simulate the growth of perturbations shows (see Fig. 2.4) that the growth in relative density would in any case be non-linear. For example, when $R = R_{max} = 2GM/c^2$ and $t = \pi GM/c^3$, the enhancement is a factor of $(3\pi/4)^2 = 5.6$, compared with 1.06 for the linear case. However, the conclusion still stands that the observed level of fluctuations in the microwave radiation would have been too small to account for the observed structures in terms of growth of fluctuations in the baryonic component alone.

5.8   Evolution of fluctuations during the radiation era

In the early stages of the Big Bang, the universe was radiation-dominated and the velocity of sound was $v_s = (\partial P/\partial \rho)^{1/2} \approx c/\sqrt{3}$—see Table 2.2. This means that, using (2.36), with $\rho_r c^2 = (3c^2/32\pi G)/t^2$, the Jeans length was

$$\lambda_J = c \left[\frac{\pi}{(3G\rho_r)} \right]^{1/2} = ct \left(\frac{32\pi}{9} \right)^{1/2} \tag{5.37}$$

Thus, the horizon distance and the Jeans length are both of order ct during the radiation era.

We now trace the development with time of the mass inside the horizon during the radiation-dominated era, and consider whether, during that period, initial density fluctuations could survive. The actual baryonic mass inside the horizon during this era would be

$$M_H(t) \sim \rho_b(t)(ct)^3 \propto \frac{1}{T^3} \tag{5.38}$$

where the T dependence arises from the fact that $\rho_b \propto 1/R^3 \propto T^3$ and from (2.38), $t \propto 1/T^2$ during the radiation-dominated era. (This dependence will flatten off near the decoupling temperature because of the increasing effect of the baryons in reducing the sound velocity.) At the time of baryon–photon decoupling, $z_{dec} \sim 1000$, $\rho_b = \rho_c \Omega_b (1 + z_{dec})^3$, and $t_{dec} = t_0/(1 + z_{dec})^{3/2} \sim 10^{13}$ s (see Example 2.6). Inserting the value of ρ_c from (2.23) we find

$$M_H(t_{dec}) \sim 10^{18} \Omega_b M_\odot \sim 10^{17} M_\odot \tag{5.39}$$

for $\Omega_b = 0.05$ (see (2.28)). The Jeans mass (5.31) will be an order of magnitude larger. This demonstrates that fluctuations on the scale of galaxies ($M \sim 10^{11} M_\odot$) and clusters ($M \sim 10^{13} M_\odot$) come inside the horizon during the radiation era, at redshifts $(1+z) \sim 10^5$ and $\sim 20,000$, respectively. The variation of M_H and M_J with T is shown in Fig. 5.4.

In discussing the survival of fluctuations during the radiation era, we should point out that there are several different possible types of fluctuation. **Adiabatic** fluctuations behave like sound waves, with baryon and photon densities fluctuating together, while for **isothermal** fluctuations the matter density fluctuates but the photon density does not, so that matter is in a constant temperature photon bath. Or it could be that both matter and photon densities fluctuate but with opposite phases, in so-called **isocurvature** fluctuations.

We consider here only the adiabatic fluctuations, which are likely to be the most limiting from the point of view of growth. What happens depends on the scale of distance considered. While the matter is non-relativistic, photons travel at light velocity and through radiation pressure can stream away from regions of higher density to ones of lower density, so ironing out any fluctuations. During most of the radiation era, the photon energy density is much larger than that of the baryons, so that if the photons diffuse away, the amplitude of the fluctuation will be severely reduced, a process called **Silk damping**.

This loss of photons will be prevented if they are locked in to the baryonic matter by 'Thomson drag', that is by Compton scattering by electrons of the

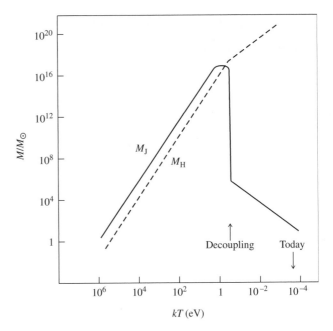

Fig. 5.4 The variation with radiation temperature T of the (baryonic) mass M_H inside the horizon (i.e. inside the largest distance over which causal effects are possible) and of the Jeans mass M_J (i.e. the smallest mass which can overcome the pressure of radiation and contract under gravity). After decoupling of matter and radiation, the Jeans mass falls abruptly as the velocity of sound reduces by a factor of 10^4, while the mass inside the horizon continues to increase (as $1/T^{1.5}$).

baryon–electron plasma. It is found, as shown in Example 5.3, that fluctuations containing baryonic masses well above $10^{12} M_\odot$—that is, the size of galaxies or larger objects—will survive without significant reduction of amplitude to the era of decoupling, after which they can grow. Fluctuations on smaller scales will, in contrast, be ironed out, as indicated in Fig. 5.5.

Example 5.3   *Estimate the minimum mass associated with a primordial density fluctuation, which could survive to the era of decoupling.*

The scattering mean free path of photons through ionized baryonic matter will be $l = 1/(n_e \sigma)$, where the electron number density $n_e \sim \rho_b N_0$, N_0 is Avogadro's number and σ is the Compton cross section for $\gamma e \to \gamma e$, equal at these energies to the Thomson cross section (1.26). Since the scattering is isotropic, the result of N successive scatters is that the photon travels a bee-line distance D where $\langle D^2 \rangle = (\mathbf{l}_1 + \mathbf{l}_2 + \mathbf{l}_3 + \cdots + \mathbf{l}_N)^2 = N \langle l^2 \rangle$, since the cross-terms cancel in the square (this is an example of the famous 'drunkard's walk' problem).

The time taken for the photon to cover a bee-line distance D is therefore $t = Nl/c$, so that $D = (ctl)^{1/2}$ is the geometric mean of the horizon distance and the scattering mean free path. Hence, the time required for a photon to diffuse out of a fluctuation of scale-length D is of order D^2/lc, and if this is much less than the time t since the onset of the Big Bang, the photons will stream away and the fluctuation will be damped out. In order for the fluctuation to survive until $t = t_{dec}$ (and thereafter grow), we therefore need

$$D^2 > lct_{dec} = \frac{ct_{dec}}{\rho_b N_0 \sigma}$$

For a lower limit to D we take $\rho_b = \rho_b(\text{dec}) = \rho_b(0)(1 + z_{dec})^3$, giving $M > D^3 \rho_b \sim 10^{12} M_\odot$.

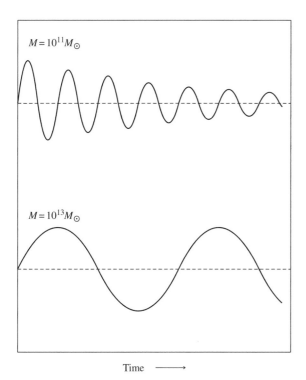

Fig. 5.5 During the radiation era, adiabatic fluctuations encompassing baryonic masses below $10^{12}M_\odot$ are damped out by the leakage of the photon component, while those of masses above $10^{13}M_\odot$ remain at a practically constant amplitude until the epoch of matter–radiation decoupling (see Example 5.3).

As indicated in Example 5.4, as soon as matter and radiation decoupled and neutral atoms formed, the velocity of sound and hence the Jeans length decreased by over 10 000 times. Growth of inhomogeneities on galactic and smaller scales then became possible.

Example 5.4   *Estimate the value of the Jeans mass just after the epoch of decoupling of matter (baryons) and radiation.*

After decoupling of baryonic matter from radiation and the recombination of protons and electrons to form atoms, the velocity of sound is given by (5.32)

$$\frac{v_s^2}{c^2} = \frac{5kT}{3m_Hc^2}$$

Taking $kT = 0.3\,\text{eV}$ from (2.55) and the mass of the hydrogen atom $m_Hc^2 = 0.94\,\text{GeV}$, we find

$$\frac{v_s}{c} = 2 \times 10^{-5}$$

Thus, compared with the pre-decoupling epoch, the sound velocity and the Jeans length have fallen by a factor of 10^4, and the Jeans mass from $10^{18}M_\odot$ to $10^6M_\odot$. This last is the typical mass of **globular clusters** of the order of 10^5 stars, which are some of the oldest objects in the sky. Obviously, even larger objects such as galaxies and galaxy clusters would have had no problem in condensing under gravity at this epoch.

We can see from the above discussion that only after the universe became matter-dominated did perturbations really have a chance to grow. In this respect, dark matter, and specifically cold dark matter, plays a vital role in the development of structures at the galactic and super-galactic level. It turns out in fact that the calculated increase in density contrast with time, as described above and starting out from values of $\Delta\rho/\rho \sim 10^{-5}$ following inflation, is not enough to account for the observed growth of galaxies and clusters, relying simply on the gravitational collapse of the baryonic component (with $\Omega_b(0) \sim 0.05$), once it has decoupled from radiation at a redshift $z \sim 1000$. As previously stated, one requires substantial amounts of cold (i.e. non-relativistic) dark matter (with $\Omega_m(0) \sim 0.30$) as a component of the primordial universe. Unlike normal (baryonic) matter, this will not interact with radiation via Thomson scattering and it also begins to dominate over radiation at an earlier epoch ($z \sim 10^4$, see Example 2.6), and thus is both more efficient at achieving gravitational collapse and has more time to achieve it. Of course, once the dark matter agglomerations have formed gravitational potential wells, baryons will fall into them and indeed their increase in density contrast will follow that of the dark matter.

Figure 5.6 shows the values of the root-mean-square amplitude of density fluctuations plotted against the distance scale λ introduced in (5.24). On very large scales (typically angular ranges of $10°–100°$) these were deduced from the temperature fluctuations of the microwave background, as observed by the COBE satellite experiments. On smaller scales and angles, the density fluctuations have been deduced from large-scale galaxy surveys (LSS), such as those at infrared wavelengths using the IRAS satellite, which analysed almost 20 000 galaxies in the early 1990s.

At large scales, the spectrum of fluctuations does seem to follow the $1/\lambda^2$ variation predicted by inflation in (5.24), the universe becoming progressively smoother over the largest distances, while the spectrum flattens off at the smaller scales of galaxy clusters. The curves show the expected amplitude for a cold dark matter scenario, and one where hot and cold dark matter are mixed. Cold dark matter alone produces clumping on small (galaxy cluster) scales that is too big. Hot dark matter alone would lead to fluctuations which are much too small, since they get ironed out by the relativistic dark matter particles, streaming from regions of high density as described above. The mixed dark matter model (70% CDM, 30% HDM) seems to be the best compromise fit, and implies that structure evolves from the 'bottom up', clumping occurring first on the smaller scales (galaxies and clusters) and the larger super-clusters following later (and still apparently forming today).

5.9   Temperature and density fluctuations in the cosmic microwave background: 'acoustic peaks' at small angles

As already mentioned in Chapter 2, the cosmic microwave background radiation exhibits small deviations in the value of the temperature, as recorded by the COBE satellite observations in Fig. 5.6. The largest effect is in fact in the form of a dipole term which corresponds to the Doppler formula $[1 + (v/c)\cos\theta]$, arising from the velocity of 370 km s^{-1} of the Earth relative to the Hubble flow. In addition to this dipole term, the COBE satellite observations detected higher multipoles, which corresponded to variations in the temperature over angular scales of $7°$ or more, the scale being set by the detector resolution. The

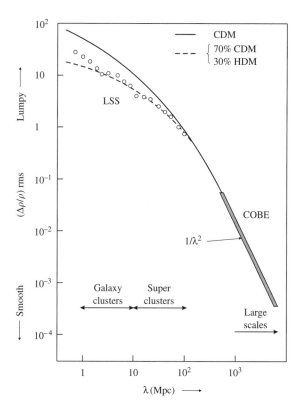

Fig. 5.6 Plot of density fluctuations against the scale λ, from the COBE satellite observations on the microwave background at the largest angular scales, and from galaxy surveys (e.g. the IRAS satellite experiments, labelled LSS) on small angular scales. Such surveys consist of counting the number of galaxies contained in each of many adjacent volumes λ^3 of sky, determining the r.m.s. fractional fluctuation about the average number, and repeating the procedure for different values of λ. The curves show typical predictions from cold and mixed dark matter models (after Kolb 1998).

temperature fluctuations observed were of the order of 10^{-5}. Although observed today, these fluctuations should also be those that the photons exhibited at the 'surface of last scatter' when they decoupled from matter at a temperature of $kT \sim 0.3$ eV many billions of years ago.

More recently, measurements shown in Fig. 5.8 have been carried out with detectors having much higher angular resolution—typically $10'$ of arc. Two were flown in high altitude balloons—the BOOMERANG (de Bernardis *et al.* 2002) and MAXIMA (Lee *et al.* 2001) experiments—and one was a ground-based interferometer—the DASI (Halverson *et al.* 2001) experiment—all at the South Pole. These variations in temperature on small scales have been of prime importance in providing information on the parameters of the early universe, such as the curvature and the contributions of matter, radiation, and vacuum terms to the overall energy density. Of course, not all the effects observed will be due to 'primordial' density and temperature fluctuations in the early universe: some Thomson (elastic) scattering of the microwave photons by free electrons could take place **en route** to the detectors from distant parts of the universe, but the amount of angular smearing from this cause is quite small and can be neglected in this discussion (see Problem 5.3).

Variations in the microwave background temperature will mirror those in the primordial energy density. There are two effects, going in opposite directions. First, in a more dense region, local compression will cause heating. But, on the other hand, the larger gravitational potential in the denser region means that the photons are red-shifted as they climb out of the local potential well. There is some cancellation of these competing factors, but the net effect is that the denser regions appear cooler, because as a region condenses, the potential well that the photon has to climb on the way out is deeper than the one it fell into on the way in. For a radiation-dominated universe, $\Delta T/T = -\Delta\Phi/2c^2$, while for a matter-dominated universe $\Delta T/T = -\Delta\Phi/3c^2$, where $\Delta\Phi$ is the local increase in the gravitational potential due to the over-density.

In the discussion of the horizon distance in Section 5.1, we concluded that, for a flat universe ($K = 0$), the angle subtended today by the (optical) horizon at the time of decoupling of matter and radiation was of the order of one or two degrees, as in (5.3). A pressure wave can arise from density inhomogeneities and the interplay of gravitational attraction and compressional effects of non-relativistic matter, on the one hand, opposed by photon pressure on the other. The propagation of such a pressure wave depends on the velocity of sound v_s, and the acoustic horizon is v_s/c times the optical horizon distance. If the cosmic fluid is radiation dominated at this epoch, then from Table 2.2 this ratio is approximately $1/\sqrt{3}$. The acoustic horizon at the time of last scatter of the microwave radiation in this case subtends today an angle at the Earth of approximately

$$\theta_{\text{acoustic}} \sim \frac{2t_{\text{dec}}(1 + z_{\text{dec}})}{\sqrt{3} \times (t_0 - t_{\text{dec}})} \sim 1° \tag{5.40}$$

A more detailed calculation, given in Appendix C, comes up with essentially the same angle.

Coming now to experiment, the fluctuations in the temperature of the microwave radiation can be measured as a function of position in the sky, and the correlation determined between two points separated by a particular angle θ. Suppose a measurement of the radiation temperature in a direction specified by a unit vector **n**, as compared with the whole-sky average T, indicates a deviation $\Delta T(\mathbf{n})$, while in a direction **m** it is $\Delta T(\mathbf{m})$. The correlation between pairs of points in the sky is given by the average quantity

$$C(\theta) = \left\langle \left(\frac{\Delta T(\mathbf{n})}{T}\right) \left(\frac{\Delta T(\mathbf{m})}{T}\right) \right\rangle \quad \text{with } \mathbf{n} \cdot \mathbf{m} = \cos\theta$$

$$= (1/4\pi) \sum (2l + 1)C_l P_l(\cos\theta) \tag{5.41}$$

where the average is taken over all pairs of points in the sky separated by angle θ. In the second line, the distribution $C(\theta)$ has been expanded as a sum of Legendre polynomials $P_l(\cos\theta)$ running over all values of the integer l. The coefficients C_l describe the fluctuation spectrum, which depends not only on the initial spectrum of density fluctuations as discussed in Section 5.5, but on several other parameters, such as the baryon–photon ratio, the amount of dark matter, the Hubble constant and so on. As had been predicted over thirty years ago, the measurement of the values of C_l for l-values in the hundreds should determine these parameters (the coefficient C_l for $l = 1$ corresponds to the dipole term mentioned above and is disregarded).

The Legendre polynomial $P_l(\cos\theta)$ in (5.41) oscillates as a function of θ between positive and negative values, having l zeros between 0 and π rad, with approximately equal spacing

$$\Delta\theta \approx \frac{\pi}{l} \approx \frac{200°}{l} \tag{5.42}$$

The sum $\sum(2l + 1)P_l(\cos\theta)$ from $l = 1$ to $l_{max} \gg 1$ has a strong maximum in the forward direction ($\theta = 0$), where the amplitudes of all the different l values add, while at larger angles the various contributions largely cancel out, the amplitude falling off to practically zero within the angular interval $\Delta\theta = 200°/l_{max}$. As (5.40) indicates, fluctuations over an angular range of a degree or less are relevant to the acoustic horizon distance at decoupling and will therefore be concerned with polynomial contributions of $l > 100$.

Suppose, now, that an initial 'primordial' perturbation in the density of the 'photon–baryon fluid' occurs. This can be decomposed into a superposition of modes of different wavenumbers k and wavelengths $\lambda = 2\pi/k$. If the wavelength becomes larger than the horizon size L_H—as it certainly would in the course of inflation—then the amplitude of that mode of the perturbation will become frozen; as previously stated, there could no longer be a causal connection between the troughs and the crests. However, as time evolves, $\lambda(t)$ will increase with the scale parameter $R(t) \propto t^n$, where $n = \frac{2}{3}$ for a matter-dominated situation and $n = \frac{1}{2}$ for radiation domination. Hence, $\lambda(t)/L_H(t) \sim 1/t^{(1-n)}$ as in (5.2) and, since $n < 1$, it follows that in time, λ will come inside the acoustic horizon and the amplitude of that mode will then start to oscillate as a **standing acoustic wave** in the cosmic fluid. Modes of smaller wavelength will enter the horizon earlier and oscillate more quickly (since the frequency varies as $1/\lambda$). The effect of having components of different wavelengths and phases is that one obtains a series of **acoustic peaks** when the amplitude is plotted as a function of l, as shown in Fig. 5.7. The quantity $l(l + 1)C_l$ is plotted against l (since it turns out that this is a constant, independent of l, for a scale-invariant spectrum of fluctuations).

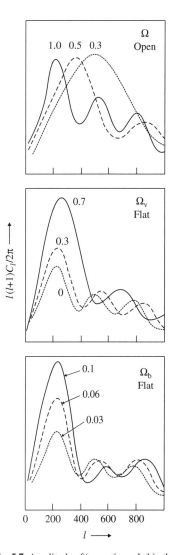

Fig. 5.7 Amplitude of 'acoustic peaks' in the fluctuations in temperature expected in cosmic microwave radiation, plotted against the value l of the multipole involved. The **position** of the first peak (top panel) depends principally on the total density Ω, while its height—that is, the strength of the acoustic oscillation—is seen to depend on the baryon density (bottom panel). The second and subsequent peaks depend on other cosmological parameters, as described in the text. Notice that, provided $\Omega = 1$, the results (middle panel) are insensitive to the division of density between matter and vacuum contributions, Ω_m and $\Omega_v = 1 - \Omega_m$ (after Kamionkowski and Kosowsky 1999).

Example 5.5   *Estimate the magnitudes of the physical objects corresponding to the 'acoustic peaks' in the angular power spectrum of the microwave radiation.*

The amount of material contained inside the horizon at the epoch under consideration, that is, when matter decoupled from radiation and protons started to recombine with electrons to form atoms and molecules, would in fact be somewhat larger than the typical mass of a supercluster. The baryonic mass inside the horizon (since $\rho_b \sim \rho_r$ at $t = t_{dec}$) would be

$$M_{hor} \sim (ct_{dec})^3 \rho_r(dec) \sim 10^{17} M_\odot \qquad (5.43)$$

using the fact that $z_{dec} \sim 1000$ (see (2.57)), $t_{dec} \sim 10^{13}$ s (see Example (2.6)), and $\rho_r(dec) = \rho_r(0)(1 + z_{dec})^4 \sim 10^{-19}$ kg m^{-3}. So the acoustic peaks observed today in the angular spectrum of the microwaves correspond to regions of space, which, at the time of decoupling, contained extremely large masses. Providing that recombination of protons and electrons into atoms and molecules had started so that the velocity of sound had fallen to $10^{-4}c$ or less, fluctuations on smaller scales, as in (5.24), would have grown first. As pointed out above, this suggests a 'bottom-up' scenario, with galaxies forming first, followed by galaxy clusters and then superclusters. The precise hierarchy would depend on details of the model, for example, the amount and nature of the dark matter and the form of the fluctuation spectrum.

Coming to the observations, the first peak in Fig. 5.8, at $l \sim 200$, corresponds to the mode that has just come inside the horizon and compressed only once,

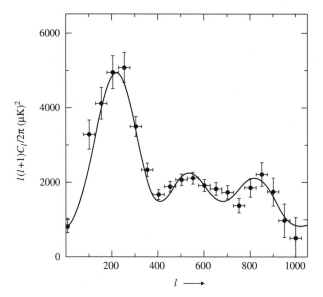

Fig. 5.8 Observed amplitude of 'acoustic peaks' in cosmic microwave radiation as a function of the order l of the polynomial $P_l(\cos\theta)$, from the BOOMERANG experiment, which employed high angular resolution bolometric (total energy) detectors flown in a balloon over the Antarctic (after de Bernardis *et al.* 2002).

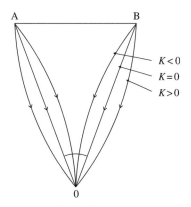

Fig. 5.9 The effect of the curvature of space (the gravitational deflection of photons) on angular measurements of distant objects. For curvature $K > 0$, that is, a closed universe, the angle is increased, while for $K < 0$ (open universe) it is decreased, in comparison with the value for a spatially flat universe ($K = 0$).

the second corresponds to a shorter wavelength mode that has undergone two oscillations, and so on.

The **position** (angle or l-value) of the first peak depends very strongly on the **total density** Ω, or equivalently through (2.24) and (2.26) on the curvature K or $\Omega_k \equiv 1 - \Omega$. For a flat universe, the value of the angle θ of the peak will be approximately as in (5.40), with the corresponding value of l in (5.42). However, for a 'closed' universe of positive curvature ($K > 0$), the angle will be increased, the gravitational field between the source and the detector acting as a converging lens. Thus, the peak will move to lower l values, while if the curvature is negative—the case of an 'open' universe—the angle will be decreased as for a diverging lens and the peak will move to higher l (see Fig. 5.9).

As shown in Appendix C, when one is, as in this case, looking back to large z-values of order 1000, this lensing effect has a simple form. In the first place, the distance to the 'surface of last scatter' of the microwave radiation for the case of an open universe (with $\Omega_k = 1 - \Omega$) varies as $1/\Omega$. Second, the horizon distance at the time of matter–radiation decoupling varies as $\Omega^{-1/2}$. So the angle subtended by the acoustic horizon scales as $\Omega^{1/2}$ and the l-value of the first acoustic peak is proportional to $1/\Omega^{1/2}$. Provided the total value of $\Omega \approx 1$, the universe is flat ($\Omega_k \sim 0$) and the peak position varies little, irrespective of how the matter and vacuum energy density is shared (we remind ourselves from (2.25) that $\Omega = \Omega_m + \Omega_v$ and that these symbols refer to the quantities today). The variation of the angle with closure parameter is shown in Fig. 5.10 for the two important cases, of an open universe with no vacuum energy, and a flat universe with variable vacuum energy, as calculated in Appendix C. The main and very important result from these experiments is that **the universe turns out to be remarkably flat.**

As indicated in Fig. 5.7, the **height** of the first peak measures the intensity of the acoustic oscillation, which is sensitive to the baryon/photon ratio (a greater baryon density will help gravitational collapse and enhance the oscillation

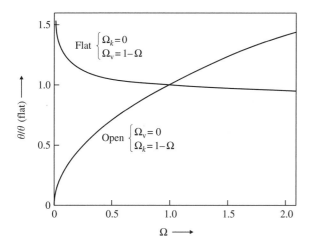

Fig. 5.10 The calculated angle subtended by the cosmic microwave 'acoustic peaks' relative to the angle expected for a flat, matter-dominated universe, plotted against the matter density parameter $\Omega = \Omega_m$. One curve is for an open universe with no vacuum energy, that is, a curvature term $\Omega_k = 1 - \Omega$, giving a variation proportional to $\Omega^{1/2}$. The other curve is for a flat universe ($\Omega_k = 0$) with a vacuum energy $\Omega_v = 1 - \Omega$. Note that the angle in this case depends only weakly on the partition between matter and vacuum energy. For details of the calculation, see Appendix C.

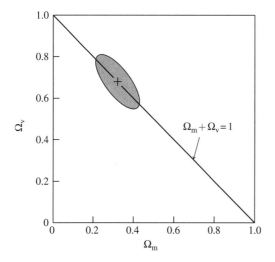

Fig. 5.11 Fits to the data from a combination of cosmic microwave observations on small angle structure (BOOMERANG, MAXIMA, and DASI experiments), supernovae Ia data at large redshift (Fig. 4.11), and analysis of large-scale structures from infrared and X-ray surveys. The ellipse is the 95% confidence contour within which the solutions (5.44) are confined (after de Bernardis *et al.* 2002, Harun-or-Rashid and Roos 2001).

amplitude, whereas the pressure of the photons will oppose collapse and tend to iron out inhomogeneities). The positions and heights of the other peaks are sensitive to other cosmological parameters.

Before the microwave radiation decouples from matter, it will undergo a last act of Thomson scattering from free electrons. Because of the anisotropies in the radiation described above, this scattering should result in polarization—just as sunlight becomes polarized when scattered in the atmosphere. The degree of polarization measured with the DASI interferometer is exactly of the magnitude (parts per million) expected. As stated above, another source of photon polarization, due to gravitational radiation accompanying inflation, will be much more difficult to detect.

With relatively weak assumptions on the age of the universe and the value of the Hubble constant, and in combination with the results of measurements of large-scale structure from infrared and X-ray surveys and of Type Ia supernovae observations (see Figs 2.3 and 4.11), the microwave data then yield as typical fits the following values for the various parameters:

$$\text{Total closure parameter } \Omega_{\text{tot}} = 1.02 \pm 0.05$$

$$\text{Vacuum energy contribution } \Omega_{\text{v}} = 0.70 \pm 0.05$$

$$\text{Total matter contribution } \Omega_{\text{m}} = 0.30 \pm 0.05 \tag{5.44}$$

$$\text{Baryon density contribution } \Omega_{\text{b}} = 0.05 \pm 0.01$$

The total closure density is consistent with the value of unity predicted by the inflationary scenario. The percentage error is principally associated with, and just twice, the fractional error in the *l*-value of the peak. The value of the vacuum energy density or cosmological constant contribution deduced from the supernova measurements of the Hubble constant and its variation with time, described in Chapter 4, are in good agreement with estimates from combining the microwave data with galaxy redshift surveys, and of course that result, as shown in Example 2.3, gives an estimate for the age of the universe in excellent agreement with independent estimates from radioactive isotope analysis, stellar ages in globular clusters, etc. Furthermore, the baryon density in (5.44) is in fair agreement with the (less precisely) known value from Big Bang nucleosynthesis described in Chapter 2. The bulk of the matter contribution is obviously accounted for by dark matter. Typical fits to the data are shown in Fig. 5.11.

5.10 Summary

- The conventional Big Bang model, while successfully accounting for the redshift, abundance of light elements and the microwave background radiation, suffers from the horizon and flatness problems.
- The horizon problem arises on account of the observed isotropy of the microwave background out to the largest angles. The horizon at the time of the decoupling of matter and radiation, at $z \sim 1000$ (when the radiation had the last opportunity of interaction and achieving thermal equilibrium), now only subtends about $1°$ at the Earth. It is, therefore, impossible to understand how the large-angle uniformity in temperature could have been achieved by causal processes.
- The flatness problem arises because the fractional difference between the observed density ρ and the critical density ρ_c should be proportional to t in a radiation-dominated universe, and $t^{4/3}$ in the matter-dominated case. Thus, at very early times, ρ must have been very finely tuned to ρ_c (only one part in 10^{52}, if we go back to the Planck era).
- The postulate of a preliminary inflationary stage of exponential expansion, when the radius of the initial micro-universe expanded from 10^{-26} to 1 m, solves both these problems, and also accounts for the absence of magnetic monopoles.
- Quantum fluctuations at the commencement of inflation may account for the observed perturbations, of the order of 10^{-5}, in the temperature (and hence density) of the microwave radiation. The quantum fluctuations would become classical, frozen fluctuations in density when they were inflated beyond the causal horizon.
- Large-scale structures in the universe—galaxies, galaxy clusters, voids, etc.—were seeded by these primordial perturbations in the density (or metric curvature).

- Initially, the density perturbations $\delta = \Delta\rho/\rho$ grew linearly with the expansion parameter R as they collapsed under gravity in the era of matter-domination. The collapse became possible on all scales larger than the Jeans length, in turn determined by the speed of sound v_s in the cosmic fluid. As v_s decreased abruptly when atoms began to form, more dramatic, and non-linear, collapse on smaller and smaller scales became possible.

- The spectrum of density fluctuations predicted by inflation is a power law, with no preferred scale, and $\Delta\rho/\rho \sim \lambda^{-2}$, where λ is the length involved. The spectrum observed in COBE microwave measurements at large angular scales fits this prediction. At smaller scales the fluctuation spectrum from galaxy surveys such as IRAS is flatter and consistent with expectations from mixed dark matter models.

- The tiny (10^{-5}) variations in the observed microwave temperature between pairs of points in the sky separated by angle θ can be described by a sum of Legendre polynomials $P_l(\cos\theta)$, where l-values of 100–1000 are relevant to investigations at separation angles of the order of $1°$. The fluctuations in temperature, consequent on the density fluctuations, appear as a series of so-called 'acoustic peaks' when plotted against the l-value.

- The position of the first acoustic peak, at $l \approx 200$, provides a measure of the total density parameter Ω, indicating a value close to unity, and a consequent curvature parameter $\Omega_k = 1 - \Omega$ of less than 0.05: the universe is practically flat. The height of the first peak and the heights and positions of the other peaks provide estimates for other cosmological parameters, such as the baryon density, the amount of dark matter, Hubble parameter, etc.

Problems

More challenging problems are denoted with an asterisk.

(5.1) Show that the free-fall time (5.26) is of the same order of magnitude as the period of a satellite in close orbit about a spherical cloud of density ρ.

(5.2) Calculate the Jeans mass and Jeans length for a mass of air at NTP ($T = 273$ K, $\rho = 1.29$ kg m^{-3}).

*(5.3) If the intergalactic medium is appreciably ionized, the cosmic microwave radiation may have been scattered (Thomson scattering) since the 'epoch of last scattering' at redshift $z \sim 1100$. Estimate the resulting mean free path of the microwave radiation, using the baryon densities given in Chapter 2. Is this important in the interpretation of the angular anisotropies of the radiation?

*(5.4) Verify the expressions (5.34)–(5.36) for the growth of the density contrast with time in a closed ($K = +1$)

matter-dominated universe. Derive the corresponding expressions for an open universe, with $K = -1$, and show that in this case the density contrast will decrease with time.

(5.5) Assume that a density fluctuation occurs during the inflation process, and that at the end of inflation, at $t_i \sim 10^{-32}$ s, this has become 'frozen out' at a length scale λ, when the universe has reached a radius ~ 1 m. Calculate at what subsequent time the perturbation will come inside the horizon and start to oscillate as an 'acoustic' wave, for $\lambda = 1$ mm and 1 cm. Estimate the masses inside the horizon in the two cases and identify them with large-scale structures.

(5.6) Calculate the Jeans length at time t in the early, radiation-dominated universe, and show that it is approximately equal to the horizon distance at that time.

Cosmic particles

The particles circulating in the cosmos include the cosmic rays, which have been intensively studied ever since their discovery by Hess in 1912. Karl K. Darrow, a past president of the American Physical Society, caught some of the atmosphere of this research when he described their study as remarkable "for the delicacy of the apparatus, the minuteness of the phenomena, the adventurous excursions of the experimenters and the grandeur of the inferences".

Cosmic rays consist of high energy particles incident on the Earth from outer space, plus the secondary particles, which they generate as they traverse the atmosphere. Their study has a special place in physics, not only in its own right, but because of the pioneering role that cosmic ray research has played in the study of elementary particles and their interactions. Let us recall that the discovery in cosmic rays of antimatter, in the form of the positron in 1932, and of pions and muons and strange particles in the late 1940s, really kick-started the building of large particle accelerators and the development of their associated detection equipment, developments which were essential in putting the subject of elementary particle physics on a sound quantitative basis.

Equally, in the 1990s, the study of the interactions of solar and atmospheric neutrinos, on distance scales far larger than anything that had been attempted at accelerators or reactors, revealed the first cracks in the Standard Model, with evidence for neutrino flavour mixing and for finite neutrino masses, as described in Sections 6.9–6.11. This has led to a revival, in the new millenium, of lepton physics in fixed-target experiments at accelerators, and to the development of radically new proposals such as the building of muon storage rings to serve as sources of high energy electron- and muon-neutrinos.

At the highest energies, studies of γ-rays in the TeV range and above have indicated point sources in the skies where it seems the most violent events in the universe have taken place, and intensive studies of both γ-rays and ultra-high energy protons and heavier nuclei will certainly shed new light on mechanisms for particle acceleration, as well, perhaps, as revealing new fundamental processes taking place at energies far in excess of what could ever be achieved on the Earth.

6.1 The spectrum and composition of cosmic rays

The charged primary particles of the cosmic rays consist principally of protons (86%), α-particles (11%), nuclei of heavier elements up to uranium (1%), and electrons (2%). While these come from primary sources, there are also very small proportions of positrons and antiprotons, which we believe are of secondary origin and generated by interactions of the primary particles with interstellar gas. The above percentages are for particles above a given **magnetic rigidity** $R = pc/z|e|$, where p is the momentum and $z|e|$ the particle charge; that is, for particles with the same probability of penetrating to the atmosphere through the geomagnetic field. Neutral particles consist of γ-rays and of neutrinos and antineutrinos. Some of these can be identified as coming from 'point' sources in the sky; for example, neutrinos from the Sun and from supernovae, and gamma rays from sources such as the Crab Nebula and active galactic nuclei.

The nature of the charged primary particles was first established by means of nuclear emulsion detectors flown in high altitude balloons—see Fig. 6.1. At low energies, primary energies may be estimated from the range in the absorber. For the GeV–TeV energy region, the calorimetric method has been employed. This involves measuring the ionization energy in electromagnetic showers that develop as a result of the nuclear cascade which the primary generates as it traverses great thicknesses of absorber (see Section 6.5). Detectors flown in satellites have employed scintillation counters to measure the primary nuclear charge from the pulse height, and gas-filled Cerenkov counters to measure the particle velocity and hence the energy (see Fig. 6.10).

The energy density in the cosmic rays is about 1 eV cm^{-3} when calculated for deep space outside the influence of solar system magnetic fields, and is therefore quite comparable with the energy density in starlight of 0.6 eV cm^{-3}, that in the cosmic microwave background radiation of 0.26 eV cm^{-3}, and that in the galactic magnetic field of 3 μG or 0.25 eV cm^{-3}. The bulk of the primary radiation is of galactic origin: however, the fact that the spectrum extends to very high energies (10^{20} eV) indicates that some at least of the radiation must be of extra-galactic origin, since the interstellar magnetic field could not contain such particles inside our local galaxy.

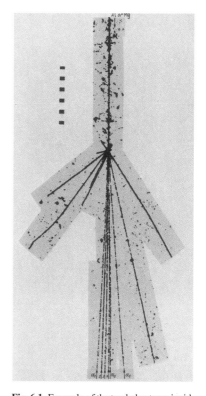

Fig. 6.1 Example of the track due to an incident high energy primary aluminium nucleus ($Z = 13$) which interacts in a photographic nuclear emulsion flown on a balloon in the stratosphere. Charged particles ionize the atoms they traverse and the electrons liberated form a latent image as they are trapped in the microcrystals of silver halide. Upon processing, the unaffected halide is dissolved out and the crystals with latent images are reduced to black metallic silver, so forming the tracks. In this example, the incident nucleus undergoes fragmentation into a 'jet' of six α-particles ($Z = 2$) in the forward direction. The charges of primary and secondary nuclei were established by measurement of the frequency along the tracks of the hair-like δ-rays (knock-on electrons), which varies as Z^2. The tracks at wider angles are due to protons ejected from the struck nucleus. The mean free path for interaction of heavy primary nuclei with those of the atmosphere is of the order of 10 gm cm^{-2} so that they do not penetrate much below an altitude of 25 km. The scale on the left-hand side of this micro-photograph is 50 μm.

Example 6.1 *Calculate the radius of curvature of the trajectory of a proton of energy 10^{20} eV in a galactic magnetic field of 3 μG (3×10^{-10} T). Compare this with the typical disc radius of a spiral galaxy of 15 kpc.*

Referring to a book on electromagnetism or the formula in Appendix A, the radius of curvature ρ in metres of a singly charged particle of momentum p GeV/c in a magnetic field of B T is given by $\rho = pc/(0.3B)$. Substituting, one finds $\rho = 10^{21}$ m or 30 kpc.

Galactic magnetic fields may be detected from the fact that paramagnetic dust grains would be partly aligned in the field, and the scattering of polarized light by the grains would give rise to a rotation of the plane of polarization (called the Faraday effect). This can be used to show that the magnetic fields are trapped in the spiral arms of the galaxy.

Figure 6.2 shows the energy spectrum of cosmic ray protons. Above an energy of a few GeV, the spectrum up to the so-called 'knee' at 10^{14} eV (100 TeV) follows a simple power law

$$N(E) \, dE = \text{const.} \, E^{-2.7} \, dE \qquad (6.1)$$

Above the 'knee' the spectrum becomes steeper with an index of about -3.0, before apparently flattening off again above 10^{18} eV. These ultra-high energy

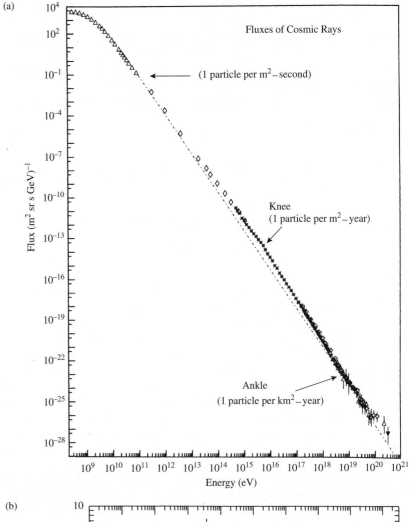

(a)

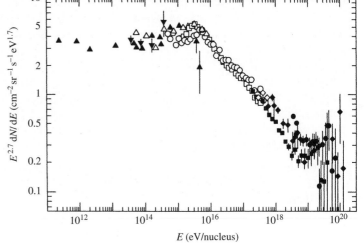

(b)

Fig. 6.2 (a) The primary cosmic ray differential energy spectrum showing the power law $E^{-2.7}$ dependence at energies below the 'knee', steepening to $E^{-3.0}$ at energies above it, followed by indication of a flattening above the 'ankle' at 10^{18} eV. Arrows show the integral fluxes of particles above certain energies (graph by S. Swordy, reproduced courtesy of J. Cronin (1999)). (b) The primary spectrum multiplied by $E^{2.7}$, showing the knee in more detail (reprinted from Review of particle properties, by Barnett, R.M. *et al.* (1996). *Phys. Rev.*, **D54**, 1).

cosmic rays are discussed in more detail in Section 6.6. At energies above 30 GeV, where effects due to the magnetic fields of the Earth or the Sun are unimportant, the radiation appears to be quite isotropic. This is in any case to be expected at all but the very highest energies, since the galactic magnetic fields would destroy any initial anisotropy.

The chemical composition of the cosmic ray nuclei exhibits remarkable similarities to the solar system abundances, which are deduced from absorption lines in the solar photosphere and from meteorites, but it also shows some significant differences, as seen in Fig. 6.3. The cosmic and solar abundances both show the odd–even effect, associated with the fact that nuclei with Z and A even are more strongly bound than those with odd A and/or odd Z, and therefore are more frequent products in thermonuclear reactions in stars. The peaks in the

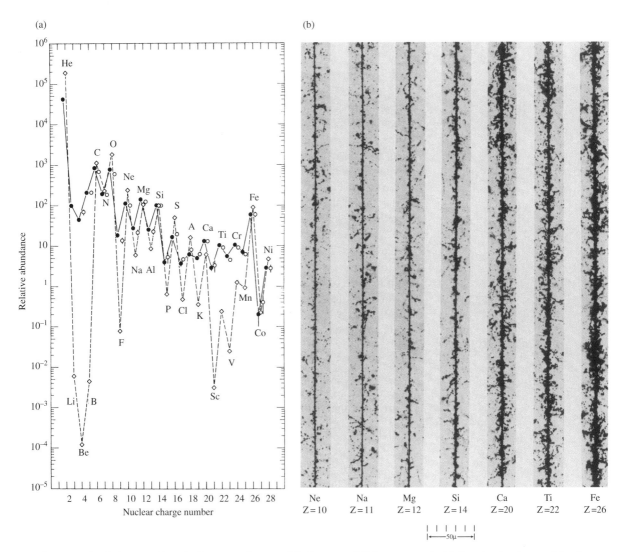

Fig. 6.3 (a) The chemical composition of primary cosmic ray nuclei, shown as a full line, compared with the solar abundances of the elements, shown as a dashed line (from Simpson (1983) with permission from *Annual Reviews of Nuclear and Particle Science* **33**); (b) tracks of various relativistic primary cosmic ray nuclei recorded in nuclear emulsions flown on a high altitude balloon.

normalized abundances for C, N, and O and for Fe are also closely similar, suggesting that many of the cosmic ray nuclei must be of stellar origin.

The big differences between the cosmic and solar abundances are in those of Li, Be, and B. The abundance of such elements in stars is very small, since they have low Coulomb barriers and are weakly bound and rapidly consumed in nuclear reactions in stellar cores. Their comparative abundance in cosmic rays is due to **spallation** of carbon and oxygen nuclei as they traverse the interstellar hydrogen (see Fig. 6.1). In fact the amount of these light elements determines the average thickness of interstellar matter that the radiation traverses and indicates an average lifetime of the cosmic rays in the galaxy of about three million years. It is found that the energy spectra of Li, Be, and B are somewhat steeper than those of carbon or oxygen, indicating that at the higher energies nuclei do not undergo so much fragmentation, presumably because they leak out of the galaxy sooner than those of lower energy. In a similar way, the abundance of Sc, Ti, V, and Mn in the cosmic rays is due to spallation of the abundant Fe and Ni nuclei.

6.2   Geomagnetic and solar effects

The primary radiation, for charged particles below 10 GeV energy, does show directional effects and also time dependence. The charged primaries are affected by the Earth's magnetic field, which approximates that due to a simple magnetic dipole, and also by modulation in time due to the solar wind, which follows the eleven-year solar cycle.

We first discuss the geomagnetic effects. The axis of the dipole is at an angle to the axis of the Earth's rotation. The geographical coordinates of the poles varies slowly with geological time, the present N pole being located at longitude $101°$ W, latitude $75°$ N. The calculation of the actual orbits of particles incident on the Earth as they spiral in the dipole field is rather tedious and complex, and most easily accomplished using a computer program. However, some of the main features of the geomagnetic effects can be understood analytically.

Consider first a particle of charge $z|e|$, velocity v and momentum $p = mv$ travelling in a circular equatorial path of radius r around a short dipole of moment M. Equating centrifugal and magnetic forces we obtain

$$z|e| \, |\mathbf{B} \times \mathbf{v}| = \frac{mv^2}{r}$$

where the equatorial field due to the dipole is

$$B = \left(\frac{\mu_0}{4\pi}\right) \frac{M}{r^3}$$

The radius of the orbit is, therefore,

$$r_S = \left[\left(\frac{\mu_0}{4\pi}\right) \frac{Mz|e|}{p}\right]^{1/2} \tag{6.2}$$

known as the **Størmer unit**, after the physicist who first treated the problem. A significant value of the particle momentum is that which makes the Earth's radius r_E equal to one Størmer unit, that is,

$$\frac{pc}{z} = \left(\frac{\mu_0}{4\pi}\right)\frac{Mc|e|}{r_E^2} = 59.6 \text{ GeV} \tag{6.3}$$

where we have inserted the values in SI units of $\mu_0/4\pi = 10^{-7}$, $M = 8 \times 10^{22}$ A m, $r_E = 6.38 \times 10^6$ m, $|e| = 1.6 \times 10^{-19}$ C and 1 GeV $= 1.6 \times 10^{-10}$ J. In fact, it turns out that no proton of momentum less than the above value can reach the Earth from the eastern horizon at the magnetic equator. Størmer showed that the equation of motion obeyed by a particle has the form

$$b = r \sin\theta \cos\lambda + \frac{\cos^2\lambda}{r} \tag{6.4}$$

where r is the distance of the particle from the dipole centre in Størmer units, λ is the geomagnetic latitude and θ is the angle between the velocity vector $\mathbf{v}$ and its projection in the meridian plane OAB co-moving with the particle—see Fig. 6.4. The angle θ is called positive for particles travelling from east to west, as that shown, while it is negative for particles travelling in the opposite direction. The quantity b is the impact parameter or closest distance of approach to the dipole axis by a tangent to the particle trajectory at infinity (again in Størmer units). Since we must have $|\sin\theta| < 1$, (6.4) places restrictions on the values of b, r, and λ for the 'allowed' trajectories of particles reaching the Earth. The condition $b \leq 2$ is found to be critical in determining which momenta are cut off by the Earth's field. Inserting $b = 2$ in (6.4) the equation for the cut-off momentum at any λ and θ is given by

$$r = \frac{\cos^2\lambda}{1 + (1 - \sin\theta\cos^3\lambda)^{1/2}} \tag{6.5a}$$

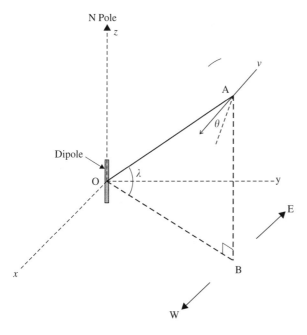

Fig. 6.4 Coordinate system and variables describing a particle A with velocity $\mathbf{v}$ in the field of a dipole M at O. θ is the angle between the velocity vector $\mathbf{v}$ of the particle and the meridian plane OAB rotating with the particle.

where, from (6.3),

$$\frac{pc}{z} = 59.6r^2 \text{ GeV} \tag{6.5b}$$

since we are concerned with particles arriving at the Earth, so that $r = r_E/r_S$. For example, for particles incident from the vertical, $\theta = 0$ and $r = \frac{1}{2}\cos^2\lambda$ so that the cut-off momentum is

$$(pc)_{\min}(\theta = 0) = 14.9\, z \cos^4\lambda \text{ GeV} \tag{6.6}$$

In NW Europe for example, with $\lambda \sim 50°$ N, the vertical cut-off momentum would be $pc/z = 1.1$ GeV, or a minimum kinetic energy for a proton of 0.48 GeV.

From (6.5) and (6.6) we see that at the magnetic equator, the vertical cut-off is 14.9 GeV c^{-1}. That for particles from the eastern horizon ($\sin\theta = +1$) is 59.6 GeV c^{-1} while that for particles from the western horizon ($\sin\theta = -1$) is only $59.6/(1 + \sqrt{2})^2 = 10.2$ GeV c^{-1}. This results in the so-called **east–west effect**, namely that at all latitudes, more (positively charged) particles arrive from the west than from the east, because of the lower momentum cut-off. The effect arises essentially because all positively charged particles are deflected in a clockwise spiral, as viewed from above the N pole. Fig. 6.5 shows a map of the vertical cut-off rigidities.

The azimuthal and latitude dependences of the primary particles are promulgated in the secondaries they produce in traversing the atmosphere. Such effects are observed, for example, in the interactions of atmospheric neutrinos, coming from the decay of secondary pions and muons, and were important in establishing the credibility of the experiments and their interpretation in terms of neutrino flavour oscillations, as discussed in Section 6.9.

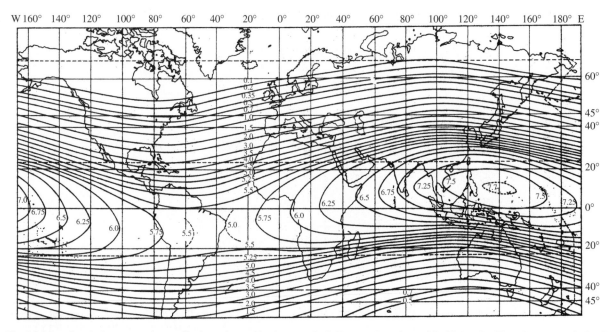

Fig. 6.5 Map of vertical geomagnetic cut-off values, given as kinetic energy in GeV per nucleon, for nuclei with $A = 2Z$. The values were calculated for a displaced dipole field. The maximum cut-off is about 7.7 GeV, or a momentum cut-off of 8.6 GeV/c, per nucleon. For protons the momentum cut-off would then be 17.2 GeV/c, to be compared with the value (6.6) for an undisplaced dipole field (from Webber 1958).

Example 6.2 *Estimate the ratio of the intensity of primary protons incident from the eastern horizon, as compared with that from the west, at magnetic latitude 45° N.*

From equation (6.5) the cut-off momentum is

$$\frac{59.6 \cos^4 \lambda}{[1 + (1 - \cos^3 \lambda)^{1/2}]^2} = 4.58 \text{ GeV}/c$$

from the east and

$$\frac{59.6 \cos^4 \lambda}{[1 + \{1 + \cos^3 \lambda)^{1/2}\}]^2} = 3.18 \text{ GeV}/c$$

from the west. Assuming a power law momentum spectrum of the form $dp/p^{2.7}$, the ratio of eastern over western intensities is $(3.18/4.58)^{1.7} = 0.54$.

In reality, the Earth's magnetic dipole (formed by ring currents deep in the Earth) is offset by some 400 km from the Earth's centre, and there are also higher order (quadrupole) components to the field. Furthermore, at distances beyond a few Earth radii, the trajectories are strongly distorted by the effects of the **solar wind**, which is a plasma of low energy protons and electrons ejected from the Sun. Variations in this wind follow the 11 yr sunspot cycle. The counting rate of sea-level neutron monitors has been measured for many decades and is in exact anti-correlation with the sunspot number, the difference between maximum and minimum counting rates being of the order of 20%. Although the protons and electrons in the solar wind are of low energy (with proton kinetic energies of the order of 0.5 keV) they have high intensities, with a kinetic energy density of the order of 3 keV cm^{-3} and an associated magnetic field of about 10^{-8} T. If the Earth happens to be in the path of this wind, it experiences phenomena called **solar flares**. For example, dramatic aurora phenomena are observed in latitudes near the magnetic poles. These arise because charged particles from the flare become scattered and trapped in the Earth's field (which acts as a sort of magnetic mirror), spiralling to and fro from pole to pole around the lines of force and producing excitation of the air molecules in the stratosphere, with the resultant optical display.

6.3 Acceleration of cosmic rays

How do cosmic rays attain their colossal energies, up to 10^{20} eV, and how do we account for the form of the energy spectrum? Many years ago, it was remarked that the energy density in cosmic rays, coupled with their lifetime in the galaxy, required a power supply somewhat similar to the rate of energy generation in supernova shells. Our own galaxy has a radius $R \sim 15$ kpc and disc thickness $D \sim 0.2$ kpc. The total power requirement to accelerate the cosmic rays in the disc, for an average energy density of $\rho_E = 1$ eV cm^{-3} is thus

$$W_{CR} = \rho_E \pi R^2 \frac{D}{\tau} = 2 \times 10^{41} \text{ J yr}^{-1} \tag{6.7}$$

where $\tau \sim 3 \times 10^6$ y is the average age of a cosmic ray particle in the galaxy, before it diffuses out or is depleted and lost in interactions with the interstellar gas. A Type II supernova (see Section 7.11) typically ejects a shell of material of about $10\,M_\odot$. $(2 \times 10^{31}\,\text{kg})$, with velocity of order $10^7\,\text{m s}^{-1}$ into the interstellar medium, approximately once per century in our galaxy. This gives an average power output per galaxy of

$$W_{\text{SN}} = 5 \times 10^{42}\,\text{J yr}^{-1} \tag{6.8}$$

Although the galactic supernova rate is quite uncertain, it appears, therefore, that an efficiency for the shockwave to transmit energy to cosmic rays of a few per cent would be enough to account for the total energy in the cosmic ray beam.

In the 1950s, Fermi had considered the problem of cosmic ray acceleration. He first envisaged charged cosmic ray particles being reflected from 'magnetic mirrors' provided by the fields associated with massive clouds of ionized interstellar gas in random motion. It turns out, however, that such a mechanism is too slow to obtain high particle energies in the known lifetime of cosmic rays in the galaxy. Fermi also proposed that acceleration could occur due to **shock-fronts**. Consider, in a simplified one-dimensional picture (Fig. 6.6) a relativistic particle travelling in the positive x-direction, which traverses a shock-front moving with velocity $-u_1$ in the negative x-direction. Suppose that the particle is back-scattered by the field in the gas behind the front; it will have a velocity component in the direction of the shock of

$$u_2 = \frac{2u_1}{C_p/C_v + 1} = \frac{3u_1}{4} \tag{6.9}$$

where the ratio of specific heats $C_p/C_v = \frac{5}{3}$ for an ionized gas. Thus, the particle travels back across the shock-front, to be scattered by magnetized clouds upstream of the front. If these scatter the particle backwards again (i.e. in the direction of positive x), the particle can re-cross the front and repeat the cycle of acceleration once more. Because the front is planar (i.e. unidirectional) a straightforward application of the Lorentz transformations (see Appendix B) shows that the fractional energy gain is of the order of the shock-front velocity (see Problem 6.14):

$$\frac{\Delta E}{E} \sim \frac{u_1}{c} \tag{6.10}$$

There are many possible sources of shocks, but as indicated above, Type II supernovae shells seem to be good candidates, with shock velocities of order $10^7\,\text{m s}^{-1}$.

Fig. 6.6 Diagram depicting the acceleration of a charged particle on crossing a shock-front, and being scattered back across the front by the upstream gas.

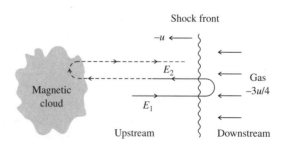

Suppose that, in each cycle of acceleration at the shock-front, the particle gets an energy increment $\Delta E = \alpha E$. After n cycles its energy becomes

$$E = E_0(1 + \alpha)^n$$

Thus, in terms of the final energy E, the number of acceleration cycles is

$$n = \frac{\ln(E/E_0)}{\ln(1 + \alpha)} \tag{6.11}$$

At each stage of the acceleration the particle can escape further cycles. Let P be the probability that the particle stays for further acceleration, so that after n cycles the number of particles remaining for further acceleration will be

$$N = N_0 P^n$$

where N_0 is the initial number of particles. Substituting for n we get

$$\ln\left(\frac{N}{N_0}\right) = n \ln P = \frac{\ln(E/E_0) \ln P}{\ln(1 + \alpha)} = \ln(E_0/E)^s$$

where $s = -\ln P/\ln(1 + \alpha)$. The number N will be the number of particles with n or more cycles, thus with energy $\geq E$. Hence, the differential energy spectrum will follow the power law dependence

$$\frac{dN(E)}{dE} = \text{const.}\left(\frac{1}{E^{(1+s)}}\right) \tag{6.12}$$

For shock-wave acceleration, it turns out that $s \sim 1.1$ typically, so that the differential spectrum index is -2.1, compared with the observed value of -2.7. The steeper observed spectrum could be accounted for if the escape probability $(1-P)$ was energy dependent. As we have already seen, the spallation spectrum of Li, Be, and B indeed falls off more rapidly with energy than that of the parent C and O nuclei, suggesting that the escape probability does increase with energy.

The shock-wave acceleration from supernovae shells appears capable of accounting for the energies of cosmic ray nuclei of charge $Z|e|$ up to about $100Z$ TeV ($10^{14}Z$ eV), but not beyond this. Other, largely unknown, acceleration mechanisms must be invoked for the very highest energy cosmic rays.

6.4   Secondary cosmic radiation: hard and soft components

The term 'cosmic rays' properly refers to particles and radiation incident from outside the Earth's atmosphere. These primary particles will produce secondaries (mesons) in traversing the atmosphere, which plays the same role as a target in an accelerator beam. The situation is shown schematically in Fig. 6.7. The most commonly produced particles are pions, which occur in three charged and neutral states π^+, π^-, and π^0. Since the nuclear interaction mean free path in air is about 100 gm cm^{-2} for a proton (and much less for a heavy nuclear primary), compared with a total atmospheric depth of $X = 1030$ gm cm^{-2}, the pions will

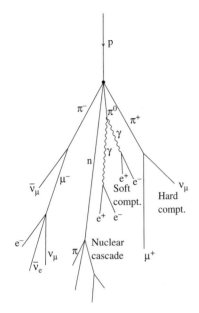

Fig. 6.7 Diagram (not to scale) indicating the production and decay of pions and muons in the atmosphere.

be created mostly in the stratosphere. The **charged pions** decay to muons and neutrinos: $\pi^+ \rightarrow \mu^+ + \nu_\mu$ and $\pi^- \rightarrow \mu^- + \bar{\nu}_\mu$, with a proper lifetime of $\tau = 26$ ns and a mean free path before decay of $\lambda = \gamma c\tau$ where $\gamma = E_\pi/m_\pi c^2$ is the time dilation factor. With $m_\pi c^2 = 0.139$ GeV, $\lambda = 55$ m for a 1 GeV pion. Now to a fair approximation the upper atmosphere is isothermal and the depth x (gm cm^{-2}) varies exponentially with height h (km), according to the formula

$$x = X \exp(-h/H) \quad \text{where } H = 6.5 \text{ km}$$

By differentiating this expression one sees that in an interval Δh of $\lambda = 55$ m $\sim 0.01H$, the depth will change by only 1%. Thus, nuclear absorption will only become important for charged pions with $\lambda \sim H$ or energies of order 100 GeV or more, and at GeV energies practically all charged pions decay in flight (rather than interact).

At high (TeV) energies, on the contrary, the pion decay probability is of order $100 \sec\theta/E_\pi$ (GeV), where θ is the zenith angle, and the majority of pions in this case undergo nuclear interaction before they have a chance to decay. Note that because of this $1/E$ factor, the power law energy spectrum of the daughter muons at TeV energies will have a (negative) index greater by one unit than the parent pions or protons. Fifty years ago, the position of this 'knee' in the muon spectrum was in fact used by Greisen to get one of the first estimates of the charged pion lifetime.

The daughter muons are also unstable, undergoing the decay $\mu^+ \rightarrow e^+ + \nu_e + \bar{\nu}_\mu$, with a proper lifetime of $\tau = 2200$ ns. Since the muon mass is 0.105 GeV, a 1 GeV muon has a mean decay length of 6.6 km, about equal to the scale height H of the atmosphere. Muons of energy 1 GeV or less will decay in flight in the atmosphere (there is no competition with nuclear interaction since muons do not have strong interactions). However, a 3 GeV muon, for example, has a mean decay length of 20 km, of the same order as the typical distance from its point of production to sea-level. Moreover, with an ionization energy loss rate of 2 MeV gm^{-1} cm^2 of air traversed (see equation (6.13)), muons with 3 GeV or more energy can get through the entire atmosphere without being brought to rest or decaying. Still higher energy muons can reach deep underground, and for this reason they are said to constitute the **hard component** of the cosmic radiation. The remaining products of charged pion and muon production, the neutrinos, are discussed in Section 6.8.

So much for the charged pions and their decay products. The **neutral pions** undergo electromagnetic decay, $\pi^0 \rightarrow 2\gamma$, with an extremely short lifetime of 8×10^{-17} s. As described below, the photons from the decay develop electron–photon cascades, mostly in the high atmosphere, since the absorption length of these cascades is short compared with the total atmospheric depth. The electrons and photons of these cascades constitute the easily absorbed **soft component** of the cosmic radiation.

Among the products of the nuclear interactions of primary cosmic rays in the atmosphere are radioactive isotopes, of which an important one is ^{14}C formed, for example, by neutron capture in nitrogen: n $+^{14}$N $\rightarrow$ ^{14}C$+^1$H. The ^{14}C atoms produced in this way combine to form CO_2 molecules and thus participate, like the more common, stable ^{12}C atoms, in the circulation of this gas in the atmosphere, and through rainfall into the oceans, and in absorption in organic matter. Since carbon-14 has a mean lifetime of 5600 yr, its abundance relative to carbon-12 in organic matter can be used to date the sample. This, of course, assumes that the carbon-14 production rate by cosmic rays has been constant with time. In fact, comparison of the age from the isotope ratio with that from ancient tree ring counts shows that the cosmic ray intensity did vary in the past and was some 20% larger 5000 yr ago. This variation was presumably due to long-term fluctuations in the value of the Earth's magnetic field, which is known from rock samples to have changed its sign and magnitude many times over geological time.

6.5   Electromagnetic cascades and air showers

6.5.1   Ionization and radiation losses

First of all we discuss the **ionization energy loss** of high energy charged particles, and the radiation loss of electrons, in traversing an absorber. Charged particles lose energy as a result of collisions with atomic electrons, leading to ionization of the atoms. The rate of energy loss is given by the Bethe–Bloch formula

$$\left(\frac{dE}{dx}\right)_{ion} = \left(\frac{4\pi N_0 z^2 e^4}{mv^2}\right)\left(\frac{Z}{A}\right)\left\{\ln\left[\frac{2mv^2\gamma^2}{I}\right] - \beta^2\right\} \tag{6.13}$$

where m is the electron mass and v and ze are the velocity and charge of the incident particle, $\beta = v/c$, $\gamma^2 = 1/(1 - \beta^2)$, N_0 is Avogadro's number, Z and A are the atomic and mass numbers of the atoms of the medium, and x is the path length in the medium, usually measured in gm cm^{-2}. The quantity I is the mean ionization potential of the medium, averaged over all electrons in the atom, and is approximately $I = 10Z$ eV. Notice that dE/dx is a function of velocity v and is independent of the mass M of the incident particle. It varies as $1/v^2$ at low velocity. After passing through a minimum value at an energy of about $3Mc^2$, the ionization loss increases logarithmically with energy. At higher energies, polarization effects set in and the ionization loss reaches a plateau value of about 2 MeV gm^{-1}cm^2, as in Fig. 6.8. Note also that, since $Z/A \sim \frac{1}{2}$ in most materials (except hydrogen and the very heavy elements) the energy loss, expressed per gm cm^{-2} of material traversed, depends little on the medium.

In addition to suffering ionization energy loss, high energy electrons undergo Coulomb scattering by the atomic nuclei of the medium as well as **radiation loss** with the emission of photons, a process known as **'bremsstrahlung'** or braking radiation. The average rate of radiative energy loss of an electron in traversing a thickness dx of medium is

$$\left(\frac{dE}{dx}\right)_{rad} = -\frac{E}{X_0} \tag{6.14}$$

where the **radiation length** X_0 is given by

$$\frac{1}{X_0} = 4\alpha\left(\frac{Z}{A}\right)(Z+1)r_e^2 N_0 \ln\left(\frac{183}{Z^{1/3}}\right) \tag{6.15}$$

where $r_e = e^2/4\pi mc^2$ is the classical electron radius and $\alpha = \frac{1}{137}$. Note that because the radiation probability is proportional to the square of the acceleration, $X_0 \propto 1/r_e^2 \propto m^2$ and the radiation length for a muon will be of order

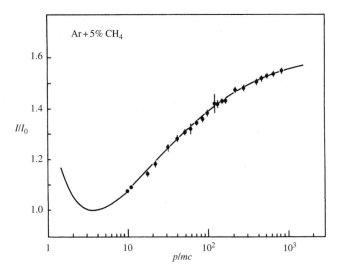

Fig. 6.8 Mean ionization energy loss of charged particles in an argon–methane gas mixture, plotted as a function of momentum in mass units, p/mc. The measurements were made by multiple ionization sampling, and show the relativistic rise from the minimum to a plateau (after Lehraus *et al.* 1978).

$(m_\mu/m_e)^2$ times that for an electron. If ionization losses are neglected, the mean energy of an electron of initial energy E_0 after having traversed a thickness x of medium will be

$$\langle E \rangle = E_0 \exp\left(\frac{-x}{X_0}\right) \tag{6.16}$$

We see from (6.15) that the radiation length varies approximately as $1/Z$. For example it is 40 gm cm^{-2} in air, compared with 6 gm cm^{-2} in lead. While the rate of ionization energy loss is practically constant for high energy electrons, the rate of radiation loss is proportional to energy E. The energy at which $(\mathrm{d}E/\mathrm{d}x)_\mathrm{ion} = (\mathrm{d}E/\mathrm{d}x)_\mathrm{rad}$ is known as the **critical energy** E_c. Electrons above the critical energy lose their energy principally through radiation processes, while for those below the critical energy it is mostly through ionization. Roughly, $E_c \sim 600/Z$ MeV.

Provided they have energy $E_\gamma > 2mc^2$, the photons radiated by an electron can themselves transform into e$^+$e$^-$ pairs, again in the Coulomb field of a nucleus to conserve momentum. The mean distance travelled by a photon before converting to a pair in a medium is called the **conversion length**. This is energy dependent, but at high energies (GeV) has an asymptotic value of approximately $(9/7)\,X_0$.

6.5.2   Development of electromagnetic showers

We are now in a position to discuss the longitudinal development of an electromagnetic shower in simple terms. Let us consider an electron of initial energy E_0 traversing a medium. In the first radiation length, suppose the electron radiates one photon, of energy $E_0/2$. In the next radiation length, suppose the photon converts to an electron–positron pair, each with energy $E_0/4$, and that the original electron radiates a further photon, also of energy $E_0/4$. Thus, after two radiation lengths, we have one photon, two electrons, and one positron, each of them with the energy $E_0/4$. Proceeding in this way, it follows that after t radiation lengths, we shall have electrons, positrons, and photons in approximately equal numbers, each with energy $E(t) = E_0/2^t$. Here, we have completely neglected ionization losses. We assume that this cascade multiplication process continues until the particle energy falls to $E = E_c$, the critical energy, when we suppose that the ionization loss suddenly becomes dominant and that no further radiation or pair conversion processes are possible. Thus, the cascade reaches a maximum and then ceases abruptly. The main features of this model are

1. The shower maximum is at a depth $t = t(\max) = \ln(E_0/E_c)/\ln 2$, that is, it increases logarithmically with primary energy E_0.
2. The number of particles at shower maximum is $N(\max) = 2^{t(\max)} = E_0/E_c$, that is, proportional to the primary energy.
3. The number of shower particles above energy E is equal to the number created at depths less than $t(E)$, that is,

$$N(>E) = \int \exp(t \ln 2)\, \mathrm{d}t = \frac{(E_0/E)}{\ln 2}$$

so that the differential energy spectrum of the particles is

$$\frac{\mathrm{d}N}{\mathrm{d}E} \propto \frac{\mathrm{d}E}{E^2}$$

4. The total track-length integral (of charged particles) in radiation lengths is

$$L = \left(\frac{2}{3}\right)\int 2^t\, \mathrm{d}t \sim \left(\frac{2}{3 \ln 2}\right)\frac{E_0}{E_c} \sim \frac{E_0}{E_c}$$

This last result also follows from energy conservation: since the ionization loss per particle is E_c per radiation length, essentially all the incident energy is finally dissipated as ionization energy loss. Thus, we obtain the important result that the track-length integral gives a measurement of the primary energy.

Example 6.3 *A cosmic-ray photon of energy* 10 TeV *is incident vertically on the atmosphere. Estimate the height of the maximum of the ensuing electron–photon shower in km. The critical energy in air is* 100 MeV *and the radiation length is* 37 gm cm^{-2}. *Assume that the depth x and height h are related by x = X (exp −h/H) where H = 6.5 km and X = 1030 gm cm^{-2} is the total depth of the atmosphere.*

Referring to the above formulae, the depth of the maximum is $x = \ln(E_0/E_c)/\ln 2 = 16.6$ radiation lengths or 615 gm cm^{-2}. Using the expression for the exponential atmosphere results in a height for the maximum of 3.4 km.

In practice, of course, the effects of both radiation loss and ionization loss are present throughout the shower process, and an actual shower consists of an initial exponential rise, a broad maximum and a gradual decline thereafter, as shown in Fig. 6.9. Nevertheless, the above simple model reproduces many of the essential quantitative features of actual electromagnetic cascades. (Our model has also treated the shower as one-dimensional. Actual showers spread out laterally, due mostly to Coulomb scattering of the electrons as they traverse the medium. The lateral spread of a shower, in radiation lengths, is a few times the so-called **Moliere unit**, equal to $21/E_c$ (MeV).)

6.5.3 Extensive air showers

If the primary particle is a high energy proton rather than an electron, a nuclear cascade will develop through the atmosphere. The longitudinal scale is the

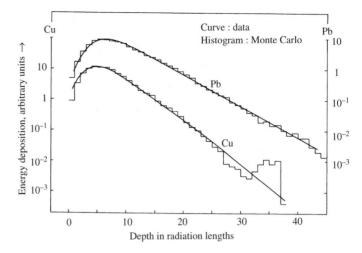

Fig. 6.9 Longitudinal development of electromagnetic showers due to 6 GeV electrons in CERN experiments (after Bathow *et al.* 1970).

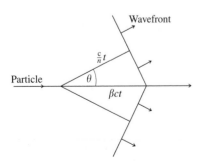

Fig. 6.10 Huyghens construction for emission of Cerenkov light by a relativistic particle.

nuclear interaction length in air, of about 100 gm cm^{-2}. The proton (or heavier nucleus) generates mesons in these interactions, and they can in turn generate further particles in subsequent collisions. While, in the electron–photon shower, the electrons lose the bulk of their energies in a radiation length, the nucleons can in general penetrate through several interaction lengths, losing only a fraction of their energy in each encounter, to struck nucleons as well as mesons. Coupled with the fact that, in air, the nuclear interaction length is 2.5 times the radiation length, cascades initiated by nucleons are much more penetrating than purely electromagnetic cascades. The lateral spread of nuclear showers is determined mostly by the transverse momentum of the secondaries in nuclear interactions, typically 0.3 GeV/c, and is much larger than for an electromagnetic shower of the same primary energy. Such **extensive air showers** will contain a high energy core, predominantly of nucleons, with a more widely spread electron–photon component that is continually fed by fresh neutral pion production and decay, $\pi^0 \to 2\gamma$, and fresh electromagnetic cascades. Apart from a small amount of energy in the form of neutrinos from pion decay, the great bulk of the energy as well as the overwhelming majority of the particles in a proton-initiated extensive air shower will end up in electron–photon cascades. So the track-length integral will again give a measure of the primary energy of the shower.

The detection of extensive air showers is accomplished by a variety of techniques. The oldest technique, pioneered by Auger 75 yr ago, uses an extended array of detectors in coincidence. These sample the charged particles in the shower, usually with scintillator or Cerenkov counters. Such showers only become detectable at sea-level for primary energies $E_0 > 1000$ TeV (10^{15} eV), when the maximum occurs near the ground. At mountain altitudes, the threshold is typically 100 TeV. The particles in such showers all have $v \sim c$, so that the shower front is quite well defined and the direction of the primary particle can be measured by timing the different parts of the shower front as it crosses the array.

As relativistic particles traverse the atmosphere, part of the energy loss appears in the form of a coherent wavefront of **Cerenkov radiation** (somewhat akin to the bow wave at the stem of a ship) as shown in Fig. 6.10. This radiation is mostly in the ultraviolet or blue region of the spectrum. The

Huyghens construction in the figure gives

$$\cos\theta = \frac{(ct/n)}{\beta ct} = \frac{1}{\beta n}, \quad \beta > \frac{1}{n} \tag{6.17}$$

where the refractive index n of the air at ground level is given by $\varepsilon = n - 1 = 3 \times 10^{-4}$, a quantity which is proportional to air pressure. The threshold energy for an electron with $mc^2 = 0.51$ MeV will be $mc^2/(1-\beta^2)^{1/2} = mc^2/(2\varepsilon)^{1/2} = 21$ MeV, while for a muon of $mc^2 = 106$ MeV it is 4.3 GeV. Most of the components of air showers have much greater energies, so they will produce abundant Cerenkov light. Typically a relativistic particle above threshold generates about 10 000 photons per km of path near ground level (and less at high altitudes where the pressure is lower). This light can be detected by means of large spherical mirror arrays which direct it on to photomultipliers placed at the focus (see Fig. 6.11).

The ionizing particles in a shower can also excite **fluorescence** from nitrogen molecules in the atmosphere, with typically 5000 photons per km of track

Fig. 6.11 Photograph of the 10 m mirror array at the Whipple Observatory, Arizona, for detection of Cerenkov light from air showers. A second nearby (11 m) dish allows construction of stereo images of extensive air shower profiles (courtesy T.C. Weekes, 1998).

length, again in the blue wavelength region. We note that the Cerenkov light is emitted in a narrow cone of angle $\theta \sim (2\varepsilon)^{1/2}$ ($=1.4°$ at ground level, although Coulomb scattering of the electrons will considerably broaden this), so that the light appears in a restricted radius of 100 m or so around the shower axis, and the axis must be fairly close to the mirror system in order to record any signal. In contrast, the fluorescent light is emitted isotropically. This means that distant showers several kilometers away, not aimed towards the mirror/photomultiplier system, can be detected, and therefore the sensitivity to the highest energy—and rarest—events is greatly increased. The first system employing these techniques was the large mirror array at the Whipple Observatory (see Fig. 6.11). It employs two spatially separated arrays of mirrors to give stereo images of the showers from the light output. In this way, it is possible to reconstruct the shower profile and to distinguish showers initiated by primary photons, which develop early and are contained in the upper atmosphere, from those generated by nucleons, which develop more slowly and are more penetrating. This feature is valuable in identifying point sources of γ-rays, as described in Section 6.7.

The weakness of these techniques is that they have a poor duty cycle. The problem of stray background light can only be overcome by operating on cloudless, moonless nights.

6.6   Ultra-high energy cosmic ray showers

As shown in Fig. 6.2, the spectrum of charged primary cosmic rays extends to at least 10^{20} eV. Most of the data from this plot comes from very extensive counter arrays (at sea level or on mountains). The largest of these to date is the AGASA array covering 100 km^2 in Japan. A larger array, called AUGER in Argentina is planned to cover 3000 km^2 and should yield about 50 events per year above 10^{20} eV. Although there is a 'knee' followed by an increasing slope of the spectrum above 10^{15} eV, above the 'ankle' at 10^{18} eV there is some indication for a flattening of the spectrum. Although at these energies the primaries should suffer only a minor deflection in the galactic magnetic field, the measured directions of the primaries do not indicate any localized sources. However, the most surprising feature of the spectrum is that there is no sign of it fading away above 10^{19} eV. Many years ago Greisen, Zatsepin and Kuzmin pointed out that the universe could become opaque at such energies through photopion production, on account of the Δ^+ resonance (see Fig. 1.13) in collisions of primary protons with photons of the microwave background radiation:

$$\gamma + p \rightarrow \Delta^+ \rightarrow p + \pi^0 \qquad (6.18)$$
$$\rightarrow n + \pi^+$$

If the proton has mass M, momentum $\mathbf{p}$ and energy E, and the microwave photon has momentum $\mathbf{q}$ and energy qc then the square of the total centre-of-momentum energy in the collision will be given (in units $c = 1$) by (see Appendix B on relativistic kinematics)

$$s = E_{\text{cms}}^2 = (E + q)^2 - (\mathbf{p} + \mathbf{q})^2$$
$$= M^2 + 2q(E - |\mathbf{p}| \cos \theta)$$

where θ is the angle between the proton and photon directions. The above quantity must be at least equal to the square of the sum of proton and pion masses, so that

$$M^2 + 2q(E - p\cos\theta) > M^2 + m_\pi^2 + 2Mm_\pi$$

Since the proton is extreme relativistic, $E = p$ and for a head-on collision $(1 - \cos\theta) = 2$, so the threshold proton energy is

$$E_{th} = m_\pi \frac{(M + m_\pi/2)}{2q}$$

The microwave background has $T = 2.725$ K, $kT = 2.35 \times 10^{-4}$ eV and the threshold energy is $(4.26/y) \times 10^{20}$ eV, where the photon energy is $qc = ykT$. So even for 0.1% of the photons in the extreme tail of the black-body distribution at $y > 10$, the threshold proton energy is still almost 10^{20} eV. Such protons are too energetic to be confined to our local galaxy by the galactic fields, which are of the order of microgauss.

The cross section for the reaction (6.18) near threshold is of order $\sigma = 2 \times 10^{-28}$ cm^2, while the total microwave photon density is $\rho = 400$ cm^{-3} (see equation (2.41)). Hence, the mean free path for collision $\lambda = 1/\rho\sigma \sim 10^{25}$ cm or some 5 Mpc, comparable to the size of a galaxy cluster (see Table 2.1). Thus, protons with energies well above 10^{20} eV at much greater distances than this would have their energies attenuated below 10^{20} eV. The fact that the AGASA spectrum shows no sign of such a cut-off could suggest that even the most energetic cosmic rays are produced in the local cluster (still containing dozens of galaxies, but with no known sources or mechanisms that could generate such tremendous energies). New counter arrays with much larger acceptance, as well as detection based on air fluorescence, are coming into operation and will hopefully shed light on this problem.

6.7   Radio galaxies and quasars

The electromagnetic radiation from a galaxy such as the Milky Way encompasses a vast spectral range, from radio wavelengths (centimetres to kilometres) to γ-rays with energies up to many TeV. In our own galaxy, less than 1% of the total electromagnetic output is at radio wavelengths, but so-called 'radio galaxies' are observed in which the radio emission can far exceed the optical output (from stars). The most dramatic radio emission is from **quasars** (standing for quasi-stellar radio sources), which are by far the brightest optical and radio sources in the sky, far exceeding the total light output from their host galaxies. Quasars invariably have large redshifts (up to $z \sim 5$) and they correspond to the most distant events known, occurring at times $t \sim t_0/(1 + z)^{3/2}$—see (2.6) and (2.11)—in the development of the universe. Quasars are indeed very largely an ancient phenomenon, occurring billions of years ago, at an early stage in the evolution of galaxies. We observe them today purely because of their great distance and the finiteness of the velocity of light or radio waves.

Quasars are often associated with galaxies so distant that the optical signal is hardly detectable, and their original discovery was made with radio telescopes, which incorporate giant receiving dishes. These have several advantages over optical telescopes: the radio signal does not suffer appreciable absorption by gas and dust, so one can probe deep into galactic centres; the signal can be amplified electronically, and its phase can easily be measured, so that the amplitudes of signals from different telescopes on a very long baseline L can be combined coherently to yield an effective aperture of L and a very high angular resolution $\Delta\theta = \lambda/L$. For example if $L \sim 1000$ km, $\Delta\theta$ is of the order of microradians.

When the optical signal is detected, it is found that the radio emission is primarily from a pair of lobes on either side of the galactic centre, and these may be separated by huge distances (even of the order of Mpc). The belief is that quasars are associated with very massive black holes, of typically 10^6–$10^8 M_\odot$ at galactic centres. Black holes are discussed in Chapter 7: they are objects with such strong gravitational fields that even relativistic particles such as photons are trapped inside them. A massive black hole would be surrounded by a spinning pancake-like accretion disc of galactic material—gas, dust, and stars—which feeds its growth. In this process, the accretion material will undergo violent oscillations, and a result is that charged particles are somehow accelerated to very high energies and punch their way through the minor axis of the accretion disc, giving two jets in opposite directions. These jets create enormous lobes of plasma as they traverse the intergalactic medium, and it is this plasma that generates the radio emission. Since magnetic fields will be associated with jets of charged particles, radio emission would be part of the **synchrotron radiation** produced. The name comes from particle accelerators called synchrotrons, in which electrons are confined and accelerated in circular paths by strong magnetic fields, and radiate quanta as a result of the acceleration.

Figure 6.14 shows a sketch of the two-jet process, and Figure 6.15 a picture of a typical radio galaxy, Cygnus A.

6.8   Point sources of γ-rays; γ-ray bursts

The majority of cosmic γ-rays form a random background of secondary origin, coming, for example, from the decay of neutral pions produced when primary protons interact with interstellar matter. Nevertheless, using instruments such as EGRET on the GRO (Gamma Ray Observatory) satellite launched in 1991 (see Fig. 6.12), point sources have been detected in the range of quantum energies from 3 keV to 30 GeV, and using the ground-based air Cerenkov method, to energies of 10 TeV.

Several known pulsar sources (see Section 7.10) such as Crab, Geminga, and Vela have been detected in this way by different laboratories (see Fig. 6.13). The main mechanism producing the γ-rays is believed to be that of radiation by electrons in the intense magnetic fields of the pulsars, that is, in the form of synchrotron radiation as described above. The rate of emission is proportional to the square of the acceleration. We may note here that if, instead of being due to magnetic bending, the acceleration is due to nuclear Coulomb scattering in traversing a medium, it is called **bremsstrahlung**. In either case, the power radiated in photons varies as m^{-4}, where m is the particle mass. Hence, it is important for the lightest charged particles, that is, electrons.

The energy spectrum of the radiated photons is roughly of the form dE/E, that is peaked to low energies. However, the electron and photon intensities near such sources are so high that very energetic photons can be produced via the

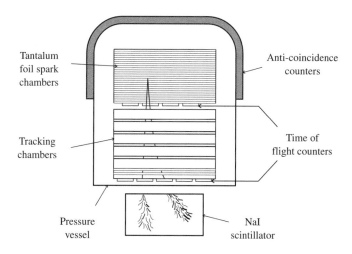

Fig. 6.12 Diagram of the EGRET (Energetic Gamma Ray Experimental Telescope) instrument on the GRO satellite. Gamma rays are detected when they materialize into e^+e^- pairs in the upper stack of tantalum sheets and spark chambers, the total energy of the ensuing electron–photon shower (Section 6.5.2) being measured from pulse heights in the lower array of NaI scintillators (after Ong 1998).

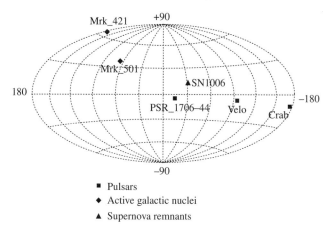

Fig. 6.13 Map of γ-ray point sources detected by the EGRET (Fig. 6.12) experiment, for gamma energies above 100 MeV. The coordinates are the longitude and latitude relative to the plane of our galaxy (from Ong 1998).

inverse Compton effect, that is, low energy photons being boosted by collisions with the very energetic electrons (which, in turn, are the products of shock acceleration).

The sources described above are (relatively) steady sources of γ-rays. The source in the Crab, for example, originated nearly one thousand years ago (the AD 1054 supernova), and produces γ-rays of energies up to several TeV. However, many of the γ-ray sources have been identified with the so-called **active galactic nuclei** (AGNs) with redshifts up to $z \sim 2.5$ (about 1% of all galaxies are 'active'). These sources do vary with time. They, just like the quasars—with which they may even be identified—are considered to be associated with very massive black holes located at galactic centres, because it seems that black holes are the only compact sources capable of generating such enormous energies and intensities of radiation, principally concentrated in 'jets' as described above. Presumably, like the radio emission, the γ-rays originate from these jets as electrons spiral in the magnetic field of the jet and emit synchrotron radiation (Fig. 6.14).

Indeed, it has become clear that the nature of the phenomena associated with massive galactic black holes depends to a large extent on the angle between

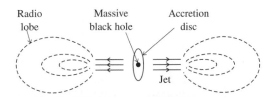

Fig. 6.14 Sketch of a possible two-jet mechanism involved in radio emission from a quasar.

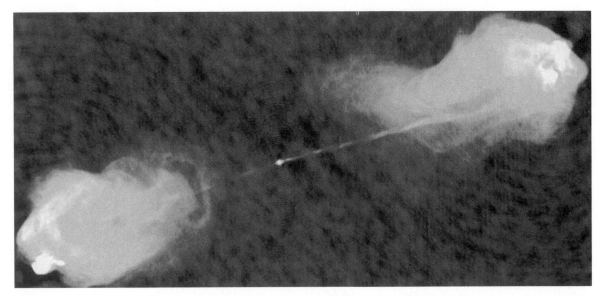

Fig. 6.15 Radio image of the radio galaxy Cygnus A. The galactic centre is the small dot midway between the massive radio lobes. (Courtesy Chris Carilli, NRAO.)

the jet axis and the line of sight to the Earth. If it is large, one obtains two jets of comparable size as in Fig. 6.15 in the case of Cygnus A. However, if the jet velocities are extreme relativistic and the angle happens to be small, so that one jet is approaching and the other is receding, the Doppler shift of the frequency can mean that the approaching jet is very bright while the receding one is below the threshold for detection. There are many interesting effects associated with the jets. For example, the observed transverse velocity may apparently exceed the velocity of light (see Problem 6.16). As another example, if the photons observed are of high energy, they would have been radiated by electrons of very large Lorentz factor γ, and hence are confined to a narrow angular spread of order $1/\gamma$ (see Problem 6.17). Thus, a small fluctuation in the jet angle could deflect the beam away from the observer, and the observed intensity could vary on short timescales. This may be a possible cause of the extreme variability of some of the γ-ray sources.

Even more intriguing than the relatively steady γ-ray sources are the **γ-ray bursts**, which last typically from 10 ms to 10 s. The quantum energy of this radiation is in the region of 0.1–100 MeV, and the sources seem to be distributed more or less isotropically over the sky, suggesting an extra-galactic origin. When finally an X-ray burst was pinpointed with an optical signal, the redshift of $z \sim 0.8$ measured from the optical spectrum confirmed this. The amount of energy in these bursts is comparable with or even larger than that emitted in neutrinos in the course of a Type II supernova explosion (10^{46} J or about one tenth of the mass energy of the Sun)—see Section 7.8.

AGN sources from which the γ-ray bursts are in the TeV energy region, are referred to as **blazars**, the most famous example being Markarian 501 (see Fig. 6.13). From this source the γ-ray flux can vary by an order of magnitude from night to night, and during the year 1997 it increased by a factor of 50.

γ-ray bursts are among the most intense and oldest sources of radiation known, with the redshifts indicating in some cases an origin when the universe was only one or two billion years old. How they originate is presently unclear, however the short burst lengths indicate an origin from very compact objects. They might arise, for example, from orbiting neutron stars that are components of a binary system, as they collapse under energy loss due to gravitational radiation to form black holes. Or they could be due to exceptionally massive stars with very short lives, which finally explode as supernovae with extreme violence. The location of these bursts seems to be widely distributed in both space and time throughout the cosmos. The rate is about ten thousand times less than that of supernova explosions resulting in neutron stars, which is of the order of one per galaxy per century. However, since the sources are located at such vast distances, which encompass millions of galaxies, the overall observed rate of about 1 burst per day is compatible with the estimated rate of neutron star mergers.

As stated above, most of the high redshift quasars we observe today, actually occurred at very early times in the universe, and represent long-past phenomena. It is believed that, at an early stage in their evolution, galaxies were dominated by extremely massive stars. These would have evolved very rapidly, since as shown in Chapter 7, the lifetime of a star of mass M on the main sequence varies as $M^{-2.5}$. Thus a star of $100 M_\odot$ would have a lifetime some 10^{-5} of that of the Sun, or less than a hundred thousand years, passing through the supernova stage and thereafter forming a black hole, which could grow rapidly in mass by absorbing gas, dust, and other nearby stars, and giving rise, when the mass became large enough, to the quasar phenomena described above.

Two-jet phenomena, on a much smaller scale, and termed **microquasars** are also observed in local galaxies. These are assumed to be due to black holes of the order of a few solar masses only, with accretion from a nearby companion star. The distance scale of the radio lobes is of the order of parsecs in this case, rather than the megaparsec scale of quasars.

Example 6.4 *High energy γ-rays from very distant sources may encounter a cut-off in energy due to collisions with photons of the cosmic microwave background, or of starlight, through formation of electron–positron pairs. Estimate the threshold energies involved, and the relevant absorption lengths.*

If E_{th} denotes the threshold photon energy, E_0 the quantum energy of the target photon, the condition for pair-production, $\gamma\gamma \rightarrow e^+e^-$ is $s = (E_{th} + E_0)^2 - (\mathbf{p_{th}} + \mathbf{p_o})^2 > 4m^2$, where m is the electron mass. For a head-on collision, $E_{th} = m^2/E_0$. For microwave photons $T = 2.73$ K, $kT = 2.35 \times 10^{-4}$ eV, and for one of energy $E_0 = y(kT)$, the value of $E_{th} \sim 10^3/y$ TeV. Most of the collisions are not head-on, but we can take this as a typical threshold energy.

The absorption length for the γ-rays is $\lambda = 1/(\rho\sigma)$, where ρ is the density of microwave target photons and σ is the interaction cross section. At the threshold $s_{th} = 4m^2$, the cross section is zero, and it rises with energy to a maximum of $\sim 0.25\sigma_{Thomson}$ at $s \sim 8m^2$ (see equations (1.26)), before falling off at higher energies. We take $\sigma \sim \pi\alpha^2/m^2 = 2.5 \times 10^{-25}$ cm^2 as indicative of an upper limit. Inserting $\rho \sim 400$ cm^{-3} for the total microwave photon number density in (2.41) results in $\lambda \sim 10^{22}$ cm ~ 4 kpc. Although this is only a rough lower limit, it shows that for photons of energies above 10^3 TeV (10^{15} eV), the universe on the scale of megaparsecs will be quite opaque. On the other hand, less than 10^{-6} of the microwave photons have $y > 20$ in the high energy tail of the Planck spectrum, and for these the threshold energy is only 50 TeV while the mean free path is over 100 Mpc, and such photons could be received from point sources spread over a large fraction of the universe (one must remember here, that at high redshifts, the microwave photon energies, and hence threshold energies, are increased by a $(1 + z)$ factor)).

For starlight photons, we can take the solar photosphere temperature of 6000 K, that is, a quantum energy of 6000/2.73 times that of the microwave photons, with a correspondingly reduced threshold energy of typically 1 TeV. The local starlight energy density is of the order of 1 eV cm^{-3} or $\rho \sim 1$ photons cm^{-3}, hence $\lambda > 1$ Mpc, which is large compared with the galactic radius. So there would be no problem in detecting γ-rays of any energy from sources in the local galaxy.

6.9   Atmospheric neutrinos: neutrino oscillations

Neutrinos and antineutrinos are constituents of the secondary cosmic rays generated in the Earth's atmosphere by interactions of the primary particles, as discussed in Section 6.4. The primaries generate mesons (pions and kaons) in collisions with air nuclei. These in turn decay in flight into neutrinos and muons: for example, $\pi^+ \rightarrow \mu^+ + \nu_\mu$. The muons in turn decay to muon- and electron-neutrinos: for example, $\mu^+ \rightarrow e^+ + \nu_e + \bar{\nu}_\mu$. The neutrino energy spectrum peaks at around 0.25 GeV, and falls off as $E^{-2.7}$ at high energy. As the above decay modes suggest, the expected ratio of fluxes $\phi(\nu_\mu)/\phi(\nu_e) \approx 2$ in the GeV energy region, and this ratio should be reflected in the relative numbers of interactions with secondary muons or electrons. Several underground experiments in the 1990s discovered on the contrary that the observed ratio of numbers of interactions containing muons as compared with electrons was nearer to 1 than to 2, signalling the possibility of a new phenomenon such as **neutrino flavour oscillations**.

Although the absolute flux of atmospheric neutrinos is low (about 1 cm^{-2} s^{-1} at sea-level) and the interaction cross sections are feeble (of the order of 10^{-38} cm^2 per nucleon at a typical energy of 1 GeV), their interactions have been recorded in substantial numbers using large (multikiloton) underground detectors originally intended to search for proton decay. The reaction rate is of the order of one per kiloton per day, which is about the same as the rate predicted for proton decay in the early (and incorrect) models of grand unification.

In the Standard Model, neutrinos are massless and exist in only one (left-handed) helicity state. However, the assumed masslessness of neutrinos was questioned many years ago, in connection with the possibility of flavour oscillations (the first proposal, by Pontecorvo, related to neutrino–antineutrino oscillations, for which there is no evidence and which is not considered further). It was proposed that, while neutrinos are created or destroyed as **flavour eigenstates**, they propagate through space as **mass eigenstates**. A particular flavour eigenstate denoted by the amplitude v_e, v_μ, or v_τ is therefore expressed, as regards its time evolution, as a linear combination v_1, v_2, and v_3 of mass eigenstates, which propagate through space with slightly different frequencies due to their different masses, and between which different phases develop with distance traversed, corresponding to a change or oscillation in the neutrino flavour.

The 3×3 matrix connecting neutrino flavour and mass eigenstates is analogous to the CKM matrix (3.38) connecting quark flavour (strong interaction) eigenstates with the weak decay eigenstates. However, in order to simplify the treatment we shall consider the case of just two flavours. In fact, it turns out that the actual effects observed to date can all be accounted for in terms of twofold mixing only. Using neutrino symbols to denote the wave amplitudes of the particles involved, let us for example consider the mixing of v_e and v_μ in terms of v_1 and v_2:

$$\begin{pmatrix} v_e \\ v_\mu \end{pmatrix} = \begin{pmatrix} \cos\theta & \sin\theta \\ -\sin\theta & \cos\theta \end{pmatrix} \begin{pmatrix} v_1 \\ v_2 \end{pmatrix} \tag{6.19}$$

where θ denotes some arbitrary mixing angle. The wave amplitudes

$$v_e = v_1 \cos\theta + v_2 \sin\theta$$
$$v_\mu = -v_1 \sin\theta + v_2 \cos\theta \tag{6.20}$$

are orthonormal states. If E denotes the neutrino energy, the amplitudes of the mass eigenstates as a function of time will be

$$v_1(t) = v_1(0) \exp(-iE_1 t)$$
$$v_2(t) = v_2(0) \exp(-iE_2 t) \tag{6.21}$$

where we have used units $\hbar = c = 1$, so the angular frequency $\omega = E$. The mass eigenstates will have a fixed momentum p, so that if the masses are $m_i \ll E_i$ (where $i = 1, 2$)

$$E_i = p + \frac{m_i^2}{2p} \tag{6.22}$$

Suppose that we start off at $t = 0$ with electron-type neutrinos, that is, $v_\mu(0) = 0$ and $v_e(0) = 1$. Inverting (6.19) we have

$$v_1(0) = v_e(0) \cos\theta$$
$$v_2(0) = v_e(0) \sin\theta \tag{6.23}$$

and

$$v_e(t) = v_1(t) \cos\theta + v_2(t) \sin\theta$$

From (6.21) and (6.23) the amplitude of the electron neutrinos becomes

$$A_e(t) = \frac{v_e(t)}{v_e(0)} = \cos^2\theta \exp(-iE_1 t) + \sin^2\theta \exp(-iE_2 t)$$

so that the intensity is

$$\frac{I_e(t)}{I_e(0)} = A_e A_e^* = 1 - \sin^2 2\theta \sin^2\left[(E_2 - E_1)\frac{t}{2}\right]$$

We use (6.22) and write the difference of the squares of the masses as $\Delta m^2 = m_2^2 - m_1^2$, where here and in what follows we assume $m_2 > m_1$. The probability of finding one or the other flavour after a time $t = L/c$, where L is the distance travelled, is

$$P(\nu_e \to \nu_e) = 1 - \sin^2 2\theta \sin^2\left(1.27\Delta m^2 \frac{L}{E}\right)$$

$$P(\nu_e \to \nu_\mu) = 1 - P(\nu_e \to \nu_e)$$

$$(6.24)$$

Here, the numerical coefficient is just $1/(4\hbar c)$ if we retain all the factors of $\hbar$ and c, and equals 1.27 if L is expressed in km, Δm^2 in $(eV/c^2)^2$ and E in GeV. Figure 6.16 shows how the flavour amplitudes oscillate for the case $\theta = 45°$. The oscillation wavelength is $\lambda = 4\pi E/\Delta m^2$. For example, for a value of $\Delta m^2 = 3 \times 10^{-3}$ eV^2 found from the atmospheric data as discussed below, $\lambda = 2400$ km for $E = 2$ GeV.

As stated previously, experimental studies of atmospheric neutrinos started in earnest in the late 1980s in massive (kiloton) detectors originally intended to search for proton decay, in which the neutrino interactions were at first considered to be an annoying background incapable of eradication. In fact, a handful of atmospheric neutrino interactions had first been observed in small detectors placed deep underground in the early 1960s, but at that time they were considered to be of little interest. So the discovery of neutrino oscillations with atmospheric neutrinos, like many other discoveries in science, has been largely accidental, as a by-product of an investigation that failed in its original aim. By good fortune it turns out that the typical energies of atmospheric neutrinos, of the order of 1 GeV and determined by the effect of the Earth's magnetic field on the primary cosmic-ray nuclei, combined with the accessible neutrino path-lengths determined by the Earth radius, are exactly matched to the relevant scale of neutrino mass differences. (Since there are three mass eigenstates, there will be two independent mass differences. The larger one is associated with atmospheric neutrinos, and the smaller one with solar neutrinos, as described below.)

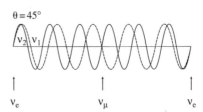

Fig. 6.16 Two neutrino ($\nu_e \to \nu_\mu$) oscillations, showing amplitudes of the mass eigenstates for the case $\theta = 45°$. They are in phase at the beginning and end of the plot, separated by one oscillatory wavelength, and thus from (6.19) corresponding at these points to pure electron-neutrino flavour eigenstates. In the centre of the plot the two amplitudes are 180° out of phase, corresponding to the muon-neutrino flavour eigenstate.

It is also relevant to remark here that searches for neutrino flavour oscillations had been carried out at both reactors and accelerators for the last thirty years, without success, presumably because the baselines employed were too short in comparison with the oscillation wavelength to achieve a measurable effect (see Fig. 6.21).

Figure 6.17 shows the zenith angle distribution of events attributed to muon- and electron-neutrinos in the Superkamiokande detector, containing 50 000 tons of water viewed by 11 000 photomultipliers (see also Fig. 3.13). Charged current reactions of the electron- and muon-type neutrinos result in production of charged electrons or muons. These emit Cerenkov radiation as they traverse the water (see Fig. 6.10), and this appears as a ring of light at the water surface, which is detected by the photomultiplier array. Muons give clean Cerenkov rings, while those for electrons are more diffuse, due to bremsstrahlung and multiple scattering as the electron traverses the water. The direction of the charged lepton is found from the timing of the pulses, and at the energies involved, this gives a fair indication of the zenith angle of the incident neutrino.

As shown on the x-axis in Fig. 6.17, the typical path length of the neutrinos through the atmosphere and the Earth is a strong function of the zenith angle θ, ranging from about 20 km for neutrinos from directly overhead, to 200 km for those coming in horizontally, to 12 000 km for those coming vertically upwards from the atmosphere on the far side of the Earth. The points in this graph show the ratio of the observed event rate to that expected for the case of no oscil- lations. The ratio for electrons, and hence electron-neutrinos, shows no zenith angle dependence, while the upward-travelling muons from muon-neutrinos are strongly suppressed relative to the downward ones, the factor being 0.5 for those moving vertically upwards. In comparing these results to the expecta- tions from (6.24), it is clear that because the events are integrated over a very broad energy spectrum as well as a range of path-lengths, no actual oscilla- tions will be observed and the mean value of the factor $\sin^2(1.27 \; \Delta m^2 L/E)$ for large L will be just $\frac{1}{2}$. Thus, the fact that the observed suppression is $\frac{1}{2}$ implies that $\sin^2(2\theta) \approx 1$, that is, the mixing is maximal as in Fig. 6.16 (see also Fig. 6.21). For the muon events, the curve shows the distribution expected for maximal mixing and for the best-fit value of the squared mass difference $\Delta m^2 \approx 3 \times 10^{-3} \; (eV/c^2)^2$. Since the electron events show no zenith angle effect, the results are ascribed to $\nu_\mu \rightarrow \nu_\tau$ oscillations.

Using a man-made muon-neutrino beam of a few hundred MeV energy gen- erated at a proton accelerator at KEK near Tokyo, oscillatory effects have also been observed over a long baseline of 250 km between the KEK laboratory and the Superkamiokande detector, by comparing the muon rate in this distant detector with that in a detector near the accelerator.

6.10   Solar neutrinos

Anomalously low rates, which have also been interpreted in terms of neut- rino oscillations, were first observed for solar neutrinos some thirty years ago. Section 7.6 describes the reactions involved in the so-called pp cycle of thermo- nuclear fusion of hydrogen to helium in the solar core. Neutrinos are produced from a number of reactions (see equations (7.5)–(7.11)).

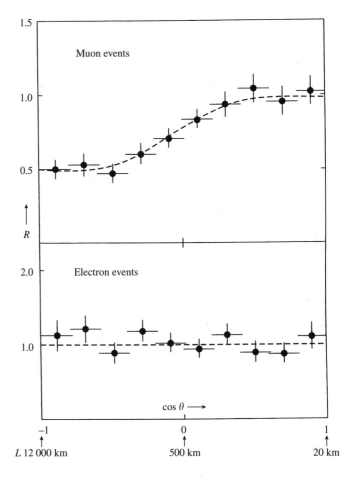

Fig. 6.17 Observed zenith angle distribution of electron and muon events with lepton momenta above 1.3 GeV/c in the Super-kamiokande detector, plotted as a ratio to the rate expected for no oscillations (Fukuda *et al* (1998)). The dashed curve in the upper plot shows the predicted variation for maximum $\nu_\mu \rightarrow \nu_\tau$ mixing with $\Delta m^2 = 3 \times 10^{-3}$ eV^2. The lower plot indicates no oscillations of electron-neutrinos. The horizontal scale indicates the neutrino path lengths through the Earth and the atmosphere, at different zenith angles.

Figure 6.18 shows the calculated fluxes of neutrinos at the Earth as a function of energy. Although the fluxes at the higher energies, notably from ^{8}B decay, are very small compared with those from the pp reaction, they make substantial contributions to the total event rate since the cross sections for the detectors employed vary approximately as E_ν^3. Table 6.1 shows the results from several experiments, giving the ratio of the observed event rate to that calculated by Bahcall *et al.* (2001) in the absence of oscillations. The first two entries are for radiochemical experiments, detecting the accumulated activity of the product nuclei after fixed time periods. They offered formidable experimental challenges, requiring the detection of less than one atom of the product element per day, in a mass of 50 tons (in the case of gallium) or 600 tons (in the case of chlorine). The gallium experiments SAGE and GNO have a threshold energy of 0.2 MeV and are therefore sensitive to pp neutrinos, which from Fig. 6.18 extend up to an energy of 0.4 MeV.

The remaining experiments have higher thresholds and are not sensitive to the pp neutrinos. The SNO and SUPER-K experiments employ 1 kiloton of heavy water and 30 kilotons of light water, respectively, and both measure events in real time, detecting the Cerenkov light emitted by the relativistic electrons traversing the water using large photomultiplier arrays (see Fig. 3.13). The electrons originate from elastic scattering reactions (rows 4 and 5 Table 6.1)

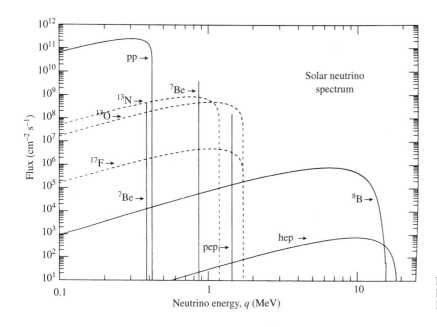

Fig. 6.18 Fluxes of solar neutrinos at the Earth from various reactions in the Sun (from Bahcall 1989).

Table 6.1 Solar neutrino experiments

Experiment		Reaction	Threshold (MeV)	Observed/expected rate
SAGE + GNO	CC	$^{71}Ga(\nu_e, e)^{71}Ge$	0.2	0.58 ± 0.04
HOMESTAKE	CC	$^{37}Cl(\nu_e, e)^{37}Ar$	0.8	0.34 ± 0.03
SNO	CC	$\nu_e + {}^2H \rightarrow p + p + e$	~5	0.35 ± 0.03
SUPER-K	ES	$\nu + e \rightarrow \nu + e$	~5	0.46 ± 0.01
SNO	ES	$\nu + e \rightarrow \nu + e$	~5	0.47 ± 0.05
SNO	NC	$\nu + {}^2H \rightarrow p + n + \nu$	~5	1.01 ± 0.12

CC = charged current (W exchange); NC = neutral current (Z exchange); ES = electron scattering (via NC for ν_μ, ν_τ, and via NC and CC for ν_e).

or from charged current reactions (row 3). The SNO experiment also detects neutrons from disintegration of deuterium in a neutral current reaction (row 6 of Table 6.1). The higher threshold of 5 MeV for the SNO and SUPER-K experiments is determined by the radioactive background levels in the water, photomultipliers, etc. The Homestake, SUPER-K, and SNO experiments are largely sensitive to the boron-8 neutrinos.

The SUPER-K experiment measures neutrino–electron elastic scattering, from the magnitude of the forward peak in the angular distribution of the scattered electrons relative to the Sun's direction (see Fig. 6.19). This can proceed through either charged current (CC) scattering (for ν_e only) **or** through neutral current (NC) scattering (via Z^0 exchange) which can apply to all flavours of neutrino, that is, to ν_e, ν_μ, or ν_τ.

The significant facts about Table 6.1 are, first, that the CC experiments measure a rate well below that expected, while the SNO neutral current reaction is consistent with expectations. Since the NC cross section is the same for all neutrino flavours, the rate is independent of any oscillations the electron-neutrinos may undergo, and bears out the correctness of the solar model.

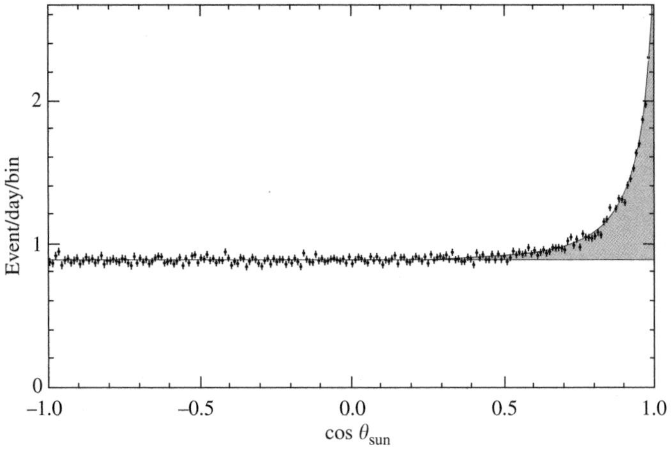

Fig. 6.19 Angular distribution of electrons relative to the Sun in the SUPER-K experiment. The flat distribution in $\cos\theta$ is due to the background of secondary cosmic rays, while the forward peak is due to scattering by solar neutrinos, $\nu + e \rightarrow \nu + e$ (courtesy M. Smy, for Superkamiokande collaboration 2002).

Second, the ratio of observed to expected CC rates is less for the experiments insensitive to the low energy (pp) neutrinos, implying an energy-dependent suppression factor. Finally, the SUPER-K and SNO results on electron scattering (ES) are in excellent agreement with the Homestake and SNO results on the CC reactions, if one accepts that these latter results indicate that only about 35% of the total neutrino flux arriving at the Earth is in the form of electron-neutrinos and that the remaining 65% has been transformed from ν_e to ν_μ and/or ν_τ, which then scatter from electrons through Z^0 exchange. The rate for these neutral current events calculated from the known value of the Weinberg angle ($\sin^2\theta_w = 0.23$) should then be about one-third of that for the ν_e, bringing the expected total ES rate into excellent agreement with the measurement.

The fact that the suppression factor for the CC reactions depends on the neutrino energy range, as evidenced by the first two entries in the table, led to the possibility that mechanisms other than vacuum oscillations were involved. First Wolfenstein (1978) and later Mikhaev and Smirnov (1986) pointed out that the oscillations could be considerably modified by matter effects, namely by what is called the **MSW mechanism**, after the initials of these three physicists.

6.11   Neutrino oscillations in matter

We start the discussion of the MSW mechanism by writing out the time evolution of the mass eigenstates (6.21), given by the Schroedinger equation $i\,d\psi/dt = E\psi$ for the time dependence of the wavefunction in terms of the energy eigenvalue E. In matrix form, using (6.22) this appears as

$$i\frac{d}{dt}\begin{pmatrix} \nu_1 \\ \nu_2 \end{pmatrix} = \begin{pmatrix} E_1 & 0 \\ 0 & E_2 \end{pmatrix}\begin{pmatrix} \nu_1 \\ \nu_2 \end{pmatrix} = \begin{pmatrix} m_1^2/2p & 0 \\ 0 & m_2^2/2p \end{pmatrix}\begin{pmatrix} \nu_1 \\ \nu_2 \end{pmatrix} + \begin{pmatrix} p & 0 \\ 0 & p \end{pmatrix}\begin{pmatrix} \nu_1 \\ \nu_2 \end{pmatrix}$$

$$(6.25)$$

The term on the extreme right is a constant phase factor which affects ν_1 and ν_2 equally and can, therefore, be omitted. If we substitute the expression for ν_e and ν_μ in terms of ν_1 and ν_2 in (6.20) we find after a little straightforward algebra that, for vacuum oscillations,

$$i\frac{d}{dt}\begin{pmatrix} \nu_e \\ \nu_\mu \end{pmatrix} = M_V \begin{pmatrix} \nu_e \\ \nu_\mu \end{pmatrix} \tag{6.26}$$

where

$$M_V = \left[\frac{(m_1^2 + m_2^2)}{4p}\right]\begin{pmatrix} 1 & 0 \\ 0 & 1 \end{pmatrix} + \left(\frac{\Delta m^2}{4p}\right)\begin{pmatrix} -\cos 2\theta & \sin 2\theta \\ \sin 2\theta & \cos 2\theta \end{pmatrix}$$

In interactions with matter, electron-neutrinos in the MeV energy range can undergo both charged ($W^\pm$ exchange) and neutral current (Z^0 exchange) interactions, while muon- or tau-neutrinos have only the neutral current option, as their energies are too low to generate the charged lepton. Hence, electron-neutrinos suffer an extra potential V_e affecting the forward scattering amplitude, which leads to a change in the effective mass:

$$V_e = G_F\sqrt{2}\,N_e$$
$$m^2 = E^2 - p^2 \rightarrow (E + V_e)^2 - p^2 \approx m^2 + 2EV_e \tag{6.27}$$
$$\Delta m_m^2 = 2\sqrt{2}\,G_F N_e E$$

where N_e is the electron density, $E = pc$ is the neutrino energy, G_F is the Fermi constant, and Δm_m^2 is the shift in mass squared. (For antineutrinos, which are the *CP* transforms of neutrinos, the sign of the potential V_e is reversed.) So in the case of electron-neutrinos traversing matter, one should substitute in the vacuum expression for the average mass squared in (6.26)

$$\frac{1}{2}(m_1^2 + m_2^2)\begin{pmatrix} 1 & 0 \\ 0 & 1 \end{pmatrix} \rightarrow \frac{1}{2}(m_1^2 + m_2^2)\begin{pmatrix} 1 & 0 \\ 0 & 1 \end{pmatrix} + 2\sqrt{2}\,G_F N_e p\begin{pmatrix} 1 & 0 \\ 0 & 0 \end{pmatrix}$$
$$= \left[\frac{1}{2}(m_1^2 + m_2^2) + \sqrt{2}G_F N_e p\right]\begin{pmatrix} 1 & 0 \\ 0 & 1 \end{pmatrix}$$
$$+ \sqrt{2}\,G_F N_e p\begin{pmatrix} 1 & 0 \\ 0 & -1 \end{pmatrix}$$

Then, the matrix M_M appropriate to matter traversal is modified from the vacuum matrix M_V in (6.26) as follows:

$$M_M = \left[\frac{m_1^2 + m_2^2}{4p} + \frac{\sqrt{2}G_F N_e}{2}\right]\begin{pmatrix} 1 & 0 \\ 0 & 1 \end{pmatrix}$$
$$+ \left[\frac{\Delta m^2}{4p}\right]\begin{pmatrix} -\cos 2\theta + A & \sin 2\theta \\ \sin 2\theta & \cos 2\theta - A \end{pmatrix} \tag{6.28}$$

where $A = 2\sqrt{2}G_F N_e p/\Delta m^2$. The first term again gives the same phase factor for ν_e and ν_μ and can therefore be omitted. If we denote the mixing angle in the presence of matter as θ_m and the mass difference squared in matter as Δm_m^2, the second term in (6.28) can also be written as in (6.26):

$$\left[\frac{\Delta m_m^2}{4p}\right]\begin{pmatrix} -\cos 2\theta_m & \sin 2\theta_m \\ \sin 2\theta_m & \cos 2\theta_m \end{pmatrix} \tag{6.29}$$

and equating this with the second term in (6.28) gives

$$\tan 2\theta_m = \frac{\sin 2\theta}{[\cos 2\theta - A]} = \frac{\tan 2\theta}{[1 - (L_v/L_e)\sec 2\theta]} \tag{6.30}$$

where the vacuum oscillation length is $L_v = 4\pi p/\Delta m^2$ and the electron interaction length is defined as $L_e = 4\pi/(2\sqrt{2}G_F N_e)$, so that $A = L_v/L_e$. We note that, irrespective of the value of θ, it is possible for the matter mixing angle to go through a 'resonance' with $\theta_m = \pi/4$, provided that L_v is positive and therefore $\Delta m^2 > 0$, that is, $m_2 > m_1$. The resonance condition is clearly

$$L_v = L_e \cos 2\theta \quad \text{or} \quad N_e(\text{res}) = \Delta m^2 \frac{\cos 2\theta}{(2\sqrt{2}G_F p)} \tag{6.31}$$

where $N_e(\text{res})$ denotes the corresponding 'resonant' electron density. For example, the density in the core of the Sun is ρ (core) $\sim$ 100 gm cm^{-3}, or $N_e(\text{core}) \sim 3 \times 10^{31}$ m^{-3}, giving $L_e \sim 3 \times 10^5$ m (compared with a solar radius of 7×10^8 m). The solar density falls off roughly exponentially with radius outside the core. If at some radius the 'resonance' condition is fulfilled, electron-neutrinos could be transformed partly or entirely to muon- or tauon-neutrinos, even if the vacuum mixing angle is very small. Neutrinos will **always** pass through the resonance region if the critical electron density is less than that in the solar core, that is, if the energy exceeds a minimum value

$$E_{\min} = \Delta m^2 \frac{\cos 2\theta}{[2\sqrt{2}G_F N_e(\text{core})]} \tag{6.32}$$

This could explain the larger suppression observed in Table 6.1 for the higher energy boron-8 neutrinos.

For an intuitive picture of the MSW effect (Fig. 6.20), let us for simplicity assume that the vacuum mixing angle is small. Then, an electron-neutrino starts out in the solar core, predominantly in what in a vacuum would be called the ν_1 eigenstate of mass m_1, but the extra weak potential increases the effective mass in the appropriate region of electron density to the value m_2 (refer to equation (6.19) for $\theta_m = 45°$), which in a vacuum would be identified as consisting principally of the ν_μ flavour eigenstate. If the solar density changes fairly slowly with radius, this mass eigenstate will pass out of the Sun without further changes, and in vacuum it will be identified with the muon-neutrino eigenstate. In general, however, there will only be partial flavour conversion.

One of the most crucial aspects of the data is the observed energy spectrum of electrons from the boron-8 neutrinos in the SUPER-K and SNO experiments, which indicates an almost constant suppression factor over the range 6–14 MeV. The fits to the data indicate a fairly large vacuum mixing angle.

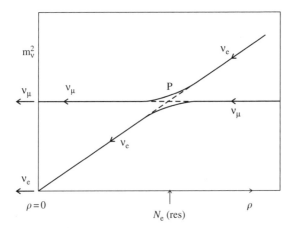

Fig. 6.20 The MSW effect. The neutrino mass squared is plotted against solar density. For muon-neutrinos, the mass is independent of density and is represented by the horizontal line. For electron-neutrinos, the mass squared is proportional to the density as in (6.27) and, if there is no flavour mixing ($\theta = 0$) is represented by the diagonal line. The two levels cross at the point P, where $\Delta m^2 \cos 2\theta = 2\sqrt{2} G_F N_e E$ (see equation (6.31)). If the electron density in the solar core is greater than the 'resonance density' at P, the electron-neutrino will be located beyond P in the upper part of the diagram. As the neutrino moves outwards into regions of lower density, it eventually reaches the resonance density, and provided the solar density varies slowly with radius, it will move along the continuous curve and emerge from the sun as a muon-neutrino.

The present atmospheric and solar neutrino results are shown in Fig. 6.21, where the fitted values of $\tan^2 \theta$ are plotted against Δm^2, and the shaded areas represent the allowed regions for the parameters. From the atmospheric results, the most massive of the three neutrino mass eigenstates will have a mass $m_3 > (\Delta m^2_{\text{atm}})^{1/2} \approx 0.05 \,\text{eV}/c^2$.

6.12 Point neutrino sources

The existence of point sources of TeV γ-rays described in Section 6.7 is generally associated with electromagnetic process (e.g. synchrotron radiation), but may also imply that of high energy neutrino point sources, since both could arise from pion decay, that is, from $\pi^0 \to 2\gamma$ and $\pi^\pm \to \mu^\pm + \nu_\mu$, respectively. Such neutrinos could be detected in underground experiments via the secondary charged muons they produce. At TeV energies, the muon range in rock would be measured in kilometres, and the object is to detect muons travelling **upwards** (since any downward flux would be completely swamped by atmospheric muons from pion decay in the atmosphere overhead). Here, the very weakness of their interactions is being exploited to detect neutrinos which have come up through the Earth and produced a signal free of background, except for the ubiquitous atmospheric neutrinos, which are however broadly distributed in angle.

The rate of such neutrino events (compared with that of γ-rays) is expected to be low on account of the weak cross section. However, this is to some extent

Fig. 6.21 Plot of $\tan^2 \theta$, where θ is the vacuum mixing angle, against neutrino mass squared difference Δm^2. For the solar and atmospheric neutrino data, the shaded areas represent the 90% confidence level limits inside which the oscillation parameters lie (i.e. the two independent mass squared differences and two mixing angles). The other curves are 'exclusion plots' from former short baseline accelerator and reactor experiments which have not observed oscillations, from which the regions to the right of and above the curves are excluded. (One laboratory short baseline experiment has reported a positive effect, but this has not been confirmed by independent experiments, and we do not include it).

First results from very long ($\sim$200 km) baseline experiments with man-made beams—the KAMLAND experiment in the Kamioka mine detecting interactions of $\bar{\nu}_e$ beams from distant reactors, and the K2K experiment in the Superkamiokande detector recording interactions of ν_μ beams from the KEK accelerator—are in excellent agreement with the results from solar and atmospheric neutrinos.

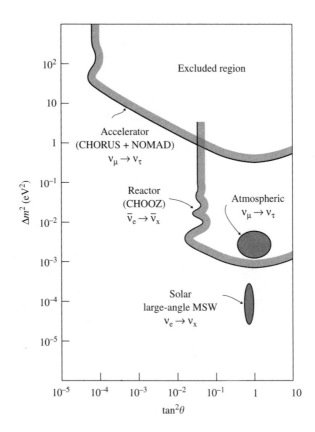

compensated, first by the fact that the cross section for the interaction of a neutrino with a quark in the target nucleus rises as the square of the CMS energy, or linearly with the laboratory neutrino energy (see equation (1.27)), and second that, as in Fig. 3.3(a), the neutrino–quark collision will have $J = 0$, so that the secondary muon will be isotropic in the CMS and thus have a uniform laboratory energy distribution extending from 0 to E_ν. Hence, the muon range in rock, increasing as the muon energy, also gives a factor proportional to neutrino energy. So although the neutrino flux falls off rapidly with energy, it is multiplied by a factor of E_ν^2 to get the event rate. This argument holds up to TeV energies, but beyond that, the cross sections flatten off because of the W propagator in (1.9), and the muon range is no longer proportional to energy because of radiation losses analogous to those of electrons discussed in Section 6.5, but coming in at a much higher energy (by a factor of order $(m_\mu / m_e)^2$).

The detection of rare high energy cosmic neutrino interactions via the secondary upward-travelling muons has to be carried out on a large scale, and uses great depths of sea water or of Antarctic ice, and again relies on the detection of the Cerenkov light radiated by the muon. Figure 6.22 and 6.23 show one such experiment, in which strings of photomultipliers collect the Cerenkov signals over a volume with dimensions of the order of several hundred metres. These arrays have successfully detected atmospheric neutrinos, but so far no events attributable to point neutrino sources have been observed.

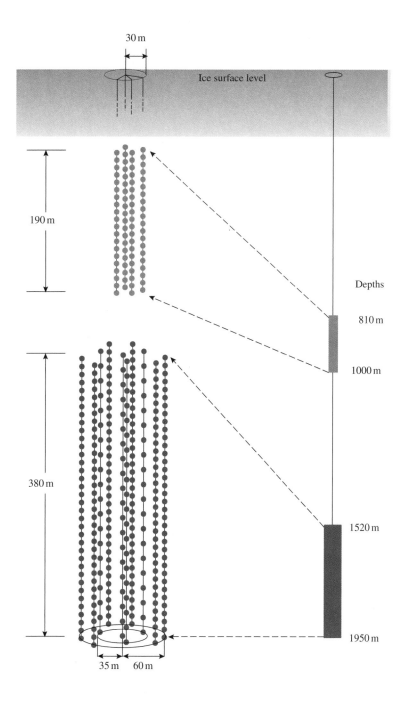

Fig. 6.22 Diagram of the strings of photomultipliers sunk in ice at the South Pole in the AMANDA experiment. They record Cerenkov light emitted by upward coming muons generated in neutrino interactions. Timing provides information on the zenith and azimuthal angles of the muon.

6.13   Gravitational radiation

Of all the radiations incident upon the Earth from the cosmos, the most elusive and most difficult to detect is surely gravitational radiation. The key equations describing gravitational radiation, its production and detection are found from the general theory of relativity, which is outside the scope of this text. Nevertheless, we can understand many of the features of gravity waves by using the analogy with electromagnetic radiation.

Fig. 6.23 Photograph of an AMANDA phototube module about to be lowered into the ice. A column of ice is melted and later freezes around the module and cables. (Reproduced by courtesy of R. Stokstad/NSF)

First of all, we remark that in scattering problems it is usual to expand the function describing a plane wave into a superposition of spherical waves with different values of the orbital angular momentum, l, with respect to the scattering centre. An oscillation can equally be represented in the form of a multipole expansion, corresponding to a superposition of oscillators with different l values. Thus dipole, quadrupole, sextupole, ... oscillator terms are associated with $l = 1, 2, 3, \ldots$ waves. Figure 6.24 depicts a simple electric dipole and two versions of an electric quadrupole. If ω represents the angular frequency of the oscillation, the power radiated is given by the formula

$$P(l) = 2cF(l)\left(\frac{\omega}{c}\right)^{2l+2}|Q_{lm}|^2 \tag{6.33}$$

where $F(l) = (l+1)/[l\{(2l+1)!!\}^2]$. Here, $n!! = 1 \cdot 3 \cdot 5 \cdot 7 \cdots\cdots$ and Q_{lm} is the lth moment of the distribution of electric charge density ρ, integrated over the volume and projected along the z-axis, along which m is the component of angular momentum:

$$Q_{lm} = \int r^l Y_{lm}^*(\theta, \phi)\rho \, dV \tag{6.34}$$

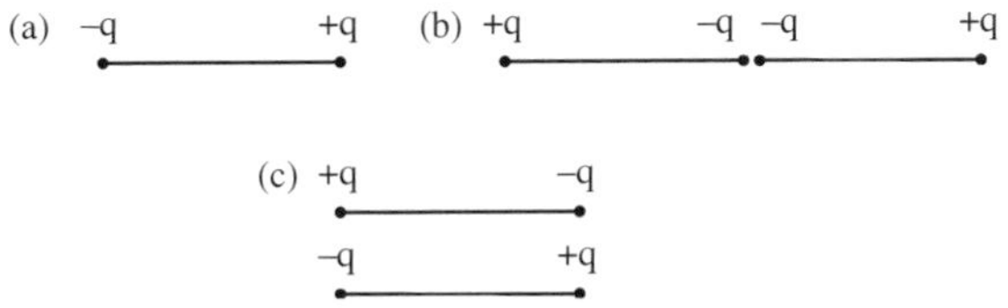

(a) $-q$ ——————— $+q$ (b) $+q$ ——————— $-q$ $-q$ ——————— $+q$

(c) $+q$ ——————— $-q$

$-q$ ——————— $+q$

Fig. 6.24 (a) An electric dipole; (b) and (c) two configurations of an electric quadrupole, for which the dipole moment is zero.

Here, $Y_{lm}(\theta, \phi)$ is a spherical harmonic. In order of magnitude, an electric dipole moment ($l = 1$) has $Q_1 \sim er$ where r is the dimension of the system and e denotes the charge. Thus, from (6.33), the power radiated is of the order of

$$P_{\text{dipole}} \sim \frac{\omega^4 e^2 r^2}{c^3} \tag{6.35}$$

The dependence on ω^4 arises because the power radiated is proportional to the square of the acceleration of the charges, varying as ω^2. The formula applies also to Rayleigh scattering of light by air molecules and by dust, and its dependence on $1/\lambda^4$ accounts for the blue of the sky and the red of the sunset. For an electric quadrupole moment ($l = 2$), $Q_2 \sim er^2$ and the power radiated is of order

$$P_{\text{quadrupole}} \sim \frac{\omega^6 e^2 r^4}{c^5} \tag{6.36}$$

The gravitational field is a tensor field (as compared with the vector, spin 1 photon field of electromagnetism), and gravitational interactions are mediated by gravitons of spin 2. As a result, the dipole emission of gravitational waves is impossible and the simplest radiator of gravitational waves is an oscillating mass quadrupole. The power emitted can be estimated by replacing e^2 in the expression e^2/r for the electric potential between charges by GM^2 in the expression GM^2/r for the gravitational potential between masses, so that

$$P_{\text{grav}} \sim \frac{\omega^6 GM^2 r^4}{c^5} \tag{6.37}$$

Here, M denotes a typical mass and r a typical dimension in the quadrupole system, and we have just quoted the order of magnitude result for orientation. It should be emphasized that a quantitative calculation using Newtonian mechanics and simply substituting from the electric quadrupole formula will in any case underestimate the power in gravitational radiation, as computed using the general theory of relativity, by a factor 4. (Recall that, in the calculation of the bending of light by a gravitational field in Chapter 4, the Newtonian approach underestimated the true deflection by a factor 2.) We quote here two examples of quantitative predictions. A rod of length L and mass M rotating about its mid-point with angular velocity ω emits gravitational radiation with the power

$$P = \left(\frac{2}{45}\right) \frac{\omega^6 GM^2 L^4}{c^5} \tag{6.38}$$

while a binary system consisting of two stars in a circular orbit of diameter D and angular frequency ω radiates power

$$P = \left(\frac{32}{5}\right) \frac{\omega^6 G \mu^2 D^4}{c^5} \tag{6.39}$$

where μ is the reduced mass, and for stars of equal mass M, $\mu = M/2$.

For an experiment on a laboratory scale, we could take $L \sim 1$ m, $M \sim 1$ kg and $\omega \sim 10^3$ s^{-1}, which from (6.37) gives $P \sim 10^{-36}$ W only. According to (6.39), even the entire Earth, in its orbit around the sun, radiates a mere 196 W. More likely objects for a measurable level of gravitational radiation may be binary stars in a late stage of evolution, just as they coalesce to black holes, when a considerable fraction of the rest energy of the stars should be emitted as gravitational radiation. Indeed, it is believed that gravitational energy loss is what leads to the collapse. Somewhat less dramatic is the radiation emitted during the formation of a neutron star, which is followed by the spectacular optical display of a Type II supernova (see Sections 7.8–7.10).

6.14   The binary pulsar

To date, the most convincing evidence for the existence of gravitational radiation, and through it, a quantitative test of general relativity, comes from the observations of the binary pulsar PSR 1913 + 16 by Hulse and Taylor in 1975. This binary consists of two neutron stars, of which one is a pulsar with a period of 0.059 s. Pulsars are neutron stars (see Section 7.10) which spin rapidly with frequencies of 10–100 s^{-1}, emitting radio waves in a beam which periodically sweeps past the observer, like a rotating beam from a lighthouse. The orbital period of this binary, $\tau = 7.8$ h, and other characteristics of the orbit were found from the Doppler shift of the pulsar signal as this neutron star revolves around its companion. The unique feature of this binary is that Hulse and Taylor observed a tiny but steady decrease in the orbital period:

$$\frac{d\tau}{dt} = -(2.423 \pm 0.006) \times 10^{-12} \tag{6.40}$$

Such a decrease is to be expected if the pair lose energy by emission of gravitational radiation. To estimate the magnitude of this effect, let us assume for simplicity that the two stars have equal masses M and execute a circular orbit of diameter D. With $\omega = 2\pi/\tau$ as the angular frequency of the orbital motion, the tangential velocity of each star is $v = \omega D/2$. The total energy of the system is the sum of kinetic and potential energies:

$$E(\text{tot}) = 2 \times \left(\frac{Mv^2}{2}\right) - \frac{GM^2}{D} \tag{6.41}$$

Balancing the gravitational and centrifugal forces on one of the masses gives

$$2\frac{Mv^2}{D} = \frac{GM^2}{D^2}$$

so that

$$E(\text{tot}) = -\frac{GM^2}{2D} \tag{6.42}$$

That is, the kinetic energy is just half the (absolute value of) the potential energy—an example of the **virial theorem**. Since from the above equations $E(\text{tot}) \propto 1/D$ and $\omega \propto 1/D^{3/2}$, the orbital period $\tau \propto E(\text{tot})^{-3/2}$. Hence,

$$\left(\frac{1}{\tau}\right) \frac{d\tau}{dt} = -\left(\frac{3}{2E}\right) \frac{dE}{dt} = -\left(\frac{3}{2}\right) P/E \qquad (6.43)$$

where P is the power radiated. The characteristics of the binary give $\omega = 2.2 \times 10^{-4}$, $M \sim 1.4 M_\odot$. Inserting these values to obtain D and $E(\text{tot})$, and using (6.39) to estimate the power radiated, one obtains $d\tau/dt \sim 10^{-13}$. A full calculation, taking into account the eccentricity of the binary orbit and the inequality of the masses yields the result

$$\frac{d\tau}{dt} = -2.40 \times 10^{-12} \qquad (6.44)$$

in astonishingly good agreement with the observed value (6.40). This result has given great confidence that the basic physics is well understood and that, despite enormous experimental difficulties, the detection of gravitational radiation in the laboratory is worth pursuing.

6.15 Detection of gravitational waves

When gravitational waves impinge on a detector, the difference in the acceleration from different parts of the wave can induce a deformation or strain, corresponding to an extension Δx in length x. The strain $h = \Delta x/x$ is given by

$$h^2 \sim \frac{GP}{(c^3 \omega^2 R^2)} \qquad (6.45)$$

where P is the power emitted by the source, R is its distance from the detector, and ω is the frequency of the radiation. Clearly, a detector with a quadrupole moment is necessary to excite a quadrupole amplitude. Inserting the value of P from (6.37), we find for the amplitude

$$h \sim \frac{GML^2 \omega^2}{(c^4 R)} \qquad (6.46)$$

where the product $ML^2 \omega^2$ is the second derivative of the quadrupole moment ML^2 of the source and is equal to the kinetic energy $E_{\text{kin}} \sim Mv^2$ associated with the source oscillations. In a violent event such as the collapse to a neutron star, gravitational energy released due to the infall is of the order of $0.1 M_\odot c^2$ (see Section 7.8). If we optimistically assume that 10% of this appears in the form of gravity waves then

$$h \sim \frac{GM_\odot}{(100 c^2 R)} \sim \frac{10^{-15}}{R} \qquad (6.47)$$

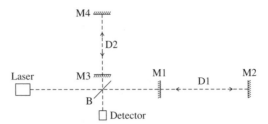

Fig. 6.25 Sketch of a Michelson interferometer layout for a gravitational wave detector.

where R is the distance of the source in parsecs. For the local galaxy, $R \sim 10$ kpc
and $h \sim 10^{-19}$, while for the Virgo cluster of galaxies $R \sim 10$ Mpc and
$h \sim 10^{-22}$. Note that, even for a bar 1 km long, $h = 10^{-19}$ corresponds to a
change in length of 10^{-16} m or one tenth of a nuclear radius! It is likely that,
by going further afield, the decrease in h with increasing R in (6.47) may be
compensated by the R^3 increase in the number of sources and the possibility
of much more violent events, such as collapse to massive black holes (AGNs),
with a gravitational wave energy far exceeding the solar mass energy.

Although the conceivable distortions to be measured by gravitational wave
detectors as a result of the most violent cosmic events will be at best of the order
of 10^{-20}, they are not considered to be totally beyond reach. The most favoured
technique is based on split laser beams and a Michelson interferometer—see
Fig. 6.25. The laser light is split into two paths at right angles by the beam
splitter B. The beams are reflected back and forth by mirrors M1–M4 (M1
and M3 being half-silvered) attached to masses and the fringes observed when
the light beams recombine and interfere. A gravitational wave will stretch one
dimension, say $D1$, and contract the orthogonal dimension $D2$, thus causing a
fringe shift. A Fabry–Perot etalon is used so that the beams make many traverses
to and fro before recombining. As an example, the values chosen for the LIGO
experiment are $D1$, $D2 \sim 4$ km, operating in a frequency range 10–1000 Hz,
typical of the collapse times for neutron stars/black holes.

The main problems for these experiments are the effects of background (seis-
mic) noise. This can be tackled by combining the signals from two or more of the
several detectors being located in different positions worldwide, using timing
information to reduce the noise and also indicate the direction of the source.

6.16 Summary

- The charged primary cosmic rays consist principally of high energy nuclei
  of the elements, their chemical composition being in general similar to
  the solar system abundances. The exception is for lithium, beryllium, and
  boron, which are abundant in the cosmic rays and produced by spallation
  of heavier nuclei in collisions with interstellar matter.
- The energy spectrum falls off as a power law, $dN/dE \sim E^{-2.7}$, which
  extends up to energies of at least 10^{20} eV.
- The charged primary radiation is affected by Solar System magnetic
  fields. The Earth's field imposes a cut-off in momentum depending on
  magnetic latitude. Cosmic rays are also moderated by solar effects (the
  solar wind), which follows the eleven-year sunspot cycle.

- The energy density in cosmic rays, at about $1 \, \text{eV} \, \text{cm}^{-3}$, is comparable with that in the cosmic microwave background, in starlight and in galactic magnetic fields. The rate at which energy needs to be injected into the cosmic rays can be accounted for in terms of shock-wave acceleration in supernova shells, provided these processes have efficiencies of a few per cent. While this mechanisms can work up to energies of $10^{14} \, \text{eV}$, the acceleration mechanism for the highest energies is unknown.

- The cosmic rays at sea level are of secondary origin, and generated by collisions of the primaries in the atmosphere. The hard component consists of muons from the decay of charged pions created in the atmosphere, while the soft component consists of electrons and photons originating from the decay of neutral pions.

- High energy cosmic ray nuclei can generate a nuclear cascade in the atmosphere, and this can lead to an extensive air shower, consisting of nucleons, muons, and electron–photon cascades extending over a large area (typically of radius 1 km). Practically all of the energy at sea-level is in the electron–photon component, and there is a linear relation between the primary energy and the shower size.

- The electron–photon showers can be detected via the Cerenkov light or scintillation light they generate in traversing the atmosphere.

- At energies above $10^{19} \, \text{eV}$, interactions of the primaries with the microwave radiation, leading to pion production, is expected to show suppression effects, but none is observed.

- Point sources of γ-rays of energies up to 30 GeV have been detected with the EGRET detector on the GRO satellite. The sources include pulsars and AGNs. Point sources involving γ-ray energies in the TeV region and above have been detected using the ground-based air Cerenkov method.

- Intermittent as well as steady sources are detected. The sporadic sources consist of bursts lasting 10 ms–10 s, which can disappear completely and reappear a year or so later. In the TeV energy region the γ-ray bursts are known as blazars. The shortness of the bursts indicates compact sources, and the blazar rate is consistent with the estimated rate of mergers of binary neutron stars to form black holes.

- Atmospheric neutrinos from the decay of pions produced in the atmosphere have been studied extensively in deep underground experiments, and show clear evidence for oscillations in the neutrino flavour (ν_μ or ν_e) over baselines comparable with the Earth radius, associated with the differences in mass of the neutrino mass eigenstates. The amplitude of the ($\nu_\mu \to \nu_\tau$) mixing is near maximal.

- Similar oscillation phenomena have been observed for neutrinos from the Sun. The suppression of ν_e events due to mixing shows an energy dependence, which can be described in terms of matter-induced oscillations inside the Sun.

- Attempts are under way to detect point sources of high energy neutrinos.

- The existence of gravitational radiation of the expected magnitude has been demonstrated from the slow-down rate of a binary pulsar. Attempts to detect gravitational radiation directly are currently under way.

Problems

More challenging problems are marked with an asterisk.

(6.1) Relativistic cosmic-ray protons are accelerated by a shock-front. Deduce the form of the differential energy spectrum of the protons, assuming that the probability that a proton will re-cross the front is 80% and that the fractional increase in energy per crossing is 20%.

(6.2) The refractive index, n, of air at sea-level is given by $n - 1 = 2.7 \times 10^{-4}$, a quantity which is proportional to pressure. Calculate the radial spread in metres of the ring of Cerenkov light at sea-level, due to an ultra-relativistic charged particle travelling vertically downwards at a depth of 100 gm cm^{-2} in the atmosphere. Assume an exponential atmosphere with density ρ and height h related by $\rho = \rho_0 \exp(-h/H)$ where $H = 6.5$ km. The total atmospheric depth is 1030 gm cm^{-2}.

(6.3) The average rate of energy loss of ultra-relativistic muons of energy E in traversing x gm cm^{-2} of material is given by the formula $dE/dx = a + bE$, where $a = 2.5$ MeV gm^{-1} cm^2 is the rate of ionization loss and the second term accounts for radiation energy loss. Calculate the average range (in km) of a 5000 GeV muon in rock of density 3 gm cm^{-3}, for which the critical muon energy is 1000 GeV.

*(6.4) Primary cosmic ray protons interact in the atmosphere with mean free path $\lambda \sim 100$ gm cm^{-2}. They produce relativistic charged pions of energy E travelling vertically downwards. These pions may subsequently decay in flight, or they may undergo nuclear interaction, again with a mean free path equal to λ. Assuming an exponential atmosphere as in Problem 6.2, show that the overall probability that a pion will decay rather than interact is $P = E_0/(E_0 + E)$ where $E_0 = m_\pi c^2 H/c\tau_\pi$. Calculate the value of E_0. How is the expression for P modified if the pion is produced at angle θ to the zenith? The total depth of the atmosphere (1030 gm cm^{-2}) can be assumed to be very large compared with λ ($m_\pi c^2 = 0.14$ GeV; $H = 6.5$ km; $\tau_\pi = 26$ ns).

*(6.5) Calculate the maximum and minimum energies of muons produced by decay in flight of high energy charged pions, $\pi^+ \to \mu^+ + \nu_\mu$. Assuming that the differential energy spectrum of the pions produced in the atmosphere follows a power law, $dN/dE \propto E^{-3} dE$, find an expression for the ratio of the numbers of muons at sea-level to the number of pions created at the same energy, that is $R = dN_\mu(E)/dN_\pi(E)$, assuming that the pion decay probability averaged over the atmosphere is $P = E_0/(E_0 + E)$ as quoted in Problem 6.4. Neglect the ionization energy loss of pions and muons in the atmosphere. Calculate the value of R for $E/E_0 = 0.1, 1.0,$ and 5.0 ($m_\pi c^2 = 139$ MeV; $m_\mu c^2 = 106$ MeV; $m_\nu \sim 0$).

(6.6) State whether you believe that CP-violating effects in a neutrino beam are possible if the mixing is between just two flavour eigenstates. What happens if matter effects are taken into account, for a beam traversing the Earth?

(6.7) In an experiment using a nuclear reactor as a source of electron-antineutrinos, the observed rate of the reaction $\bar{\nu}_e + p \to e^+ + n$ in a detector placed 250 m from the reactor core is found to be 0.95 ± 0.10 of that expected. If the mean effective antineutrino energy is 5 MeV, what limits would this place on a possible neutrino mass difference, assuming maximum mixing?

*(6.8) In the Kamiokande experiment, solar neutrinos are observed through the process of elastic scattering off electrons, and the detection of the Cerenkov light emitted by the recoiling electron as it traverses the water detector. If the incident neutrino has energy E_0, calculate the angle of scattering of the electron in terms of its recoil energy E. (Assume that the electron (and neutrino) masses can be neglected in comparison with the energies.)

*(6.9) High energy charged pions decay in flight in the atmosphere. Calculate the mean fractional energy received by the muon and by the neutrino in the decay $\pi^+ \to \mu^+ + \nu_\mu$. Estimate also the mean fractional energy of the pion, which is carried by each of the neutrinos (antineutrinos) in the subsequent muon decay $\mu^+ \to e^+ + \nu_e + \bar{\nu}_\mu$. Assume that all neutrinos are massless and neglect ionization energy losses in the atmosphere and polarization effects in muon decay (see Problem 6.5 for particle masses).

(6.10) Calculate the power radiated by the Moon in Earth orbit due to gravitational radiation (Earth's mass = 6×10^{24} kg; Moon's mass = 7.4×10^{22} kg; mean Earth–Moon distance = 3.8×10^5 km, orbital period of the moon = 27.3 days).

(6.11) Use the formulae and data in Section 6.13 to check the estimate of the decrease in orbital period of PSR 1913 + 16, assuming a circular orbit and equal masses for the members of the binary star system.

(6.12) Calculate the mean free path of high energy (1 TeV) cosmic-ray electrons in the galaxy, assuming that it is determined by collisions with the microwave background radiation. Could such electrons be extra-galactic? (Hint: Refer to equations (1.20) and (1.26) in Chapter 1.)

(6.13) Estimate the fraction of muon-neutrinos of energy 1 GeV which interact in traversing the Earth's diameter (take the mean Earth density as 3.5 gm cm^{-3}, and Earth radius as 6400 km).

(6.14) Show that, when a relativistic charged particle, travelling in the positive x direction, is scatterd backwards by the field due to a shock-front moving with non-relativistic velocity u_1 in the negative x direction, it receives a fractional energy increase of order u_1/c.

(6.15) Show that neutrinos, originating in the atmosphere at altitude h, and at zenith angle θ lying between 0 and π, have path-length L to a detector placed near the Earth's surface given by $L = (R^2 \cos^2 \theta + R^2 + 2Rh)^{1/2} - R \cos \theta$,

where R is the Earth diameter. Hence, verify the values of L quoted in Section 6.9.

*(6.16) Calculate the apparent transverse velocity of a jet travelling with velocity v at inclination θ to the line of sight to Earth, and show that this transverse velocity can appear to be superluminal, with a maximum value $\gamma \beta c$, where $\beta = v/c$ and $\gamma = 1/(1 - \beta^2)^{1/2}$.

(6.17) From the formulae for the Lorentz transformations of momentum and energy in Special Relativity (see equation (B.15), Appendix B), calculate an expression for the angle of emission θ in the laboratory system of a photon emitted at angle θ^* in the rest-frame of an electron travelling with relativistic velocity $\beta = v/c \approx 1$. Hence, show that when accelerated in a magnetic field, a parallel jet of high energy electrons will emit photons confined to a forward cone of half-angle $\theta \sim 1/\gamma$, where $\gamma = 1/(1 - \beta^2)^{1/2}$.

7 Particle physics in the stars

7.1 Introduction

In the earlier chapters we have traced the primordial development of the universe, principally through the properties and interactions of the elementary particle constituents of matter. This phase came to an end when the temperature of the fireball fell below a value $kT \sim 0.3$ eV and $z \sim 10^3$, when the radiation and (baryonic) matter decoupled, and at about the time when the universe became matter-dominated. The previously opaque universe, consisting of a plasma of electrons, hydrogen and helium nuclei, and photons, was then replaced by a relatively transparent but almost totally dark universe, consisting very largely of clouds of neutral atoms and molecules. From these clouds, stars were able to form by gravitational infall as soon as the redshift fell to $z \sim 20$–30. Then, there was light. The process of star formation has of course been continuous since that time.

The evolution of the stars, at least during most of their life, is only peripherally linked with elementary particle physics as such, so we shall discuss it rather briefly and in particular contexts, for example, with respect to solar neutrinos described in Chapter 6. However, the late stages of stellar evolution do depend very directly on particle processes at a fundamental level and at quite high energies, involving some of the most violent events in the universe, and we describe these in more detail.

7.2 Stellar evolution—the early stages

As described in Section 5.6, stars can condense out of gas clouds (predominantly hydrogen) once the mass and density of the material fulfils the Jeans criterion (5.31). The gravitational potential energy lost under contraction goes to heat up the gas. The resulting gas pressure opposes further contraction, and the so-called 'protostar' reaches a state approaching hydrostatic equilibrium (see Example 7.2). Typically, the gas density at this stage is 10^{-15} kg m^{-3} and the radius is 10^{15} m (i.e. about one million times the solar radius). As the star radiates energy from the envelope it slowly contracts further, to about a hundred times the solar radius. By then, the amount of gravitational energy that has been released is of the order of 10 eV per particle, so that collisional dissociation of hydrogen molecules (requiring 4.5 eV) and ionization of hydrogen atoms (13.6 eV) can take place. The energy generated when these atoms and molecules recombine is released as photons to the outside world, allowing still further contraction. Without any source of energy apart from gravitation, the energy

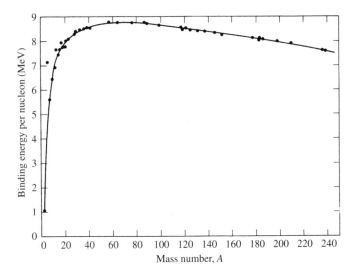

Fig. 7.1 Binding energy per nucleon as a function of mass number A, for nuclei stable against beta decay. The maximum binding is in the Fe–Ni region of the Periodic Table. (Enge 1972.)

radiated by the star must always be compensated by further contraction, and a consequent increase in pressure and temperature of the core. In fact, the kinetic (heat) energy of the star must be just equal to half its gravitational energy—an example of the **virial theorem** on the partition of kinetic and potential energy in a system in thermal equilibrium.

Further collapse of the star is eventually halted by the onset of **thermonuclear reactions**. Figure 7.1 shows the binding energy per nucleon as a function of the mass number A of the nucleus. It is clear that if two light nuclei fuse to form a heavier nucleus, energy will be released, provided the product nucleus has $A < 56$, the mass number of iron, for which the binding energy per nucleon is a maximum. The amount of energy released is substantial. For example, if as described below, helium is formed from hydrogen, the binding energy liberated is of the order of 7 MeV per nucleon.

The electrostatic potential between two nuclei of charges $Z_1 e$ and $Z_2 e$ and mass numbers A_1 and A_2 with separation r is $V = Z_1 Z_2 e^2 / 4\pi r$. When just in contact, $r = r_0(A_1^{1/3} + A_2^{1/3})$ where $r_0 = 1.2$ fm is the unit nuclear radius. The first stage of the pp fusion process in the Sun is the weak reaction

$$p + p \rightarrow d + e^+ + \nu_e + 0.32 \text{ MeV} \tag{7.1}$$

with a Coulomb barrier height $V_0 = (1/4\pi)e^2/2r_0 = (e^2/4\pi\hbar c) \times (\hbar c/2r_0)$. With $e^2/4\pi\hbar c = \alpha = \frac{1}{137}$, $\hbar c = 197$ MeV fm, and $r_0 = 1.2$ fm one finds $V_0 = 0.6$ MeV. This is very much larger than the thermal energy of protons at the core temperature of the Sun, which can be estimated from the solar luminosity to be $kT \sim 1$ keV. Although in classical terms the two nuclei cannot therefore surmount the Coulomb barrier, in quantum mechanics they can penetrate **through it** with finite probability. This effect had, in the 1920s, successfully accounted for the long lives of radioactive nuclei undergoing alpha decay. The barrier penetration probability is given by the approximate formula, valid for $E \ll E_G$:

$$P(E) = \exp\left[-\left(\frac{E_G}{E}\right)^{1/2}\right] \tag{7.2}$$

where

$$E_G = \left(\frac{2m}{\hbar^2}\right)\left(\frac{Z_1 Z_2 e^2}{4}\right)^2 \tag{7.3}$$

is the so-called Gamow energy (named for George Gamow, who first investigated the barrier penetration problem). The quantity m is the reduced mass of the two nuclei, and for the pp reaction it is half the proton mass, $m_p/2$. With $e^2 = 4\pi\alpha\hbar c$ one gets $E_G = m_p c^2 \pi^2 \alpha^2 = 0.49$ MeV. So, if the relative kinetic energy $E \sim 1$ keV as indicated above, the barrier penetrability will be of the order of $P \sim \exp(-22) \sim 10^{-10}$. In fact the protons will have a Maxwellian distribution in kinetic energy of the form

$$F(E)\,dE \sim E^{3/2} \exp\left(-\frac{E}{kT}\right) dE \tag{7.4}$$

As shown in Fig. 7.2, the penetrability factor in (7.2) increases with energy, while for $E > kT$, the number of protons in (7.4) decreases with energy. The fusion rate is the product of these two distributions. However, even if the barrier is successfully penetrated, the usual reaction between the protons will be elastic

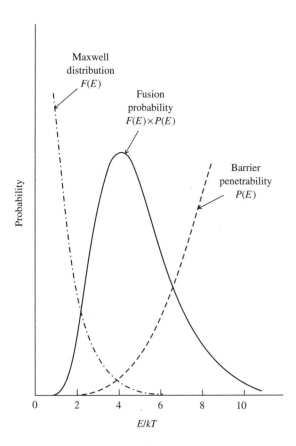

Fig. 7.2 Curves showing at left the Maxwell distribution of relative energy of colliding nuclei, and at right the barrier penetrability, for the pp reaction. The fusion rate is proportional to the product of these distributions and is shown by the solid curve.

scattering (via the strong interaction) rather than the weak reaction (7.1), which turns out to have a relative probability of the order of 1 in 10^{20}. So, although any one proton in the Sun is having millions of encounters with other protons every second, the mean lifetime for the conversion of protons to deuterium and helium is billions of years. The rate of nuclear energy generation is exactly matched to the energy radiated from the solar envelope, and while the hydrogen fuel lasts, the Sun is quite stable against any fluctuations in radius. It is also of interest to remark here that the nuclear energy generated in the core takes a very long time to work its way out to the photosphere (see Example 7.1).

Example 7.1   *Estimate the time required for energy generated by fusion in the solar core to reach the photosphere by radiative diffusion, given that the core temperature is 16×10^6 K, the surface temperature is 6000 K and the solar radius is $R = 6 \times 10^8$ m.*

At the core temperature, the energy per photon is $\sim$1 keV, and is thus in the X-ray region. The radiation is transmitted to the surface through random collisions with the plasma particles, resulting in scattering, absorption and re-emission processes. If we denote the steps between collisions by $\mathbf{d_1}, \mathbf{d_2}, \mathbf{d_3}, \ldots$ the mean square distance travelled in this 'random walk' will be $L^2(\text{av}) = \langle (\mathbf{d_1} + \mathbf{d_2} + \mathbf{d_3} + \cdots + \mathbf{d_N})^2 \rangle = Nd^2$ where d is an average step length and N is the number of steps, and we have used the fact that the steps are in random directions so that in evaluating the square, all the cross terms cancel. Thus, to reach the surface from the deep interior requires $L \sim R$ and $N \sim R^2/d^2$ steps, with an elapsed time $t_1 \sim R^2/cd$. Had the radiation been free to escape directly, the time to the surface would only have been $t_2 = R/c$, so that the process of radiative diffusion has slowed down the rate at which energy escapes the sun by a factor $t_1/t_2 = R/d$. This is the factor by which the core luminosity, of order $R^2 T_c^4$, is reduced to the surface luminosity, of order $R^2 T_s^4$. Thus, $d/R \sim (T_s/T_c)^4$ and $t_1 = (R/d)(R/c) \sim 10^{14}$ s or about a million years.

7.3   Hydrogen burning: the pp cycle in the sun

The production of energy in the Sun is via the fusion of hydrogen to helium, according to the net process

$$4p \rightarrow {}^4\text{He} + 2e^+ + 2\nu_e + 26.73 \text{ MeV} \qquad (7.5)$$

This process takes place in several stages. The first is the weak reaction (7.1) forming a deuteron:

$$p + p \rightarrow d + e^+ + \nu_e \qquad (7.6)$$

which also has a small (0.4%) contribution from the so-called 'pep' process

$$p + e^- + p \rightarrow d + \nu_e$$

The next stage is the electromagnetic reaction

$$p + d \rightarrow {}^3\text{He} + \gamma \qquad (7.7)$$

followed by two strong interaction processes. The first produces ^{4}He according to

$${}^3\text{He} + {}^3\text{He} \rightarrow {}^4\text{He} + 2p \quad (85\%) \qquad (7.8)$$

while the second produces heavier elements, generating ^{4}He in subsequent interactions:

$$^3\text{He} + {}^4\text{He} \rightarrow {}^7\text{Be} + \gamma \tag{7.9}$$

$$e^- + {}^7\text{Be} \rightarrow {}^7\text{Li} + \nu_e; \quad p + {}^7\text{Li} \rightarrow 2{}^4\text{He} \quad (15\%) \tag{7.10}$$

$$p + {}^7\text{Be} \rightarrow {}^8\text{B} + \gamma; \quad {}^8\text{B} \rightarrow {}^8\text{Be}^* + e^+ + \nu_e; \quad {}^8\text{Be}^* \rightarrow 2{}^4\text{He} \quad (0.02\%) \tag{7.11}$$

where the percentages indicate the contributions to the total helium production rate. In addition to the pp cycle, another cycle involving the elements C, N, and O accounts for about 1.6% of helium production in the Sun. Since the nuclear charges are larger, this cycle is more important for more massive hydrogen-burning stars with higher core temperatures, such as Sirius A.

The above reactions have been written out in detail, because the study of neutrinos from these reactions has been crucially important, not only in verifying our picture of solar fusion reactions but also in demonstrating the existence of neutrino flavour oscillations, as has been described in Section 6.9.

Assuming that protons in the core account for about 10% of all protons in the Sun, the observed solar luminosity (3.8×10^{26} J s^{-1}) implies through equation (7.5) that an average survival time before a proton undergoes fusion is several billion years, comparable indeed with the age of the universe. This time is determined by several factors: the Coulomb barrier penetration rate, the weak interaction cross section in the first stage of the pp cycle, and, most importantly, the opacity of the solar material, which determines the rate at which energy can escape from the Sun and hence the rate at which fusion energy is generated in the core.

As indicated above, a star first lights up when it contracts from a 'protostar' and commences hydrogen burning, which it continues for most of its life. In the Hertzsprung–Russell diagram of luminosity versus surface temperature (see Fig. 7.3(b)) a hydrogen-burning star is situated at a single point on a curve called the **main sequence**. The position on the curve depends on the mass, which determines the luminosity. Empirically, the luminosity L of main sequence stars varies with mass M as $L \propto M^{3.5}$, and so the lifetime τ of a star on the main sequence must be proportional to M/L, so $\tau \propto M^{-2.5}$. Those stars in the cluster above a certain mass, corresponding to $\tau = t_0$, the age of the cluster, have already moved off the main sequence at the main sequence turn-off (MSTO), towards the red giant branch. Here hydrogen burning proceeds in a shell outside the helium core, the envelope expands and the surface temperature falls. Eventually, the helium core ignites and the star moves over to the horizontal branch. If the carbon core in turn ignites (i.e. if the star is massive enough), the star moves back again towards the red giant branch. If M is only of the order of one solar mass, however, no fusion of the carbon/oxygen core takes place and the star moves down towards the left-hand bottom of the diagram where white dwarfs (not shown) are situated.

The age of the cluster—and hence the age t_0 of the universe for the very oldest clusters—can be found from the position of the MSTO and a stellar evolutionary model. In this way the age of the universe has been estimated at 14 ± 2 Gyr.

Fig. 7.3(a) The globular cluster M13, containing upwards of 10^5 stars. Our galaxy contains at least a hundred and fifty such globular clusters, in orbit about the centre of the galaxy. The stars in a globular cluster were formed together and they orbit about the cluster centre, to which they are gravitationally bound. Such clusters are extremely old, and as described below, can be used to set an estimate of the age of the universe.

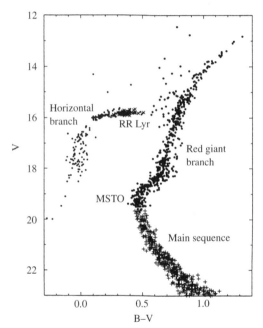

Fig. 7.3(b) The Hertzsprung–Russell diagram of stellar luminosity against surface temperature (colour), for stars in the globular cluster M15. Stars in such a cluster were all formed at essentially the same time, at a very early stage of the universe. In this graph, the magnitude (or logarithm of the luminosity) observed with a V filter ($\lambda = 540$ nm) is plotted vertically against the difference in magnitudes with a B filter ($\lambda = 440$ nm) and a V filter. Redder colours and cooler surface temperatures are to the right, and higher luminosities or lower magnitudes towards the top (from Chaboyer 1996, with permission of Elsevier Science).

7.4 Helium burning and the production of carbon and oxygen

When most of the hydrogen in the stellar core has been converted to helium, and fusion energy is no longer produced there, the core contracts and part of the gravitational energy released leads to local heating, the rest escaping from the core. The higher central temperature means that hydrogen burning now proceeds in a spherical shell surrounding the helium core, so the total mass and density of helium increases. If the star is sufficiently massive (more than about half a solar mass), the core temperature becomes high enough to ignite the helium, at a temperature just above 10^8 K, resulting in production of carbon and oxygen, the most abundant elements in the universe after hydrogen and helium.

Since the core temperature and pressure have increased, relative to the values in the pp cycle, by an order of magnitude, the outer envelope of the star expands by a large factor. It becomes a **red giant**. The luminosity increases and the surface temperature decreases, so that the star moves off the main sequence and over to the red giant branch in the Hertzsprung–Russell diagram.

Helium burning involves the following somewhat complex chain of events. In the absence of any stable nuclei with masses 5 or 8, fusion has to proceed by the so-called **triple alpha** reaction, as first discussed by Salpeter in 1952. The first stage is

$$^4\text{He} + {}^4\text{He} \leftrightarrow {}^8\text{Be} \tag{7.12}$$

The nucleus ^{8}Be is unstable by 92 keV, so that the two helium nuclei must have this relative energy in order to 'hit' the ground state, and to do this efficiently requires a temperature, read off from a curve of the same type as shown in Fig. 7.2, of $T = (1-2) \times 10^8$ K. The mean lifetime of ^{8}Be is about 2.6×10^{-16} s. The equilibrium concentration of beryllium nuclei in (7.12) is only about one-billionth of that of the helium nuclei. Nevertheless, a ^{8}Be nucleus may capture a third alpha-particle to form the nucleus ^{12}C* in an excited state at 7.654 MeV, which is just 0.3 MeV above the threshold energy for $^4\text{He} + {}^8$Be:

$$^4\text{He} + {}^8\text{Be} \leftrightarrow {}^{12}\text{C}^* \tag{7.13}$$

Hoyle (1954) had pointed out the need for, and estimated the energy of, this resonant state in carbon, and it was subsequently found in accelerator experiments. Usually, the carbon decays back to beryllium plus helium as in (7.13), but can, with only 3×10^{-4} probability, decay by radiative transitions to the ground state:

$$^{12}\text{C}^* \to {}^{12}\text{C} + 2\gamma \quad (\text{or } e^+ + e^-) \tag{7.14}$$

Once carbon nuclei have been created, the next stage of oxygen production can proceed via radiative alpha-particle capture:

$$^4\text{He} + {}^{12}\text{C} \to {}^{16}\text{O} + \gamma \tag{7.15}$$

It is fortunate that in this case there is **no** resonance in ^{16}O anywhere near the threshold energy, so that not all the carbon is consumed as soon as it is produced, and both carbon and oxygen are abundant elements in the universe. Obviously, the existence of the 7.654 MeV resonance level in the carbon 12 nucleus was vital for the development of carbon-based biological molecules and life as we know it in our particular universe.

7.5   Production of heavy elements

A massive star evolves further through fusion reactions to produce successively heavier elements, involving higher Coulomb barriers and higher core temperatures. A cross section through the star would have an onion-like appearance, as in Fig. 7.4, with the heaviest elements in the core and lighter ones in spherical shells of successively larger radius and lower temperature.

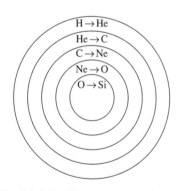

Fig. 7.4 Onion-like appearance of the cross section of a massive star at an advanced stage of nuclear fusion. The heaviest elements are in the core, where the temperature and density are greatest, and are surrounded by lighter elements in spherical layers of successively lower temperatures and density.

Table 7.1 Nuclear fusion timescales for a star of $25M_\odot$ (after Rolfs and Rodney 1988)

Fusion of	Time to complete	Core temperature (K)	Core density ($kg\,m^{-3}$)
H	7×10^6 yr	6×10^7	5×10^4
He	5×10^5 yr	2×10^8	7×10^5
C	600 yr	9×10^8	2×10^8
Ne	1 yr	1.7×10^9	4×10^9
O	0.5 yr	2.3×10^9	1×10^{10}
Si	1 day	4.1×10^9	3×10^{10}

Carbon burning commences when the core temperature and density are $T \sim 5 \times 10^8$ K and $3 \times 10^9\,kg\,m^{-3}$, respectively, and leads to the production of neon, sodium, and magnesium nuclei:

$$\begin{aligned} {}^{12}\text{C} + {}^{12}\text{C} &\rightarrow {}^{20}\text{Ne} + {}^4\text{He} \\ &\rightarrow {}^{23}\text{Na} + \text{p} \\ &\rightarrow {}^{23}\text{Mg} + \text{n} \end{aligned} \tag{7.16}$$

At still higher temperatures, of the order of 2×10^9 K, oxygen burning leads to the production of silicon:

$$ {}^{16}\text{O} + {}^{16}\text{O} \rightarrow {}^{28}\text{Si} + {}^4\text{He} \tag{7.17}$$

At such temperatures, the thermal photons have appreciable quantum energies. For example, a tiny proportion of the photons, with over 20 times the mean Planck energy, will have energies above 9 MeV and can, therefore, cause photodisintegration of silicon, with the prolific production of helium nuclei:

$$ \gamma + {}^{28}\text{Si} \rightarrow {}^{24}\text{Mg} + {}^4\text{He} \tag{7.18}$$

On account of their lower Coulomb barriers, the helium nuclei released can now, by radiative capture, induce successive fusions to form sulphur, argon, calcium, and eventually iron and nickel. These reactions proceed easily and the overall rate is really determined by the first stage of photoproduction (7.18). With the production of ^{56}Fe, however, the exothermic fusion process finally ends, since as indicated in Fig. 7.1, iron is the most strongly bound nucleus. The typical timescales, temperatures, and densities involved in nuclear fusion reactions are indicated in Table 7.1.

7.6   Electron degeneracy pressure and stellar stability

At high densities, such as those that occur in stellar cores at an advanced stage of the evolutionary path, a new form of pressure, in addition to gas pressure and radiation pressure, called **electron degeneracy pressure**, becomes important. To investigate the role of this degeneracy pressure, consider a gas of electrons at absolute zero temperature. The electrons will fall into quantum states of the lowest possible energy, and for this reason the gas is said to be degenerate. The Pauli Exclusion Principle applies to such identical fermions, so that each

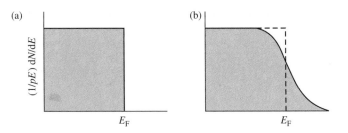

Fig. 7.5 Distribution of electron energies (a) for an electron gas at absolute zero temperature, with all levels filled up to the Fermi energy; and (b) for an electron gas at finite, low temperature, where electrons begin to spill over into states above the Fermi energy.

quantum state can be occupied by one electron only. At zero temperature, the energy is minimized if all the states are occupied, up to some maximum energy called the Fermi energy, E_F and all states of energy $E > E_F$ are unoccupied (see Fig. 7.5). The corresponding momentum is called the Fermi momentum p_F. For values of temperature T above zero, not all these quantum states are filled and the energy spectrum extends above the Fermi energy. Ultimately, when $kT \gg E_F$, the energy distribution reverts to the Fermi–Dirac distribution described by equation (2.44).

Going back to the case of the completely degenerate electron gas of Fig. 7.5(a), the number of electrons in a physical volume V with momentum $p < p_F$ will be

$$N = g_e V \int \frac{4\pi p^2 \, dp}{h^3} = g_e V \frac{4\pi p_F^3}{3h^3} \tag{7.19}$$

where $g_e = 2$ is the number of spin substates of each electron, and $4\pi p^2 \, dp/h^3$ is the number of states in phase space per unit volume. The number density of electrons will be $n = N/V$ so that

$$p_F = h \left[\frac{3n}{8\pi} \right]^{1/3} \tag{7.20}$$

If the electrons are **non-relativistic**, that is $p_F \ll m_e c$, their kinetic energy is $p^2/2m_e$ and the kinetic energy density will be

$$\frac{E_{NR}}{V} = \int 8\pi p^2 \left(\frac{p^2}{2m_e} \right) \frac{dp}{h^3} = \frac{8\pi p_F^5}{10 m_e h^3} \tag{7.21}$$

Using (2.21), the degeneracy pressure of the electron gas will, therefore, be

$$P_{NR} = \left(\frac{2}{3} \right) \left(\frac{E_{NR}}{V} \right) = \frac{8\pi p_F^5}{(15 \, m_e h^3)} = \left[\left(\frac{3}{8\pi} \right)^{2/3} \times \left(\frac{h^2}{5 m_e} \right) \right] n^{5/3} \tag{7.22}$$

However, if the electrons are mainly **relativistic**, that is if $p_F \gg m_e c$, the electron energy $E \approx pc$ and hence

$$\frac{E_R}{V} = \int \frac{8\pi p^2 \, dp(pc)}{h^3} = \frac{2\pi p_F^4 c}{h^3} \tag{7.23}$$

The corresponding pressure from (2.22) becomes

$$P_R = \left(\frac{1}{3} \right) \frac{E_R}{V} = \left[\left(\frac{hc}{4} \right) \left(\frac{3}{8\pi} \right)^{1/3} \right] n^{4/3} \tag{7.24}$$

Note that, in both relativistic and non-relativistic cases, the pressure increases with the electron density, but more rapidly in the non-relativistic case. This turns out to be crucially important in discussing stability of stars in their final evolutionary stages.

Let us now consider the effects of the gravitational pressure. The gravitational energy of a star or of a stellar core of mass M, radius R, volume V, and assumedly uniform mass density ρ is

$$E_{grav} = \left(\frac{3}{5}\right) \frac{GM^2}{R}$$

The volume-averaged gravitational pressure is (see Example 7.2)

$$P_{grav} = \frac{E_{grav}}{3V} = \left(\frac{G}{5}\right) \left(\frac{4\pi}{3}\right)^{1/3} M^{2/3} \rho^{4/3}$$

$$= \left(\frac{G}{5}\right) \left(\frac{4\pi}{3}\right)^{1/3} M^{2/3} \left(\frac{m_P A}{Z}\right)^{4/3} n^{4/3} \qquad (7.25)$$

In the second line, the mass density ρ of the material of the stellar core containing the degenerate electron gas has been expressed in terms of the electron number density n, where Z and A are the atomic and mass numbers of the nuclei of the core and m_P is the nucleon mass, so that $n = (Z/A)\rho/m_P$.

Example 7.2 *Show that, in stellar equilibrium, the volume-averaged gas pressure is one-third of the gravitational energy density.*

Let $P(r)$ be the gas pressure at radius r, and consider a spherical shell of density $\rho(r)$ and thickness dr. The outward force on the shell due to the gas pressure is

$$4\pi r^2 \left[P(r) - \left\{ P(r) + \left(\frac{dP}{dr}\right) dr \right\} \right]$$

where dP/dr is clearly negative. The inward gravitational force on the shell is $GM(<r)\,dm/r^2$ where $dm = 4\pi r^2 \rho(r)\,dr$ is the mass of the shell. Equating these forces, multiplying through by r and integrating from $r = 0$ to $r = R$, the stellar radius, we get

$$-\int 4\pi r^3 \left(\frac{dP}{dr}\right) dr = -\int GM(<r)\frac{dm}{r}$$

If we integrate the left-hand side by parts and assume that $P = 0$ at $r = R$, we get

$$3\int 4\pi r^2 \, dr \, P(r) = 3\int P(r)\,dV = -\int GM(<r)\frac{dm}{r} = E_{grav}$$

where dV denotes the volume of the spherical shell. The second term in this equation represents the gas pressure integrated over volume, while the third is the total gravitational potential energy. Hence, we find for the volume-averaged pressure

$$\langle P \rangle = -\left(\frac{1}{3}\right) \frac{E_{grav}}{V} \qquad (7.26)$$

Provided that the electrons are non-relativistic, the degeneracy pressure (7.22), varying as $n^{5/3}$, will win over the gravitational pressure, varying as $n^{4/3}$. So the star or stellar core will be stable against contraction, since any increase in density will increase the outward electron pressure relative to the inward gravitational pressure. In the case of relativistic degeneracy, however, both pressures have the same density dependence, that is varying as $n^{4/3}$, so such a core, with mass exceeding the so-called Chandrasekhar limit, is not stable and a suitable trigger can send it into a state of collapse.

The condition $p_F < m_e c$ that the electron momentum is non-relativistic implies from (7.20) that the average distance $n^{-1/3}$ between electrons exceeds the electron Compton wavelength:

$$n^{-1/3} > \left(\frac{3}{8\pi}\right)^{1/3} \left(\frac{h}{m_e c}\right) \sim 0.5 \left(\frac{h}{m_e c}\right) \qquad (7.27)$$

The number density of nucleons is An/Z, so that the critical density for stability will be

$$\rho_0 = \left(\frac{8\pi m_p A}{3Z}\right) \left(\frac{m_e c}{h}\right)^3 \qquad (7.28)$$

Equating the gravitational pressure (7.25) with the non-relativistic degeneracy pressure (7.22) gives a value for the density at which the two are equal:

$$\rho = \left(\frac{4m_e^3 G^3 M^2}{h^6}\right) \left(\frac{A m_p}{Z}\right)^5 \left(\frac{4\pi}{3}\right)^3 \qquad (7.29)$$

Identifying this with the critical density (7.28), we get a value for the maximum mass of a stellar core which is stable against collapse:

$$M_{Ch} = \left(\frac{3\sqrt{2}}{8\pi}\right) \left(\frac{hc}{G}\right)^{3/2} \left(\frac{Z}{A m_p}\right)^2 = 4.91 \left(\frac{Z}{A}\right)^2 M_\odot \approx 1.2 M_\odot \qquad (7.30)$$

where $M_\odot$ is the solar mass, and a core of helium or heavier elements, with $Z/A = \frac{1}{2}$, has been assumed. A more sophisticated stellar model, in which the density varies with radius, yields the more realistic value

$$M_{Ch} = 1.4 M_\odot \qquad (7.31)$$

The quantity M_{Ch} is called the **Chandrasekhar mass**, after the physicist who first discussed the stability of white dwarfs and obtained the above limit in 1931.

7.7   White dwarf stars

Stars of relatively low mass, such as our own Sun, after passing through the hydrogen- and helium-burning phases, will form cores of carbon and oxygen. The higher temperature of the core will lead to helium being burned in a spherical shell surrounding the core, and the stellar envelope will expand by a huge factor and eventually escape to form a planetary nebula surrounding the star. For stars in the solar mass range, the central temperature will not increase enough to lead to carbon burning, so that after the helium is finished, there is no longer any fusion energy source. Nevertheless, the star, providing its mass is less than the Chandrasekhar mass, is saved from catastrophic collapse because of the electron degeneracy pressure in the core. Such a star, bereft of its envelope and slowly cooling off, is known as a **white dwarf**.

All main sequence stars of about one solar mass will end up eventually as white dwarfs. However, these stars are limited to a fairly narrow mass range. The upper limit is determined by the Chandrasekhar mass of $1.4M_\odot$, but there is also a lower limit of approximately $0.25M_\odot$, since it turns out that the evolution of stars and their emergence as white dwarfs of masses below this limit would be on a timescale much longer than the present age of the universe (see Section 7.3). If they are partners of binary systems, the masses of white dwarfs can be measured, and those observed appear to satisfy the above limits.

> **Example 7.3**   *Using the criterion of non-relativistic electron degeneracy, estimate the radius and density of a white dwarf star of one solar mass.*
>
> From (7.29) the density needed to balance the gravitational pressure with non-relativistic degeneracy pressure is
>
> $$\rho = \left(\frac{4m_e^3 G^3 M^2}{h^6}\right) \times \left(\frac{Am_p}{Z}\right)^5 \times \left(\frac{4\pi}{3}\right)^3$$
>
> Inserting $\rho = 3M/4\pi R^3$ gives $R = 7 \times 10^6$ m $= 0.01R_\odot$ for $M = M_\odot$, $A/Z = 2$. Note that $R \propto 1/M^{1/3}$, that is, the radius of a white dwarf **decreases** as the mass increases. The average density in the case chosen is clearly 10^6 times that of the Sun, which is about 2×10^9 kg m^{-3}.

The typical radius of a white dwarf can be estimated as in the above example. It is of the order of 1% of the solar radius, corresponding to the fact that the average density is of the order of 10^6 times the mean solar density. In the above discussion, we have treated the density of the white dwarf as uniform, but as is clear from Example 7.2, both the pressure and the density must increase towards the centre of the star. For a white dwarf of about one solar mass, the central density is calculated to be in the region of 10^{11} kg m^{-3}. Since white dwarfs, as the name implies, emit white light, they have surface temperatures of the same order as that of the Sun, so that with almost a hundred times smaller radius, their luminosities are of the order of 10^{-3} of the solar luminosity. This guarantees that, even with no nuclear energy source, white dwarfs can continue shining for billions of years.

7.8   Stellar collapse: Type II supernovae

A star of mass $M > 10M_\odot$ is massive enough that it can evolve through all the stages of stellar fusion, ending up eventually with an iron core, produced by silicon burning at a temperature of the order of 4×10^9 K, as sketched in Fig. 7.4. As more silicon is burned in a shell surrounding the iron, both the mass of the iron core and its temperature will increase, until, eventually, the core mass exceeds the Chandrasekhar limit (7.31). The core is then unstable and is driven into collapse by two triggering mechanisms: photo-disintegration of iron nuclei by thermal photons, and the conversion of electrons to neutrinos by inverse beta decay. The result is a supernova explosion (rate about one per century in the Milky Way).

As the collapse proceeds, some of the gravitational energy released goes into heating up of the core to well above 10^{10} K, that is, a mean thermal photon energy above 2.5 MeV, so that a fraction of the photons can cause photo-disintegration of the iron nuclei into alpha particles (helium nuclei). With enough photons, it is clear that iron can be broken down completely into helium, thus reversing the effects of all the fusion processes since the main sequence pp cycle:

$$\gamma + {}^{56}\text{Fe} \leftrightarrow 13\,{}^{4}\text{He} + 4\text{n} \tag{7.32}$$

This equation indicates that an equilibrium between iron and helium nuclei is set up, in which the balance swings to the right-hand side as the core temperature increases. Of course, the absorption of energy in the above endothermic process (145 MeV for the complete photo-disintegration of each iron nucleus into alphas) further speeds up gravitational collapse, the core heats still further and the helium nuclei themselves undergo photo-disintegration:

$$\gamma + {}^{4}\text{He} \rightarrow 2\text{p} + 2\text{n} \tag{7.33}$$

The collapse also heralds the onset of **neutronization**, in which the electrons from the degenerate 'sea' convert free or bound protons to neutrons via inverse beta decay:

$$e^{-} + \text{p} \rightarrow \text{n} + \nu_{e} \tag{7.34}$$

with a 0.8 MeV threshold. When the radius of the core has collapsed by an order of magnitude and the density has reached the vicinity of 10^{12} kg m^{-3}, the Fermi momentum of the electrons is, from (7.20),

$$p_{F}c = hc \left[\frac{3Z\rho}{(8\pi A m_{\text{p}})} \right]^{1/3} \sim 4 \text{ MeV} \tag{7.35}$$

so that an electron of energy near the Fermi energy can trigger the above reaction, or an equivalent inverse beta decay in iron:

$$e^{-} + {}^{56}\text{Fe} \rightarrow {}^{56}\text{Mn} + \nu_{e} \tag{7.36}$$

with a threshold of 3.7 MeV. As more and more electrons are converted to neutrinos in these processes, the degeneracy pressure of the electrons will decrease steadily and the collapse will then be virtually unopposed. The free-fall time of the collapse will be given by (5.26)

$$t_{\text{FF}} = \left(\frac{3\pi}{32G\rho} \right)^{1/2} \sim 0.1 \text{ s} \tag{7.37}$$

Eventually, the collapse is halted, as the density reaches the nuclear density. The gravitational pressure is then opposed by the degeneracy pressure of the (non-relativistic) nucleons. The core contains iron nuclei, electrons, and protons as well as a preponderance of neutrons (hence the name neutron star). Very roughly, we can treat the collapsed core as a gigantic nucleus of neutrons, so that we expect the radius to be of the order of

$$R = r_{0}A^{1/3} \tag{7.38}$$

where $r_{0} = 1.2$ fm $= 1.2 \times 10^{-15}$ m is the unit nuclear radius. For a core mass $M = 1.5M_{\odot}$, the mass number $A = M/m_{\text{p}} \sim 1.9 \times 10^{57}$ and the radius

$R \sim 15$ km, for a nuclear density $\rho_N = 3m_p/4\pi r_0^3 \sim 2 \times 10^{17}$ kg m^{-3}. The repulsive nuclear force at short distances resists further compression, and it is estimated that as soon as the density exceeds nuclear density by about a factor of 2–3, the collapse will be brought to an abrupt halt, and the core material will 'bounce', producing an outgoing pressure wave that develops into a supersonic shock wave, which will traverse the infalling material of the envelope and finally give rise to the spectacular optical phenomenon of a supernova explosion. Such an event, resulting from the collapse of a massive star, is known as a Type II supernova.

As the initial collapse proceeds, the reactions (7.34) and (7.36) will result in emission of neutrinos. In particular, a short, few millisecond burst of 10^{56}–10^{57} neutrinos ν_e will accompany the outgoing shock wave. They will have energies of a few MeV and account for up to 5% of the total gravitational energy released. However, as soon as the core density exceeds about 10^{15} kg m^{-3}, it becomes effectively opaque, even to neutrinos, and they become trapped in the contracting material.

The total gravitational energy released in the collapse to a neutron star of $1.5M_\odot$ and uniform density will be

$$E_{grav} = \left(\frac{3}{5}\right) Gm_p^2 \frac{A^{5/3}}{r_0} = 3.0 \times 10^{46} \text{ J} = 1.8 \times 10^{59} \text{ MeV} \qquad (7.39)$$

which amounts, therefore, to about 100 MeV per nucleon. Note that this implies that an original uncontracted mass of $1.55M_\odot$ of nucleons will result in a neutron star with a mass of only $1.4M_\odot$. This energy release is much larger than the energy required to disintegrate the iron into its constituent nucleons (the binding energy per nucleon in iron is 8 MeV) or to convert protons and electrons to neutrons and neutrinos (0.8 MeV per nucleon). The huge amount of energy released, however, remains temporarily locked in the core, which enters a 'thermal phase' in which photons, electron–positron pairs, and neutrino–antineutrino pairs, together with the neutrons and some protons and heavier nuclei reach thermal equilibrium. All flavours of neutrinos and antineutrinos will be generated in this thermal phase:

$$\gamma \leftrightarrow e^+ + e^- \leftrightarrow \nu_i + \nu_i \qquad (7.40)$$

where $i = e, \mu, \tau$. The mean free path for neutrinos in the core material depends on both charged and neutral current scattering by nucleons, electrons, and nuclei. As an indication, we consider the charged current scattering (7.41) of ν_e by neutrons. The cross section is of the order of $G_F^2 p_f^2$ from (1.18) and (1.27), where $p_f = E \sim (E_\nu - Q)$ is the neutrino energy above the (negative) threshold ($Q = -0.8$ MeV)—see Problem 7.7. In detail, the cross section is

$$\sigma(\nu_e + n \rightarrow p + e^-) = \frac{G_F^2}{\pi}\left[1 + 3g_A^2\right] E^2 = 0.94 \times 10^{-43} E^2 \text{ cm}^2 \quad (7.41)$$

where $g_A = 1.26$ is an axial-vector coupling constant and E is in MeV. For a typical nuclear density of $\rho = 2 \times 10^{17}$ kg m^{-3}, the neutrino mean free path or diffusion length would be of the order of $\lambda = 1/\sigma N_A \rho \sim 900/E^2$ m, which for a typical value of $E = 20$ MeV, means $\lambda \sim 2m$ only. For other processes, for example neutral current scattering of muon- and tau-neutrinos, the mean free path would be several times longer.

In each scattering process (analogously to the case of photon diffusion through the Sun in Example 7.1), the neutrino will emerge in an arbitrary direction, so that after N successive scatters, this 'random walk' will carry the neutrino a root mean square straight line distance of $\lambda N^{1/2}$. Identifying this with the core radius R we obtain a diffusion time from the central region to the surface of $t \sim R^2/\lambda c$ of the order of 1 s. Since neutrinos are the only particles able to escape, the 100 MeV gravitational energy release per nucleon is divided between the six flavours of neutrino/antineutrino, and detailed computer simulations indeed indicate that neutrinos and antineutrinos of all flavours are emitted from the core in comparable numbers over a period of 0.1–10 s, with average energy $\sim$15 MeV and with an approximately Fermi–Dirac distribution as in (2.44). They are emitted from a so-called 'neutrinosphere' within a few metres of the surface. Neutrinos account for 99% of the total gravitational energy released in (7.39). The spectacular optical display of the supernova explosion accounts for only 1% of the total energy release.

7.9 Neutrinos from SN 1987A

Figure 7.6 shows a photograph of the supernova SN 1987A in the Large Magellanic Cloud, a mini-galaxy some 60 kpc from the Milky Way. It is famous because it was the first supernova from which interactions of the emitted neutrinos were observed, in fact simultaneously, in the Kamiokande and IMB water Cerenkov detectors (Fig. 7.7), originally designed to search for proton decay. The neutrino pulse was actually detected some seven hours before the optical signal became detectable.

The principal reactions that could lead to detection of supernova neutrinos in a water detector are as follows:

$$\bar{\nu}_e + p \rightarrow n + e^+ \tag{7.42a}$$

$$\nu + e^- \rightarrow \nu + e^- \tag{7.42b}$$

$$\bar{\nu} + e^- \rightarrow \bar{\nu} + e^- \tag{7.42c}$$

The secondary electrons or positrons from these reactions have relativistic velocities and part of their energy loss in traversing the water appears in the form of Cerenkov light (see Section 6.5), which is detected by an array of photomultipliers, as in Fig. 3.13.

The first reaction (7.42a) has a threshold of $Q = 1.8$ MeV and a cross section rising as the square of the neutrino energy, as in (1.27) and (7.41), with a value of 10^{-41} cm^2 per proton at $E_\nu = 10$ MeV. The angular distribution of the secondary lepton is almost isotropic. The second and third reactions go via both neutral and charged current channels for electron-neutrinos/antineutrinos, and via neutral currents only for muon- and tauon-neutrinos/antineutrinos. Although not negligible, the summed cross section for these reactions (which vary as E_ν) is only 10^{-43} cm^2 per electron at 10 MeV. So, although in water there are five electrons for every free proton, the event rate for scattering off electrons is an order of magnitude less than that for the first reaction. Moreover, in (7.42a) the secondary positron receives most of the energy ($E_e = E_\nu - 1.8$ MeV), while in (7.42b) and (7.42c) the charged lepton receives typically half the incident energy.

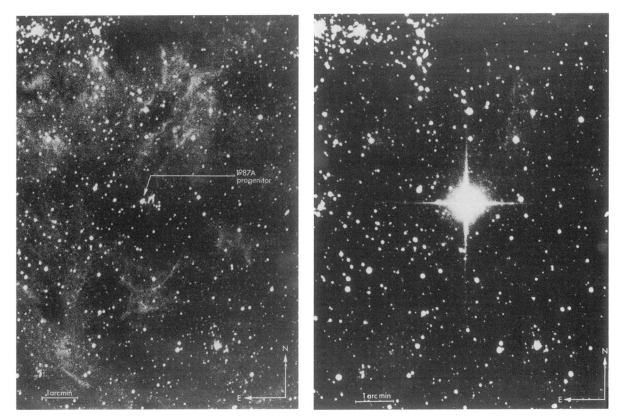

Fig. 7.6 The SN 1987A supernova. The stellar field in the Large Magellanic Cloud before (left) and two days after (right) the supernova explosion. Although such a supernova is for some time the brightest object in the local galaxy, the light emitted is only about 1% of the total energy released. The rest is accounted for by neutrinos. In this particular case, the progenitor star was a blue giant of about twenty solar masses. No neutron star (pulsar) has been detected as a remnant.

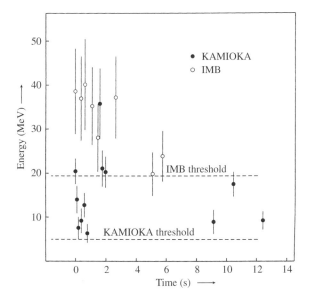

Fig. 7.7 The energies of the IMB and Kamiokande water Cerenkov events plotted against arrival time. The effective threshold energies for the two detectors were 20 MeV and 6 MeV, respectively.

The event rates recorded, together with the known distance to the supernova (60 kpc) could be used to compute the total energy flux in neutrinos and antineutrinos, assuming that the total is six times that for $\bar{\nu}_e$ alone. Both data sets, when account is taken of detection thresholds, are consistent with a mean temperature of $kT \sim 5$ MeV and, thus, a mean neutrino energy at production of $3.15kT$ appropriate to a relativistic Fermi–Dirac distribution. The integrated neutrino luminosity thus calculated from the event rates was

$$L \approx 3 \times 10^{46} \text{ J} \approx 2 \times 10^{59} \text{ MeV} \tag{7.43}$$

with an uncertainty of a factor of two, and thus in excellent agreement with the prediction (7.39). It perhaps needs to be emphasized that the neutrino flux from a supernova is indeed prodigious. Altogether some 10^{58} neutrinos were emitted from SN1987A and even at the Earth, some 170 000 light years distant, the flux was over 10^{10} neutrinos cm^{-2}.

The recording of neutrinos from SN1987A gave some information on neutrino properties. The fact that they survived a 170 000 year journey without attenuation testifies to their stability. Since the neutrino pulse lasted less than 10 s, the transit time of neutrinos of different energies was the same within 1 part in 5×10^{11}. The time of arrival on Earth, t_E will be given in terms of the emission time from the supernova, t_{SN}, its distance L and the neutrino mass m and energy E by

$$t_E = t_{SN} + \left(\frac{L}{c}\right)\left[1 + \frac{m^2 c^4}{2E^2}\right]$$

for $m^2 \ll E^2$. For two events with different energies E_1 and E_2 the time difference will be given by

$$\Delta t = |\Delta t_E - \Delta t_{SN}| = \left(\frac{Lm^2 c^4}{2c}\right)\left[\frac{1}{E_1^2} - \frac{1}{E_2^2}\right] \tag{7.44}$$

If we take as typical values $E_1 = 10$ MeV, $E_2 = 20$ MeV and $\Delta t < 10$ s, this equation gives $m < 20$ eV, a poorer limit than that from tritium beta decay (Table 1.3).

It is of interest to remark here that the neutrino burst is very probably instrumental in helping the shock wave to develop the spectacular optical display of a supernova. Early computer models suggested that the outward moving shock might stall as it met with the infalling matter from outside the core, and produce disintegration of this material into its constituent nucleons. However, when account was taken of the interactions of the neutrinos with the envelope material, the transfer of only 1% of the total neutrino energy was found to be enough to keep the shock wave moving.

So it seems possible that neutrinos of all flavours—ν_e, ν_μ, and ν_τ—interacting via both neutral and charged current processes, still play a vital part in cosmic events, while of course the corresponding charged μ- and τ-leptons disappeared by decay within microseconds of the Big Bang, and we are left with only the electrons. We may also remark here that supernovae perform a unique role in the production of very heavy elements, since they are sources of the very intense neutron fluxes which build up the nuclei in the later part of the Periodic Table, via rapid neutron capture chains. So it is worth bearing in mind that the iodine in your thyroid and the barium in your bones probably owe their

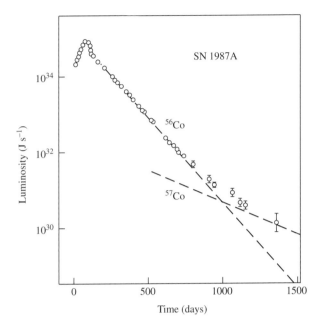

Fig. 7.8 The light curve of supernova SN 1987A. After the initial outburst, the luminosity fell rapidly over the first 100 days, being dominated by the beta decay of ^{56}Ni to ^{56}Co, with a mean lifetime $\tau = 9$ days. From time $t = 100$ to $t = 500$ days, the energy release was dominated by the beta decay of ^{56}Co to ^{56}Fe, with $\tau = 111$ days. Beyond $t = 1000$ days, the important decay is of ^{57}Co to ^{57}Fe ($\tau = 391$ days) as well as that of other long-lived isotopes. Most of the heavy nuclei would have been produced in rapid absorption reactions of neutrons with the material of the infalling envelope. Interestingly enough, no neutron star has been detected following this particular supernova (after Suntzeff *et al.* 1992).

existence to the fact that there are three flavours of neutrino and antineutrino, with both neutral and charged current couplings.

As stated above, only about 1% of the total supernova energy appears in the form of light output (although this is enough to dominate, for a time, the luminosity of the host galaxy). The light curve, at least for the first three years, is approximately exponential, being dominated by the decay of radioactive ^{56}Co, with a mean lifetime of 111 days (see Fig. 7.8).

7.10 Neutron stars and pulsars

The rump left behind after a supernova explosion is a neutron star, which contains neutrons, protons, electrons, and heavier nuclei, but with neutrons predominating. In the free state, a neutron undergoes decay with a mean lifetime of 887 ± 2 s, so we must consider the equilibrium in the reversible reaction

$$n \leftrightarrow p + e^- + \nu_e + 0.8 \text{ MeV} \qquad (7.45)$$

In the neutron star, the decay of the neutron will be prevented as a result of the Pauli principle, provided that all the quantum states that can be reached by the electron and the proton from the decay are already filled. To a very good approximation, if we neglect the Q-value in the decay, this condition is satisfied when the Fermi energies of the degenerate neutrons and electrons are equal, that is $E_F(n) \sim E_F(e)$, so that the forward and backward reactions are in equilibrium. We know from (7.39) that the neutrons and protons will be non-relativistic, while the electrons will be ultra-relativistic, so that $p_F(e) \ll p_F(n)$. Then, it is clear from (7.20) that the electron number density will be much smaller than that of the neutrons, as shown in the numerical example (Example 7.4).

Example 7.4 *Estimate the ratio of numbers of electrons (and protons) to neutrons necessary to prevent neutron decay in a neutron star of density* $\rho = 2 \times 10^{17} \text{ kg m}^{-3}$.

Assuming that practically all of the nucleons are neutrons, their number density will be $n_n = 1.2 \times 10^{44} \text{ m}^{-3}$, and their Fermi momentum from (7.20) will be

$$p_F(n)c = hc \left[\frac{3n_n}{8\pi} \right]^{1/3} = 300 \text{ MeV}$$

while their (non-relativistic) Fermi energy will be

$$E_F(n) = \frac{[p_F(n)c]^2}{2M_n c^2} = 48 \text{ MeV}$$

To prevent neutron decay, the Fermi momentum and energy of the (relativistic) electrons, therefore, need only be of the order of

$$p_F(e)c = E_F(e) = E_F(n) = 48 \text{ MeV}$$

and the electron number density, proportional to the cube of the Fermi momentum, will be $n_e = [48/300]^3 n_n \sim 0.004 n_n$. Obviously, $n_p = n_e$ by charge conservation. So a small proportion, less than 1%, of electrons and protons is sufficient to prevent neutron decay, and the equilibrium in (7.45) is very much to the left. [Note: Here, we have for simplicity neglected the small effect of the protons. If we include them, the equilibrium condition becomes $E_F(n) = E_F(e) + E_F(p)$. It is left as an exercise to solve the ensuing quadratic equation, and show that the effect of the protons is to reduce the above electron concentration by 7%.]

Although the early theory of neutron stars was developed shortly after Chadwick discovered the neutron in 1932, major experimental interest had to await the discovery of **pulsars** by Hewish *et al.* (1968). Pulsars are rapidly rotating neutron stars which emit radiation at short and extremely regular intervals, much like a rotating lighthouse beam which crosses the line of sight of an observer with a regular frequency. Over 1000 pulsars are known, with rotational periods ranging from 1.5 ms to 8.5 s. Only about 1% of pulsars can be associated with past supernova remnants, since over millions of years the neutron stars have drifted away from the remnant nebula. For a few young pulsars like that in the Crab, the nebula is still associated. This most famous example of a pulsar has a period of 33 ms, and is the remnant of the AD 1054 supernova recorded by the Chinese.

In addition to pulsars like that in the Crab which emit radio waves, some 200 **X-ray pulsars** are known. These are neutron stars which are members of binary star systems. Matter accretes from the massive companion star on to the magnetic poles of the neutron star, creating the X-ray emission ('aurora'). The X-rays are pulsed with the rotational frequency of the neutron star.

X-ray bursters are associated with neutron stars which have light main sequence companion stars. Hydrogen is accreted on to the very hot neutron star surface, and after some time, it reaches a density and temperature leading to ignition in a thermonuclear explosion lasting several seconds. The process is repeated as more material is accreted. γ**-ray bursters** have already been mentioned in Section 6.7. They are associated with the most violent events in the universe, releasing an estimated 10^{46} J in γ-rays, about the same amount of energy as in a supernova explosion. They are possibly produced as a result of neutron star binaries merging to form black holes.

The maximum angular frequency ω of a pulsar will be given by the requirement that the outward centrifugal force on the surface material should not exceed the inward gravitational attraction, that is,

$$\omega^2 R < \frac{GM}{R^2} \tag{7.46}$$

Inserting the values $R \sim 15$ km, $M \sim 1.5 M_\odot$ gives for the minimum period $\tau = 2\pi/\omega > 1$ ms, and indeed no neutron stars are observed with shorter periods than this. The very high frequencies observed for many pulsars result because much of the angular momentum of the original giant stars is retained through the white dwarf stage and the rotational frequency is enormously increased because of the dramatic contraction to the neutron star size.

The pulsar radiation is ascribed to the existence of a rotating magnetic dipole inclined at an angle θ to the axis of rotation. For a dipole of strength μ the electromagnetic power radiated is proportional to the square of the radial acceleration, that is, to ω^4:

$$P \propto \mu \omega^4 \sin^2 \theta \tag{7.47}$$

The magnetic field at the surface of a pulsar is of order 10^8 T, this high value resulting from the trapping and concentration of magnetic flux by the highly conducting plasma during stellar collapse, the field increasing inversely as the square of the radius. The energy lost through the emission of the radiation results in a small deceleration of the pulsar. If I is the moment of inertia of the pulsar, the rotational energy is $\frac{1}{2} I \omega^2$, so that the rate of change of rotational energy or power emitted is

$$P = I \omega \left(\frac{d\omega}{dt} \right) \propto \omega^4$$

so that

$$\frac{d\omega}{dt} = -A\omega^3$$

Observations on the Crab pulsar indicate that at the present time, $d\omega/dt = -2.4 \times 10^{-9}$ while $\omega = 190$ s^{-1}. If the initial angular velocity is ω_i, the time t for which it has been spinning to reach the present value of ω is

$$t = \frac{1}{2A} \left[\frac{1}{\omega^2} - \frac{1}{\omega_i^2} \right] < \frac{1}{2A\omega^2} = \frac{1}{2} \frac{\omega}{d\omega/dt} = 1255 \text{ yr}$$

in agreement with its origin in AD 1054.

7.11 Black holes

Neutrons play a similar role in supporting a neutron star as degenerate electrons do in supporting a white dwarf. The limit to which the degenerate neutron gas can do this is analogous to the Chandrasekhar limit for electron degeneracy in white dwarfs. If we forget about the strong nuclear interactions and general relativity effects in high gravitational fields, we can apply (7.28) and (7.29) with the substitution m_P for m_e and $A/Z = 1$, so that the limit (7.31) becomes

$$M_{max} = \left(\frac{3\sqrt{2}}{8\pi}\right)\left(\frac{hc}{G}\right)^{3/2} m_p^{-2} \sim 5M_\odot \qquad (7.48)$$

For neutron stars with masses $M > M_{max}$, the degenerate neutron gas becomes relativistic and gravitational collapse is inevitable. However, strong interactions between the neutrons will tend to make the neutron star matter more incompressible and increase the maximum mass. On the other hand, the gravitational binding energy of a neutron star is comparable with its mass energy (see (7.39)), so that non-linear gravitational effects associated with the mass energy of the field itself should be included and this will tend to decrease the maximum mass of the neutron star. So (7.48) should only be taken as an indication that the critical mass of neutron stars is a few solar masses.

The fate of a neutron star that undergoes gravitational collapse is a **black hole**. The crucial property associated with a black hole is the Schwarzschild radius for an object of mass M, given by the formula

$$R_{Schw} = \frac{2GM}{c^2} \qquad (7.49)$$

This result was derived by Schwarzschild, who obtained an exact solution to Einstein's field equations of general relativity, for the specific case of a gravitational field due to a large static mass M. By chance it happens that it can also be found using special relativity and the equivalence principle (see Appendix B), or by equating the radial escape velocity from a point mass M to the velocity of light.[1] As an example of (7.49), the Schwarzschild radius of a star of mass $M = 5M_\odot$ is $R_{Schw} = 15$ km. Equation (7.49) implies that when the physical radius of a collapsed star falls inside the Schwarzschild radius, there are no light paths (geodesics) to the outside world. Photons from the star cannot escape its gravitational field and it becomes black to an outside observer.

To understand how this can be, let us apply the special theory of relativity by comparing a time interval dt on a stationary clock in a remote inertial frame with that, dt', on an identical clock stationary in an inertial frame that has velocity v relative to the first frame, and is instantaneously co-moving with the surface of the collapsing star. Then $dt'^2 = dt^2(1 - v^2/c^2)$. Thus, as $v \to c$, $dt' \to 0$ and to an observer in the remote frame, the star appears 'frozen' in time. The frequency ω of light from a star collapsing inside its Schwarzschild radius will be Doppler shifted to $\omega' = \omega[(1 - v/c)/(1 + v/c)]^{1/2}$—see equations (2.1) or (B.15)—so that the quantum energy $\hbar\omega \to 0$ as $v \to c$ and the energy emitted from the star tends to zero. These phenomena are what would be recorded by an external observer. An observer **within** the Schwarzschild radius would, however, record lots of activity, but would not be able to communicate with the outside world.

Black holes are inevitable consequences of Einstein's general theory of relativity—even if Einstein could not bring himself to believe in their existence. From the general theory it follows that everything possessing energy and momentum, including photons, will be deflected by a gravitational field and will be 'turned around' if the field is strong enough.

[1] Imagine a particle approaching a point mass M along a radius vector r with velocity $v(r)$. The acceleration at r is $dv/dt = v\,dv/dr = GM/r^2$. Integrating, assuming $v = 0$ at $r = \infty$, the velocity at r is given by $v^2 = 2GM/r$.

The experimental evidence for the existence of black holes is quite convincing. It rests, for example, on observation of binary systems in which the motion of the visible star implies the existence of a compact invisible companion with mass $M > M_{max}$. Such systems are observed as compact X-ray sources, the X-rays being produced as mass from the visible star flows into the black hole. Candidates are the X-ray source Cygnus X-1, with $M = 3.4M_{\odot}$ and V404 Cygni, which includes a compact object of $M > 6M_{\odot}$. However, the most dramatic and compelling evidence is that for extremely massive (10^6–10^9 solar mass) black holes at the centres of essentially all galaxies. Our own Milky Way, for example, contains a black hole of about 4×10^6 solar masses, as deduced from the observed Keplerian orbits of one or two stars circulating it with enormous velocities (see Problem 7.8).

It is generally considered that some of the most violent events in the universe, such as γ-ray bursts (see Section 6.7) from the so-called active galactic nuclei, may be associated with massive black holes ($M \sim 10^8 M_{\odot}$) at the centres of galaxies. The infall of matter—stars, gas, and dust—towards the black hole could generate electromagnetic radiation and also be responsible for accelerating the highest energy cosmic rays. Only about 1% of present galaxies are 'active' (AGNs). In most galaxies, the mass infall of surrounding material into the black hole has been essentially completed and one is left with the 'ash' of the originally very active galactic nucleus, namely a relatively quiescent but massive black hole.

7.12 Hawking radiation from black holes

One effect of quantum fluctuations in very strong gravitational fields is that black holes are able to emit (thermal) radiation, as proved by Hawking in 1974 (Hawking 1974). The Hawking temperature for a black hole of mass M is given by

$$kT_{H} = \frac{\hbar c^3}{8\pi GM} \tag{7.50}$$

For example, for $M = 5M_{\odot}$, $T_{H} \sim 8 \times 10^{-8}$ K. Note that as the black hole loses energy and mass, it gets hotter, and thus a black hole will eventually evaporate and disappear. The lifetime can be calculated from the rate of energy loss from the surface:

$$\frac{d(Mc^2)}{dt} = 4\pi R_{Schw}^2 \sigma T_{H}^4$$

where $\sigma = \pi^2 k^4 / (60\hbar^3 c^2)$ is the Stefan constant. Substituting from (7.49) and (7.50) and integrating, one obtains for the lifetime

$$\tau_{BH} = \text{const.} \times \frac{G^2 M^3}{\hbar c^4} \sim 10^{66} \left(\frac{M}{M_{\odot}}\right)^3 \text{ yr} \tag{7.51}$$

Thus, the time for a black hole of a typical astronomical mass to evaporate is far longer than the age of the universe.

The origin of the Hawking radiation and the form of (7.50) can be made plausible by the following simple argument. Suppose that, as a result of a quantum fluctuation at a radial distance r just outside the Schwarzschild radius of a black

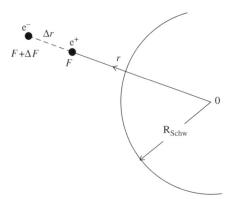

Fig. 7.9 Creation of an electron–positron pair just outside the Schwarzschild radius of a black hole.

hole of mass M, a virtual e^+e^- pair is temporarily created (see Fig. 7.9). If the pair has total energy E it can, according to the Uncertainty Principle, exist for a time $\Delta t \sim \hbar/E$. In this time the two particles can separate by a maximum radial distance $\Delta r \sim c\Delta t \sim c\hbar/E$. The difference of the gravitational field strengths at the positions of the two particles is then $(2GM/r^3)\Delta r$ and the difference ΔF of the gravitational forces upon them is this quantity multiplied by the effective mass, E/c^2, so that the tidal force is $\Delta F \sim (GM/r^3)\hbar/c$. For the gravitational field to be able to create the pair, one requires $\Delta F\,\Delta r > E$, that is $E < \hbar(GM/r^3)^{1/2}$. The largest value of E will be for the minimum value of r, namely $r \sim R_{\text{Schw}}$, giving the order-of-magnitude condition

$$E \sim \frac{\hbar c^3}{GM}$$

agreeing with (7.50) up to numerical constants. In the presence of this strong gravitational field, the pair can be separated fast enough, such that one of them gets well outside the Schwarzschild radius and escapes as a real particle, while the other is sucked back into the black hole.

7.13 Summary

- Stars form from protostars consisting of vast clouds of hydrogen in gravitational collapse, which contract until the core temperature reaches $kT \sim 1$ keV, when thermonuclear fusion of hydrogen to helium commences and the star attains hydrostatic equilibrium. This hydrogen fusion on the main sequence of the Hertzsprung–Russell diagram continues typically for billions of years.

- When the hydrogen fuel is exhausted, fusion of helium to carbon and oxygen takes place, at a higher core temperature and on a much shorter timescale. The star becomes a red giant with a bloated envelope.

- If the stellar mass is of the order of a solar mass or less, the consumption of helium marks the end of the fusion cycle and of nuclear energy release, and the star cools off slowly as a white dwarf with a degenerate electron core.

- In more massive stars, the core temperature becomes high enough for fusion to continue, with production of heavier elements up to nickel and iron.

- If the mass of the iron core exceeds the Chandrasekhar mass of $1.4M_\odot$, it is inherently unstable and suffers catastrophic collapse with the formation of a very compact neutron star of nuclear density. About 10% of the mass energy of the star is emitted in the form of a burst of neutrinos, and as a shock wave which gives rise to the optical display of a Type II supernova.
- If the core mass is much larger, around 4–$6M_\odot$, the neutron star is unstable and undergoes further collapse to a black hole. Binary systems with a black hole as one partner are compact and very intense sources of X-rays, emitted as matter flows into the black hole from the companion.
- Black holes can decay by emission of Hawking radiation, which is a manifestation of quantum fluctuations near the Schwarzschild radius of a black hole. Intense sources of X-rays from the centres of many galaxies are attributed to emission from very hot gas flowing into massive ($10^8 M_\odot$) black holes at the galactic centre. Such black holes may be identified with active galactic nuclei, associated with intense γ-ray bursts as described in Chapter 6.

Problems

More challenging problems are marked with an asterisk.

(7.1) Estimate the maximum rotational frequency of a white dwarf star, assuming that it has a mass equal to a solar mass and radius of 1% of the solar radius.

(7.2) Calculate the luminosity (in W) of the Sun, given that its surface temperature is 5780 K and its radius is 7×10^8 m. Solar energy is provided by fusion of helium from hydrogen. If 5% of the hydrogen in the Sun has so far been converted to helium, estimate the age of the Sun, assuming the luminosity to have been constant.

*(7.3) Find the maximum mass for a body containing normal atomic matter (density 10^4 kg m^{-3}) which does not require electron degeneracy pressure in order to maintain its stability against gravitational collapse. What object might such a body represent?

(7.4) The slow-down of the Crab Pulsar is assumed to be due to emission of dipole radiation as a result of its rotating magnetic field as indicated in (7.47). However, one could also ask if the slow-down could be due to gravitational quadrupole radiation, varying with rotational frequency as ω^6 as in (6.37). Show that the observed values of $\omega = 190$ s^{-1} and $d\omega/dt = -2.4 \times 10^{-9}$ would then be inconsistent with its known age.

*(7.5) Calculate the mass of a black hole with a Schwarzschild radius equal to the particle horizon distance for a universe of age t_0. If the universe had a density equal to the critical density, at what value of t_0 would the mass of the universe be equal to that of the black hole?

(7.6) Estimate the radius and mass of a black hole with a lifetime equal to that of the universe.

(7.7) Show that, in the reaction $\bar{\nu}_e + p \to e^+ + n$, the cross section is of the order of $G_F^2 p_f^2$ where the CMS momentum in the final state $p_f \approx E_\nu - Q$. Here, E_ν is the antineutrino energy, assumed to be small compared with the nucleon rest-mass but large compared with the electron rest-mass, which can be neglected, and Q is the threshold energy for the reaction.

(7.8) A star has been observed (in the infra-red) in orbit about a massive unseen object (black hole) at the centre of our galaxy, identified with the compact radio and X-ray source Sagittarius A (see *Physics Today*, Feb. 2003 for references). The elliptic orbit has a period of 15 years and eccentricity of 0.87, and the closest distance of approach (perigee) is estimated to be 17 light-hours. If necessary referring to a text on celestial mechanics, calculate the mass of the black hole and the orbital velocity of the star at perigee.

A Table of physical constants

The following table is taken from the 'Physical Constants' Table of the Particle Data Group, published in the *European Physical Journal*, **C15**, 1 (2000). Constants in the table below are quoted only to three figures of decimals.

Symbol	Name	Value
c	velocity of light (in vacuum)	$2.998 \times 10^8 \text{ m s}^{-1}$
$\hbar$	Planck's constant/2π	$1.055 \times 10^{-34} \text{ J s} = 6.582 \times 10^{-22} \text{ MeV s}$
$\hbar c$		$0.197 \text{ GeV fm} = 3.16 \times 10^{-26} \text{ J m}$
e	electron charge	$1.602 \times 10^{-19} \text{ C}$
m_e	electron mass	$0.511 \text{ MeV}/c^2 = 9.109 \times 10^{-31} \text{ kg}$
m_p	proton mass	$0.938 \text{ GeV}/c^2 = 1.672 \times 10^{-27} \text{ kg} = 1836 m_e$
m_n	neutron mass	$0.939 \text{ GeV}/c^2$
$m_n - m_p$	neutron–proton mass difference	$1.293 \text{ MeV}/c^2$
μ_0	permeability of free space	$4\pi \times 10^{-7} \text{ N A}^{-2}$
$\varepsilon_0 = 1/\mu_0 c^2$	permittivity of free space	$8.854 \times 10^{-12} \text{ F m}^{-1}$
$\alpha = e^2/4\pi\varepsilon_0\hbar c$	fine structure constant	$1/137.036$
$r_e = e^2/4\pi\varepsilon_0 m_e c^2$	classical electron radius	$2.818 \times 10^{-15} \text{ fm}$
$a_\infty = 4\pi\varepsilon_0\hbar^2/m_e e^2$	$= r_e/\alpha^2 =$ Bohr radius	$0.529 \times 10^{-10} \text{ m}$
$\lambda_c = \hbar/m_e c = r_e/\alpha$	= reduced Compton wavelength of electron	$3.861 \times 10^{-13} \text{ m}$
$\sigma_T = 8\pi r_e^2/3$	Thomson cross section	$0.665 \times 10^{-28} \text{ m}^2 = 0.665 \text{ b}$
$\mu_B = e\hbar/2m_e$	Bohr magneton	$5.788 \times 10^{-11} \text{ MeV T}^{-1}$
$\mu_N = e\hbar/2m_p$	nuclear magneton	$3.152 \times 10^{-14} \text{ MeV T}^{-1}$
$\omega/B = e/m_e$	cyclotron frequency of electron	$1.759 \times 10^{11} \text{ rad s}^{-1} \text{ T}^{-1}$
N_A	Avogadro's number	$6.022 \times 10^{23} \text{ mol}^{-1}$
k	Boltzmann's constant	$1.381 \times 10^{-23} \text{ J K}^{-1} = 8.617 \times 10^{-11} \text{ MeV K}^{-1}$
$\sigma = \pi^2 k^4/60\hbar^3 c^2$	Stefan's constant	$5.670 \times 10^{-8} \text{ W m}^{-2} \text{ K}^{-4}$
$G_F/(\hbar c)^3$	Fermi coupling constant	$1.166 \times 10^{-5} \text{ GeV}^{-2}$
$\sin^2\theta_W$	weak mixing parameter	0.2312
M_W	W-boson mass	$80.42 \text{ GeV}/c^2$
M_Z	Z-boson mass	$91.19 \text{ GeV}/c^2$
G	gravitational constant	$6.673 \times 10^{-11} \text{ m}^3 \text{ kg}^{-1} \text{ s}^{-2}$
au	astronomical unit = mean Earth–Sun distance	$1.496 \times 10^{11} \text{ m}$
$M_{PL} = (\hbar c/G)^{1/2}$	Planck mass	$1.221 \times 10^{19} \text{ GeV}/c^2 = 2.177 \times 10^{-8} \text{ kg}$

Symbol	Name	Value
pc	parsec	3.086×10^{16} m $= 3.262$ ly
$M_\odot$	solar mass	1.989×10^{30} kg
$R_\odot$	solar radius (equator)	6.961×10^{8} m
$L_\odot$	solar luminosity	3.85×10^{26} W
$M_\oplus$	Earth mass	5.975×10^{24} kg
$R_\oplus$	Earth radius (equator)	6.378×10^{6} m
H_0	Hubble expansion rate	70 (± 10) km s^{-1} Mpc^{-1}
T_0	CMBR temperature	2.725 ± 0.001 K
t_0	age of Universe	14 ± 2 Gyr

Conversion factors

1 eV $= 1.602 \times 10^{-19}$ J; 1 eV$/c^2 = 1.782 \times 10^{-36}$ kg

kT at 300 K $= 1/38.681 = 0.02585$ eV

1 erg $= 10^{-7}$ J; 1 dyne $= 10^{-5}$ N; 1 cal $= 4.18$ J; $0\,°$C $= 273.15$ K

1 atmosphere $= 760$ Torr $= 101\,325$ Pa $= 1013$ gm cm^{-2}

1 barn $= 10^{-28}$ m^2; $\pi = 3.141592$; $e = 2.718281828$

pc $= 0.3B\rho =$ momentum in GeV of singly-charged particle with radius of curvature ρ in metres in a magnetic field of B tesla.

B.1 Coordinate transformations

The special theory of relativity, proposed by Einstein in 1905, involves transformations between inertial frames of reference. An inertial frame is one in which Newton's law of inertia holds: a body in such a frame, not acted on by any external force continues in its state of rest or of uniform motion in a straight line. Although an inertial frame is, strictly speaking, an idealized concept, a reference frame far removed from any fields or gravitating masses approximates to such a frame, as does a lift in free fall on Earth. On the scale of experiments in high energy physics at accelerators, gravitational effects are negligibly small and the laboratory can, to all intents and purposes, be treated as an inertial frame. However, on the scale of the cosmos, gravity is the most important of the fundamental interactions.

We summarize here the coordinate transformations between inertial frames in special relativity. These are obtained from two assumptions: that the coordinate transformations should be linear (to agree with the Galilean transformations in the non-relativistic limit); and that the velocity of light c in vacuum should be the same in all inertial frames (as observed in numerous experiments). The relation between the coordinates x', y', z', t' of an event in an inertial frame Σ' moving with velocity v along the x-axis with respect to an inertial frame Σ, where the coordinates of the event are x, y, z, t is then as follows:

$$x' = \gamma(x - vt) \quad y' = y \quad z' = z \quad t' = \gamma\left(t - \frac{vx}{c^2}\right) \tag{B.1}$$

where $\gamma = (1 - v^2/c^2)^{-1/2}$ is the so-called Lorentz factor. When $v \to 0, \gamma \to 1$, $x' \to (x - vt)$, and $t' \to t$ as in the Galilean transformation, and the transformation also makes the velocity of light invariant: $x'^2 + y'^2 + z'^2 - c^2t'^2 = x^2 + y^2 + z^2 - c^2t^2 = 0$. The invariant interval or line element is made up of the squares of the coordinate differences for two events and is defined by

$$\begin{aligned} ds^2 &= c^2\,dt'^2 - dx'^2 - dy'^2 - dz'^2 \\ &= c^2\,dt^2 - dx^2 - dy^2 - dz^2 \\ &= c^2\,d\tau^2 \end{aligned} \tag{B.2}$$

It has the same value in all inertial frames, as is easily demonstrated by substitution in (B.1). The interval is invariant not only under Lorentz transformations between inertial frames, but also under translations or rotations of the coordinate axes. In the third line above, τ refers to the 'proper time' that is the time on a clock fixed in the inertial frame (for which $dx = dy = dz = 0$). The interval ds is referred to as **timelike, null,** or **spacelike** according as ds^2 is positive, zero, or negative. Note that if the coordinate increments refer to the passage of a light ray, $ds^2 = 0$, using Pythagoras's theorem and the fact that, in inertial frames, light travels in straight lines. In the case of non-inertial frames, that is, reference frames accelerating with respect to inertial frames, light does not travel in straight lines, and space is non-Euclidean or 'curved'.

If we write the coordinates in (B.1) in a different notation, as $ct = x_0, x = x_1$, $y = x_2, z = x_3$, then, (B.2) can be set in the form of the space–time **metric**

$$ds^2 = \sum g_{\mu\nu}\, dx_\mu\, dx_\nu \tag{B.3}$$

where the summation is over $\mu, \nu = 0, 1, 2, 3$ and $g_{\mu\nu}$ is a 4×4 matrix called the **metric tensor.** For coordinate frames, in general, including those accelerating with respect to inertial frames, $g_{\mu\nu}$ will be a function of the space–time coordinates x_μ. However, for inertial frames only, it has a simple form with constant diagonal elements and all off-diagonal elements equal to zero:

$$g_{00} = 1, \quad g_{11} = g_{22} = g_{33} = -1; \quad g_{\mu\nu} = 0 \quad \text{for } \mu \neq \nu \tag{B.4}$$

or, set in matrix form,

$$g_{\mu\nu} = \begin{pmatrix} 1 & 0 & 0 & 0 \\ 0 & -1 & 0 & 0 \\ 0 & 0 & -1 & 0 \\ 0 & 0 & 0 & -1 \end{pmatrix}$$

in conformity with (B.2).

In the above we have used rectangular coordinates, but we can also set the interval in terms of spherical coordinates r, θ, ϕ where r is the radial coordinate, θ is the polar angle, and ϕ is the azimuthal angle about the z-axis, when we can write

$$ds^2 = c^2\, dt^2 - dr^2 - r^2(d\theta^2 + \sin^2 \theta\, d\phi^2) \tag{B.5}$$

That is, $dx_0 = c\, dt$, $dx_1 = dr$, $dx_2 = r\, d\theta$, and $dx_3 = r \sin\theta\, d\phi$.

The **general theory of relativity** proposed by Einstein in 1915 is concerned with providing an invariant description of physical phenomena in all conceivable reference frames, including those in accelerated motion with respect to inertial frames, the acceleration being provided by gravitational fields. The theory is summarized in Einstein's field equations, which have been solved for several situations (although no general solution has ever been found). A very important solution is the one obtained by Schwarzschild in 1916, for the metric in the neighbourhood of a spherically symmetric distribution of total mass M, far removed from other gravitating masses. In this case the elements of the metric tensor are no longer constant as in (B.4), but functions of the coordinates. The Schwarzschild line element has the form

$$ds^2 = \left[1 - \frac{2GM}{rc^2}\right] c^2\, dt^2 - \left[1 - \frac{2GM}{rc^2}\right]^{-1} dr^2 - r^2(d\theta^2 + \sin^2 \theta\, d\phi^2) \tag{B.6}$$

where the coordinates t, r, θ, and ϕ are those measured by a remote observer in a distant inertial frame. This formula can be used to calculate the deflection of light rays by the gravitational field of the mass M, and provided an early verification of Einstein's general theory of relativity, by measurement of the deflection of a light ray passing close to the Sun—the 'Einstein star shift'—in the famous 1919 eclipse expedition.

In contrast with (B.5), the coefficients in (B.6) of both dt and dr depend on M and r (however, since there is spherical symmetry, there can be no such dependence for the angular coordinates). The metric tensor is again a diagonal matrix with all off-diagonal elements (i.e. with $\mu \neq \nu$) equal to zero.

An important consequence of (B.6) applies for fixed spatial coordinates ($dr = d\theta = d\phi = 0$), when the transformation between time intervals becomes

$$ds^2 = c^2\, d\tau^2 = c^2\, dt^2 \left[1 - \frac{2GM}{rc^2} \right] \tag{B.7}$$

That is, the proper time interval $d\tau$ measured on a clock at rest in the gravitational field of the mass M is reduced in comparison with the value dt on an identical clock in the remote inertial frame. A clock in a gravitational field runs slow! When the clock is placed at the radial coordinate $r = 2GM/c^2, d\tau = 0$, that is, the time on the local clock appears frozen. The quantity $r_{Schw} = 2GM/c^2$ is called the **Schwarzschild radius** of the mass M. The formula (B.7) can also be inferred using heuristic arguments from special relativity and the equivalence principle, as shown in the next section.

B.2   The equivalence principle: clocks in gravitational fields

Suppose, an observer, initially at rest in an inertial frame Σ, is given a **small** acceleration a in the x direction. The space coordinates he records, for an event with coordinates x, y, z, t in Σ will be

$$x' = x - \tfrac{1}{2}at^2, \quad y' = y, \quad z' = z$$

With $x = x' + \tfrac{1}{2}at^2$ and $dx = (\partial x/\partial x')\, dx' + (\partial x/\partial t)\, dt = dx' + at\, dt$, the invariant interval (B.2) is

$$ds^2 = c^2\, dt^2 - dx^2 - dy^2 - dz^2$$
$$= (c^2 - a^2t^2)\, dt^2 - 2at\, dx'\, dt - dx'^2 - dy'^2 - dz'^2 \tag{B.8}$$

The second line refers to spatial coordinates measured in the accelerated frame Σ'. The time dt' elapsed on a clock fixed in this accelerating frame of reference, that is, for which $dx' = dy' = dz' = 0$, will be given by $ds^2 = c^2\, dt'^2$ and hence

$$dt'^2 = \left(1 - \frac{a^2t^2}{c^2} \right) dt^2 \tag{B.9}$$

The instantaneous velocity of the accelerating clock measured in Σ is $v = at$ so that the interval dt' of proper time measured on this clock, as compared with the value dt measured on an identical clock at rest in Σ is also given by

$$dt'^2 = \left(1 - \frac{v^2}{c^2}\right) dt^2 \qquad \text{(B.10)}$$

which is the usual formula for time dilation following from the Lorentz transformation (B.1). The interval dt' is here the same as would be measured on an identical clock at rest in an **inertial frame** Σ'' which is instantaneously co-moving with the accelerated clock and has velocity v with respect to the frame Σ. The distance which the accelerated clock has moved after time t is $H = \frac{1}{2}at^2$, so that $a^2t^2 = 2aH$ and (B.9) can be written as

$$dt'^2 = \left(1 - \frac{2aH}{c^2}\right) dt^2 \qquad \text{(B.11)}$$

The **principle of equivalence** states that a frame Σ' accelerating with respect to an inertial frame Σ is exactly equivalent to a system at rest in Σ but subject to a homogeneous gravitational field. This comes about because of the equivalence of gravitational and inertial mass. The inertial mass of a body is defined as the ratio of the force F_I applied to the acceleration g produced, that is,

$$F_I = M_I g$$

The gravitational mass is defined by the force on the body in a gravitational field, due, for example, to a point mass M at distance r:

$$F_G = M_G \left(\frac{GM}{r^2}\right)$$

If F_I is the gravitational force F_G it follows that the 'gravitational acceleration' is

$$g = \left(\frac{M_G}{M_I}\right) \frac{GM}{r^2} \qquad \text{(B.12)}$$

so that g will be the same for all bodies provided they have the same ratio of gravitational to inertial mass (and if they have, an obvious convention is to set the ratio equal to unity). The principle of equivalence has been checked experimentally to a precision of better than 1 part in 10^{12}, using very precise torsion balance experiments.

In (B.11) we can therefore replace the **accelerated, moving** clock by an identical, **stationary** clock in a gravitational field providing a gravitational acceleration a, so that

$$dt'^2 = \left(1 + \frac{2\Delta\Phi}{c^2}\right) dt^2 \qquad \text{(B.13)}$$

where dt' is the time interval on the clock in the field, dt is that on an identical clock in an inertial frame remote from any gravitational field, and $\Delta\Phi = -aH$ is the difference in the gravitational potential. For the remote clock, $\Phi = 0$, while for that in the field, $\Phi < 0$. Thus, a clock at low (negative) gravitational potential, such as one at sea-level, should run slower than an identical clock at a higher (less negative) potential on a mountain top. This predicted gravitational shift was verified experimentally by Pound and Snider in 1965. In their experiment, the very small (10^{-15}) increase in frequency of ^{57}Fe γ-rays falling down a vertical 22 m tube was measured by means of an ^{57}Fe absorber at the bottom, utilizing the Mossbauer effect. The photons from the emitter at the higher potential are 'blue-shifted' compared with the absorption frequency at the lower potential, and this was compensated using the Doppler effect, by slowly moving the absorber downwards at the appropriate velocity $v/c = \Delta v/v \sim 10^{-15}$. Since that time, atomic clocks have been carried on aircraft to directly verify the above formula by comparing with similar clocks at ground level.

Relative to our remote clock at $\Phi = 0$, a clock in the field of a point mass M at distance r is at a potential $\Phi = -GM/r$ and

$$dt'^2 = \left[1 - \frac{2GM}{rc^2}\right] dt^2 \tag{B.14}$$

The analysis here has assumed **small** values of acceleration, that is $\Delta\Phi \ll c^2$. It happens by chance, however, that (B.14) is correct even for strong fields, and gives the same result as (B.7), which follows from the full analysis using the general theory of relativity.

B.3   Relativistic kinematics

The transformations of energy and momentum of a particle between inertial frames in special relativity are easily found from the coordinate transformations (B.1) by replacing x, y, z by the Cartesian components p_x, p_y, p_z of the three-momentum $\mathbf{p}$ of the particle, and the time component t by the total energy E (using units $c = 1$ for brevity). Then, the transformations between an inertial frame Σ and another frame Σ' moving at velocity $\beta = v/c$ in the x-direction with respect to Σ are

$$\begin{aligned}
p'_x &= \gamma(p_x - \beta E) \\
p'_y &= p_y \\
p'_z &= p_z \\
E' &= \gamma(E - \beta p_x)
\end{aligned} \tag{B.15}$$

where $\gamma = (1 - \beta^2)^{-1/2}$, and the quantity

$$E'^2 - p'^2 = E^2 - p^2 = m^2 \tag{B.16}$$

where m, the rest-mass of the particle, is invariant under the transformation.

We already saw in (B.2) that the expression for the square of the interval ds was an invariant under a transformation to another inertial frame. The quantity ds is called a **four-vector**, since it has four space and time components, and the invariance under Lorentz transformations is analogous to the invariance in three space dimensions, of the length squared of a three-component vector under translations or rotations of the coordinate axes. The general scalar product of two four-vectors A_μ and B_ν is

$$\sum g_{\mu\nu}A_\mu B_\nu = A_0 B_0 - \mathbf{A} \cdot \mathbf{B}$$

with $g_{\mu\nu}$ from (B.4).

So, in kinematics, we express the quantities in (B.15) as components of four-vectors called **four-momenta** p_μ ($\mu = 0, 1, 2, 3$), where $p_0 = E$, $p_1 = p_x$, $p_2 = p_y$, $p_3 = p_z$. The four-momentum squared is a Lorentz scalar with the value

$$p^2 = \sum g_{\mu\nu} p_\mu p_\nu = m^2 \tag{B.17}$$

In scattering experiments in high energy physics, the result of the scattering of a particle by the interaction can be expressed in terms of the invariant **four-momentum transfer** $q_\mu = p_\mu - p'_\mu$, so that $q^2 = (E - E')^2 - (\mathbf{p} - \mathbf{p'})^2$, where the unprimed and primed quantities refer to the values for a particle before and after the interaction. It is left as an exercise to show that q^2 is always negative in a scattering process.

In kinematic problems, it is advantageous to evaluate quantities in the centre-of-momentum system (CMS), that is, in a coordinate frame where the total three-momentum $\mathbf{p}$ of the colliding particles is zero. Then, the invariant four-momentum squared (which can be evaluated in any inertial frame) is just equal to the square of the total CMS energy, conventionally denoted by the symbol $s = E_{\text{CMS}}^2$ (but not to be confused with the space-time interval, also called s).

B.4 Fixed target and colliding beam accelerators

As an example of the use of four-vectors and the centre-of-momentum frame of reference, we consider fixed target and colliding-beam accelerators. Suppose a beam of particles of mass m_a, energy E_a, and momentum $\mathbf{p_a}$ collides with a target particle of mass m_b, energy E_b, and momentum $\mathbf{p_b}$. Then the total four-momentum squared is given by

$$s = (E_a + E_b)^2 - (\mathbf{p_a} + \mathbf{p_b})^2 = m_a^2 + m_b^2 + 2(E_a E_b - \mathbf{p_a} \cdot \mathbf{p_b}) \tag{B.18}$$

The energy available for new particle creation is $\varepsilon = \sqrt{s} - m_b - m_a$. If $E_a \gg m_a$ and $E_b \gg m_b$, then,

$$\varepsilon^2 \approx 2(E_a E_b - \mathbf{p_a} \cdot \mathbf{p_b}) \tag{B.19}$$

(a) Fixed target
If the beam of particles a collides with a stationary target b, so that $E_b = m_b$ and $\mathbf{p_b} = 0$, then,

$$\varepsilon \approx (2m_b E_a)^{1/2} \tag{B.20}$$

and the available energy rises with the square root of the incident energy. Examples of accelerators which have used fixed targets are the CERN PS (Proton Synchrotron) accelerating protons to 28 GeV, and the CERN SPS (Super Proton Synchrotron) for protons up to 400 GeV.

(b) Colliding beams
If the beams of particles a and b collide head-on, then $\mathbf{p_a} \cdot \mathbf{p_b} = -|\mathbf{p_a}||\mathbf{p_b}|$ and assuming both beams are extreme relativistic, we obtain

$$\varepsilon^2 \approx [2(E_a E_b + |\mathbf{p_a}||\mathbf{p_b}|)] \approx 4E_a E_b$$

In many colliders, the two beams are of particles of equal masses and equal energies, that is, $E_a = E_b = E$, in which case

$$\varepsilon \approx 2E \tag{B.21}$$

so that the available energy rises in proportion to the beam energy. An example is the LEP II e^+e^- collider at CERN, accelerating electrons and positrons in opposite directions in the same vacuum tube. The beam energies were $E = 100$ GeV, so that the 200 GeV available CMS energy was sufficient to investigate the reactions $e^+e^- \rightarrow W^+W^-$ and Z^0Z^0, where the threshold energies are $2M_W \sim 160$ GeV and $2M_Z \sim 180$ GeV. The HERA machine is an example of an asymmetric ep collider, accelerating electrons or positrons to 28 GeV energy in one vacuum ring, and protons to 820 GeV in the other direction in a second ring above the first, the two beams being brought into collision in two intersection regions. In this case, the square of the CMS energy is $s = 93\,000$ GeV^2, and the useful maximum value of four-momentum transfer squared between the particles is $|q^2(\mathrm{max})| \sim 20\,000$ GeV^2, about a hundred times the maximum useful value attainable using secondary muon or neutrino beams from fixed target machines.

Colliding-beam machines have the obvious advantage of providing much higher CMS energies for a given beam energy, and have been essential in identifying the more massive fundamental particles—the $W^\pm$ and Z^0 mediators of the electroweak interactions and the bottom and top quarks during the 1980s and 1990s. They are, however, limited to beams of stable or nearly stable particles, namely electrons, positrons, protons, antiprotons, heavy ions, and, possibly in the future, muons (in $\mu^+\mu^-$ colliders).

Fixed target machines achieve lower CMS energies but have the advantage that they can produce a range of intense high energy beams of secondary particles, for example, $\pi, K^\pm, K^0, \mu, \nu_\mu$. These were important, historically, in laying the quantitative experimental foundations of particle physics, including the establishment of the quark substructure of matter and of CP violation in weak interactions in the 1960s, and of quantum chromodynamics of the strong interquark interactions in the 1970s. They still have very important applications today (2003) both as injectors for colliders and as sources of neutrino beams for the study of neutrino oscillations.

The Robertson–Walker line element: distances and angles in cosmology

C.1 The line element

For the FLRW model of an isotropic and homogeneous expanding universe, the line element (B.5) appropriate to inertial frames in special relativity is modified in general relativity as follows:

$$ds^2 = c^2\,dt^2 - R^2(t)\left(\frac{dr^2}{1 - Kr^2} + r^2\,d\theta^2 + r^2 \sin^2\theta\,d\phi^2 \right) \qquad (C.1)$$

where $R(t)$ is the universal expansion parameter, r the radial space coordinate in a reference frame co-moving with the expansion, and the parameter K describes the curvature of space. $K > 0$ corresponds to positive curvature, $K < 0$ to negative curvature, and $K = 0$ to the flat Euclidean space of special relativity.

Equation (C.1) follows from the solution of the field equations of general relativity, and we have simply stated the result. However, the form of the curvature term can be understood from a two-dimensional analogy. The accompanying figure (Fig. C.1) shows a section through a sphere of radius ρ, centre O, A and B being two points on the surface of the sphere. The shortest distance between A and B along the surface is the arc AB of a great circle, of length l. Denoting the angle subtended at the centre by $\alpha = l/\rho$ and the chord AB through the sphere by $r = 2\rho \sin(\alpha/2)$, and defining the curvature parameter K by $\rho = 1/(2\sqrt{K})$, it follows that

$$l = \frac{1}{\sqrt{K}} \sin^{-1}(r\sqrt{K})$$

and

$$dl = \frac{dr}{\cos(l\sqrt{K})} = \frac{dr}{\sqrt{1 - Kr^2}} \qquad (C.2)$$

expressing the element of arc length dl along the surface of the sphere in terms of the curvature K and the corresponding element of arc length dr for the limit $K \to 0$. Note however that this two-dimensional analogy applies for $K > 0$ only and that there is no two-dimensional analogue for $K < 0$.

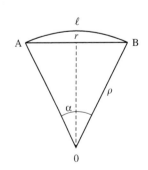

Fig. C.1

C.2   Distances in cosmology

Consider a galaxy G, of fixed diameter d, fixed angular coordinates θ, ϕ, and redshift z. To reach a telescope on the Earth at time t_0, light pulses must leave G at time t, say. Those leaving opposite ends P and Q of the diameter will subtend an angle at the Earth of $\Delta\theta$, which can be found by putting $d\phi = dt = dr = 0$ and $d\theta = \Delta\theta$ in (C.1), when we obtain (see Fig. C.2)

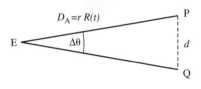

Fig. C.2

$$ds^2 = -[R(t)r\Delta\theta]^2$$

where r is the co-moving and time-independent space coordinate of G relative to the Earth, as in (2.5). However, $ds^2 = -d^2$ since d is the spacelike separation of P and Q in the G rest-frame. (Note that d, being a galactic quantity, is not subject to the universal expansion.) Hence,

$$\Delta\theta = \frac{d}{rR(t)} = \frac{d}{D_A} \tag{C.3}$$

where the quantity

$$D_A = rR(t)$$

is called the **angular distance** of G. The above relation simply states that the light pulses leave G at time t, when its instantaneous distance from Earth is $rR(t)$. Of course, this is not the distance of G when those pulses arrive at the Earth, at the later time t_0. That distance is larger because since the pulses left G, the universe has expanded by a factor $R(0)/R(t) = (1 + z)$. We can denote this distance by

$$D_H = rR(0) = \frac{D_A R(0)}{R(t)} = D_A(1 + z) \tag{C.4}$$

As the redshift $z \to \infty$, this quantity tends to the horizon distance discussed in Section 5.1.

A common method of estimating the distance to a pointlike source is from its apparent brightness or luminosity L, assuming that the intrinsic luminosity or power output P from the source is known. The distance to the star when the light arrives on Earth is $R(0)r$ as above. The amount of energy falling on the Earth per unit area and per unit time, that is, the flux or apparent luminosity is thus given by

$$L_{app} = \frac{P}{4\pi(R(0)r)^2} \times \frac{1}{(1+z)^2} = \frac{P}{4\pi D_L^2} \tag{C.5}$$

Here, one factor of $1/(1 + z)$ arises because pulses of light emitted from the source over a time interval Δt arrive at the detector over a dilated time interval $\Delta t(1+z)$, and the second $1/(1+z)$ factor arises because the energy per photon at emission has been red-shifted downwards. The quantity D_L defined in (2.3) and in (C.5) is called the **luminosity distance**, where, from (C.4),

$$D_L = D_A(1 + z)^2 = D_H(1 + z) \tag{C.6}$$

The distances defined in (C.4)–(C.6) are in terms of r and $R(0)$, neither of which can be observed directly. We must express these in terms of measurable quantities, namely the redshift z of the galaxy G and the Hubble parameter H_0.

First of all, we note that light travels along geodesic paths (i.e. paths of least time) given by $ds^2 = 0$. For a light ray travelling to us from galaxy G located at fixed angular coordinates θ, ϕ we therefore obtain the following differential relation from equation (C.1):

$$c(1+z)\,dt = R(0)\frac{dr}{\sqrt{1-Kr^2}}$$

Next, we express the time t in terms of the redshift z by the relation (2.32)

$$H = \frac{1}{R}\frac{dR}{dt} = -\left(\frac{1}{1+z}\right)\frac{dz}{dt} \tag{C.7}$$

Hence, we arrive at the equation

$$R(0)\int_0^r \frac{dr}{\sqrt{1-Kr^2}} = \int c(1+z)\,dt = c\int_0^z \frac{dz}{H} = \frac{c}{H_0}I(z) \tag{C.8}$$

where from (2.31) and (2.32) the integral $I(z)$ is defined as

$$I(z) = \int \frac{dz}{[\Omega_m(0)(1+z)^3 + \Omega_r(0)(1+z)^4 + \Omega_v(0) + \Omega_k(0)(1+z)^2]^{1/2}} \tag{C.9}$$

and the closure parameters Ω_m, Ω_r, Ω_v, and Ω_k refer to matter, radiation, vacuum, and curvature terms, and we recall from (2.24) that $\Omega_k(0) = -Kc^2/[R(0)H_0]^2$ and that $\Omega_m + \Omega_r + \Omega_v + \Omega_k = 1$.

The integration of the left-hand side term in (C.8) is straightforward and gives the relation (cf. (C.2) for the case $K > 0$)

$$\frac{c}{H_0}I(z) = R(0)\frac{\sin^{-1}(rK^{1/2})}{K^{1/2}} \qquad K > 0 \quad \text{Closed} \tag{C.10a}$$

$$= R(0)\frac{\sinh^{-1}(r|K|^{1/2})}{|K|^{1/2}} \qquad K < 0 \quad \text{Open} \tag{C.10b}$$

$$= R(0)r \qquad K = 0 \quad \text{Flat} \tag{C.10c}$$

Thus, the expression for the distance D_H at redshift z is

$$D_H(z) = R(0)r = \frac{R(0)}{|K|^{1/2}}S(|\Omega_k|^{1/2}I(z)) \tag{C.11}$$

where

$$S(x) = \sin x \qquad \text{for } K > 0$$
$$= \sinh x \qquad \text{for } K < 0$$
$$= x \qquad \text{for } K = 0$$

and for $K = 0$, $D_H(z) = R(0)r = (c/H_0)I(z)$.

(C.11) is the required relation expressing r in terms of z and K. We can use it to calculate the distance to the 'last scattering surface' of the cosmic microwave radiation when it decoupled from matter at $z \sim z_{dec} \sim 1100$.

First, we consider the case of an **open, matter-dominated universe with zero cosmological constant** (i.e. $\Omega_v = 0$). Since we are dealing with values of

$z < 1100$ we can also simplify things by neglecting the radiation term, so that with $\Omega \equiv \Omega_m$ and $\Omega_k = (1 - \Omega)$ we get from (C.9)

$$I(z) = \int \frac{dz}{[\Omega(1+z)^3 + (1-\Omega)(1+z)^2]^{1/2}} = \int \frac{dz}{(1+z)(1+\Omega z)^{1/2}} \tag{C.12}$$

with the limits $z = 0$ and $z = z$. This integration can be performed, for example, by making the substitution $(1 + \Omega z) = (1 - \Omega)\sec^2\theta$, when it has the value

$$I(z) = (1 - \Omega)^{-1/2} \ln \left(\frac{1 + \cos\theta}{1 - \cos\theta} \right) = \frac{1}{q} \ln \left[\frac{(p+q)(1-q)}{(p-q)(1+q)} \right]$$

where $p^2 = (1 + \Omega z)$, $q^2 = (1 - \Omega) = \Omega_k$. With $\sinh X = (e^X - e^{-X})/2$ and $|K|^{1/2} = R(0)H_0\Omega_k^{1/2}/c$, equation (C.12) gives the following expression for the distance as a function of z:

$$\begin{aligned}
D_H(z) = R(0)r &= \left[\frac{c}{H_0(1-\Omega)^{1/2}} \right] \sinh \left\{ \ln \left[\frac{(p+q)(1-q)}{(p-q)(1+q)} \right] \right\} \\
&= \frac{2c}{H_0} \left[\frac{\Omega z - (2-\Omega)((1+\Omega z)^{1/2} - 1)}{\Omega^2(1+z)} \right]
\end{aligned} \tag{C.13}$$

known as the **Mattig formula**, valid, we repeat, for an open, matter-dominated universe with zero vacuum energy and the radiation contribution neglected.

The value of the angular distance D_A is obtained by dividing D_H by the $(1+z)$ factor in (C.4). It has the interesting property that at small values of z it is proportional to z, as might be expected, but that at large z, $z \gg 1$, it varies as $1/z$. Consequently, the angle subtended by objects of fixed diameter, after first decreasing with increasing distance, actually increases at large redshifts (see Fig. 5.9 for the open universe case). If, instead, we multiply (C.13) by $(1+z)$ we get the formula for the luminosity distance D_L.

For a very distant object at high redshift (but $z < 1000$ so as to justify our neglect of the radiation term in the above), equation (C.13) gives the simple approximate result

$$D_H \approx \frac{2c}{H_0\Omega} \tag{C.14}$$

In fact, for $\Omega = 1$ this is just the expression for the horizon distance in a flat, matter-dominated universe (see equation (5.2) and Example 5.1). For an open universe, the effect is therefore just to divide the distance by a factor Ω.

The second important case is that of a **flat universe**, with the dominant contributions from **matter and from vacuum energy**, that is $\Omega_m = \Omega$ and $\Omega_v = 1-\Omega$, and $\Omega_k = \Omega_r = 0$. In this case the integral (C.9) has the form

$$I(z) = \int \frac{dz}{[\Omega(1+z)^3 + (1-\Omega)]^{1/2}} \tag{C.15}$$

No analytical solution is possible and the integration has to be performed numerically. However, for values of $z > 4$ and $\Omega > 0.05$ one can neglect the vacuum term in comparison with the matter term and perform that part of the integral analytically, so that only a few minutes work with a pocket calculator are needed to plot D versus Ω for a fixed z. We discuss the results in the next section.

C.3 Angle subtended at the earth by the horizon at the time of last scattering of the cosmic microwave radiation

The distance to the horizon at the time t_{dec}, when the microwave radiation decoupled from matter, that is, at $z_{\text{dec}} \sim 1100$, can also be found using the above type of analysis. This is relevant to the small-angle anisotropies in the cosmic microwave background discussed in Section 5.9. Also, at about that time, the energy density of matter and radiation would have been comparable, which was the reason for neglecting the radiation energy density in the above analysis for $z < 1000$. In contrast, for values of $z > 1000$, the radiation term becomes dominant and in a first approximation all the other terms in the integrand of (C.9) can be dropped. Furthermore, since the curvature term is negligible at large z values, distances are given by the flat universe formula (C.10c). Hence, integrating from $z = z_{\text{dec}}$ to $z = \infty$, the horizon distance is

$$L_{\text{H}} = \frac{c}{H_0} \int \frac{dz}{\Omega_{\text{r}}^{1/2}(1+z)^2} = \frac{c}{H_0(1 + z_{\text{dec}})\Omega_{\text{r}}^{1/2}} \qquad \text{(C.16)}$$

Note that this agrees with the value of $L_{\text{H}} = 2ct$ given in Example 5.1, taking into account that, for radiation dominance, the time elapsed between $z = \infty$ and $z = z_{\text{dec}}$, as given in (2.33) is $t = 1/[2H_0(1 + z_{\text{dec}})\Omega_{\text{r}}^{1/2}]$. Since at the present time $\Omega_{\text{r}}(0) \sim \Omega_{\text{m}}(0)/(1 + z_{\text{dec}})$, the result (C.16) can also be written as

$$L_{\text{H}} = \frac{c}{H_0(1 + z_{\text{dec}})^{1/2}\Omega^{1/2}} \qquad \text{(C.17)}$$

where $\Omega = \Omega_{\text{m}}(0)$. Dividing by D_{H} in (C.14), the angle subtended by the acoustic horizon at $z = z_{\text{dec}}$ is therefore found by including a factor 2 for the total horizon–horizon diameter and an approximate value of $1/\sqrt{3}$ for the ratio of sound to light velocity in a situation where radiation is prominent:

$$\frac{\theta}{\sqrt{\Omega}} \sim [3(1 + z_{\text{dec}})]^{-1/2} \approx 1° \qquad \text{(C.18)}$$

This result applies for the case of an **open, matter-dominated** universe with no vacuum energy term. It is based on several simplifying assumptions and the absolute value of the angle is therefore only approximate. However, the main result stands, namely that the angle subtended varies as $\Omega^{1/2}$.

For the **flat universe case with vacuum energy** (C.15), the result of numerical integration in Fig. 5.7 shows that the angle θ in this case depends little on the way that the total closure parameter $\Omega_{\text{tot}} = 1$ is divided between the matter and vacuum energy terms, Ω_{m} and $\Omega_{\text{v}} = 1 - \Omega_{\text{m}}$, so that the relative values of the quantities Ω_{m} and Ω_{v} are not well constrained by the microwave data on its own.

As indicated in Chapter 5, the era of matter–radiation decoupling specified by z_{dec} is very important, as after that era, matter became dominant, electrons and protons combined to form stable atoms, and the velocity of sound fell dramatically, allowing large-scale structures to begin forming.

The variation with Ω of the angle θ submitted by the horizon at $z = z_{\text{dec}}$, and calculated from the above formulae is shown in Fig. 5.10. Since the angle

(C.18) varies as $\Omega^{1/2}$, the precision on the closure parameter is just twice the percentage error in the measurement of the first acoustic peak at $l \sim 200$.

C.4   Hubble plot for an expanding or contracting universe

In the Hubble diagram, the apparent stellar magnitude, or equivalently the logarithm of the luminosity distance defined in (2.3), (C.5), and (C.6), is plotted against $\log z$. The value of D_L as a function of z depends on the present value H_0 of the Hubble parameter and the cosmological density parameters Ω_m, Ω_v, Ω_k (with $\Omega_m + \Omega_v + \Omega_k = 1$; for our present purposes, discussing z values of order unity or less, the radiation energy density is negligible). From (C.4) we know that

$$D_L(z) = (1+z)D_H(z)$$

Hence, the dependence of D_L on H_0 and the cosmological parameters is simply found by multiplying D_H, as computed from (C.11), by the $(1+z)$ factor. For empty, vacuum-dominated and matter-dominated universes, it is left as an exercise to demonstrate the dependence on z as given in Table 4.1 and displayed in Fig. 4.11. For $\Omega_m = 0.35$, $\Omega_v = 0.65$, a numerical integration as in (C.15) is required.

Yukawa theory and the boson propagator

The propagator term involved in the exchange of virtual bosons in the interactions between elementary particles arises in the theory of quantum exchange first proposed by Yukawa in 1935. Yukawa was seeking to describe the short-range nature of the potential between neutrons and protons in the atomic nucleus. He started with the relativistic relation between total energy E, three-momentum p, and rest-mass m as in (1.1):

$$E^2 = p^2 c^2 + m^2 c^4 \qquad \text{(D.1)}$$

We now substitute the coordinate operators $E_{\mathrm{op}} = -i\hbar \partial/\partial t$ and $p_{\mathrm{op}} = -i\hbar\nabla$, which will yield the expectation values of energy and momentum when applied to the wavefunction of a particle, so that the above equation then becomes (dividing through by $-\hbar^2 c^2$)

$$\frac{1}{c^2}\frac{\partial^2 \psi}{\partial t^2} = \nabla^2 \psi - \frac{m^2 c^2}{\hbar^2}\psi \qquad \text{(D.2)}$$

called the **Klein–Gordon wave equation** describing the propagation of a free, spinless particle of mass m. If we insert $m = 0$, (D.2) becomes the familiar wave equation describing the propagation of an electromagnetic wave with velocity c, with ψ interpreted either as the wave amplitude of the associated photons, or as the electromagnetic potential $U(\mathbf{r})$. For a static, radially symmetric potential, we drop the time-dependent term so that (D.2) assumes the form

$$\nabla^2 U(r) \equiv \left(\frac{1}{r^2}\right)\frac{\partial}{\partial r}\left(r^2 \frac{\partial U}{\partial r}\right) = \frac{m^2 c^2}{\hbar^2} U(r) \qquad \text{(D.3)}$$

As can be verified by substitution, integration of this expression yields the solution

$$U(r) = \frac{g_0}{4\pi r}\exp\left(-\frac{r}{R}\right) \qquad\qquad R = \frac{\hbar}{mc} \qquad \text{(D.4)}$$

In this expression, g_0 is a constant of integration. In the electromagnetic case, $m = 0$ and the static potential is $U(r) = Q/4\pi r$ where Q is the electric charge at the origin. Hence, Yukawa interpreted g_0 as the 'strong nuclear charge'. Inserting for R the known range of nuclear forces of about 1.4 fm, one obtains $mc^2 = \hbar c/R \sim 150$ MeV. The pion, first observed in cosmic rays in 1947 was a particle of zero spin and just this mass. However, the interpretation of nuclear forces in terms of heavy quantum exchange turns out to be much more complicated than Yukawa had envisaged seventy years ago—for example it involves spin-dependent potentials. Nor is the pion a fundamental boson but just the lightest quark–antiquark combination. Nevertheless, Yukawa's theory pointed to a fundamental relation between the range of the interaction (D.4) between two elementary particles and the mass of the associated exchange quantum, which is just as valid today as it was years ago.

Let us consider a particle of incident momentum $\mathbf{p}_i$ being scattered with momentum $\mathbf{p}_f$ by the potential $U(\mathbf{r})$ provided by a massive source, in which case no energy is transferred and the numerical value of the momentum p of the incident and scattered particles are the same. The particle will be deflected through some angle θ and receive a momentum transfer $\mathbf{q} = \mathbf{p}_i - \mathbf{p}_f (= 2p \sin[\theta/2])$. The amplitude $f(\mathbf{q})$ for scattering will be the Fourier transform of the potential $U(\mathbf{r})$, in exactly the same way that the angular distribution of light diffracted by an obstacle in classical optics is the Fourier transform of the spatial extent of the obstacle. If g represents the coupling of the particle to the potential, we can write

$$f(\mathbf{q}) = g \int U(\mathbf{r}) \exp(i\mathbf{q} \cdot \mathbf{r}) \, dV \qquad (D.5)$$

Assuming a central potential $U(\mathbf{r}) = U(r)$ and with $\mathbf{q} \cdot \mathbf{r} = qr \cos\theta$ and $dV = r^2 \, dr \, d\phi \sin\theta \, d\theta$ where θ and ϕ are polar and azimuthal angles, and introducing the Yukawa potential (D.4) we obtain

$$f(\mathbf{q}) = 2\pi g \iint U(r) \exp(iqr \cos\theta) \, d(\cos\theta) r^2 \, dr$$

$$= g g_0 \int \exp\left(-\frac{r}{R}\right) \left\{ \frac{\exp(iqr) - \exp(-iqr)}{iqr} \right\} r^2 \, dr$$

$$= \frac{g g_0}{\mathbf{q}^2 + 1/R^2} = \frac{g g_0}{\mathbf{q}^2 + m^2 c^4/\hbar^2} \qquad (D.6)$$

This result is for a massive potential source, where three-momentum but no energy has been exchanged. For an actual scattering process between two particles, the relativistically invariant four-momentum transfer squared will be $q^2 = \Delta E^2 - \Delta \mathbf{p}^2 = \Delta E^2 - \mathbf{q}^2$. So for $\mathbf{q}^2$ in (D.6), holding for $\Delta E = 0$, we should substitute $-q^2$, so that the scattering amplitude becomes, in units $\hbar = c = 1$

$$f(q^2) = \frac{g g_0}{m^2 - q^2} \qquad (D.7)$$

Thus, the scattering amplitude denoted by $|T_{fi}|$ in (1.18) consists of the product of the couplings of the two particles to the exchanged virtual boson, multiplied by the propagator term, which depends on the four-momentum transferred (where q^2 is always negative) and on the mass of the free boson. All the above expressions are for spinless particles and additional factors are required when spin is introduced.

Perturbative growth of structure in the early universe

We start with the FLRW model described in Chapter 2, which assumes a completely isotropic and homogeneous distribution of matter and radiation undergoing the Hubble expansion. We assume we are dealing, at least initially, with tiny perturbations and therefore weak gravitational fields. Further, the distances involved, although enormous, are assumed to be small compared with the horizon distance ct so that the Hubble flow is non-relativistic. We begin by using classical fluid dynamics. There are three basic equations which read as follows:

$$\frac{\partial \rho}{\partial t} + \nabla \cdot (\rho \mathbf{u}) = 0 \tag{E.1}$$

$$\frac{\partial \mathbf{u}}{\partial t} + (\mathbf{u} \cdot \nabla)\mathbf{u} = -\left[\left(\frac{1}{\rho} \right) \nabla P + \nabla \Phi \right] \tag{E.2}$$

$$\nabla^2 \Phi = 4\pi G \rho \tag{E.3}$$

In these equations, ρ is the fluid density, $\mathbf{u}$ the velocity of fluid flow, P the pressure, and Φ is the gravitational potential. The first equation (E.1) is the **equation of continuity** expressing the fact that the rate of decrease of fluid density with time is just equal to the divergence of the fluid flow, that is, the mass of fluid flowing out of the volume in unit time. Equation (E.2) is **Euler's equation**. It states that the force ∇P on a volume element is equal to the rate of change of momentum of that element, that is, $\rho \, d\mathbf{u}/dt = \nabla P + \rho \nabla \Phi$ if the gravitational pressure is included. The total rate of change of velocity, applying to a particular element of the fluid, is made up of two parts: the partial derivative $\partial \mathbf{u}/\partial t$, measuring the change in fluid velocity at a particular spatial coordinate, plus a term due to the fact that the liquid element is in motion and in time dt has travelled a distance $d\mathbf{r} = \mathbf{u} \, dt$. Thus,

$$d\mathbf{u} = dt \left(\frac{\partial \mathbf{u}}{\partial t} \right) + \left[d\mathbf{x} \left(\frac{\partial \mathbf{u}}{\partial x} \right) + d\mathbf{y} \left(\frac{\partial \mathbf{u}}{\partial y} \right) + d\mathbf{z} \left(\frac{\partial \mathbf{u}}{\partial z} \right) \right]$$

$$= dt \left(\frac{\partial \mathbf{u}}{\partial t} \right) + (d\mathbf{r} \cdot \nabla)\mathbf{u} \tag{E.4}$$

and the result (E.2) follows upon dividing by dt. The third equation (E.3) is **Poisson's equation** for the gravitational potential in terms of G and the density.

In the absence of any perturbations in density, the above equations have the following solutions:

$$\rho(t) = \frac{\rho_0}{R(t)^3}$$

$$\mathbf{u}(t, \mathbf{r}) = \frac{\dot{R}(t)}{R(t)}\mathbf{r} \tag{E.5}$$

$$\Phi(t, r) = 2\pi\frac{G\rho r^2}{3}$$

The first expresses the dependence of density on the expansion parameter $R(t)$. In equation (E.1), assuming that we are dealing with a homogeneous universe, we have $\nabla\rho = 0$, while $\rho\nabla \cdot \mathbf{u} = \rho(\dot{R}/R)\nabla \cdot \mathbf{r} = 3\rho(\dot{R}/R)$: the result follows (see footnote at left bottom). The second is the equation for Hubble flow $\mathbf{u}(t, \mathbf{r}) = H(t)\mathbf{r}$ in (2.2), and the third follows from integration of (E.3).

Now we suppose that perturbations in the values of $\mathbf{u}$ and ρ occur. It turns out to be easier to discuss the developments in a coordinate frame co-moving with the Hubble expansion. In the following, $\mathbf{r}$ denotes a position coordinate measured by a 'stationary' observer (i.e. one not moving with the Hubble flow), and $\mathbf{x}$ that in the co-moving frame. Then, $\mathbf{x} = \mathbf{r}/R(t)$. The velocity of a fluid particle defined above as $\mathbf{u}$ in the stationary frame is then

$$\mathbf{u} = \frac{d\mathbf{r}}{dt} = \mathbf{x}\dot{R} + \mathbf{v} \tag{E.6}$$

The first term on the right measures the velocity arising from the Hubble flow, and the extra term $\mathbf{v}$ (where $\mathbf{v} \ll \mathbf{u}$) is the so-called 'peculiar velocity' of the particle relative to the general expansion. In the absence of a perturbation, this would of course be zero. The perturbation in density ρ is denoted by $\Delta\rho \ll \rho$ and the fractional change, called the 'density contrast', is denoted by $\delta = \Delta\rho/\rho$. A gradient in the stationary system is denoted by ∇_s to distinguish it from that in the co-moving frame, called ∇_c, where

$$\nabla_c = R\nabla_s \tag{E.7}$$

Finally, time derivatives of any function, say F, in the two systems will be related by

$$\left(\frac{\partial F}{\partial t}\right)_s = \left(\frac{\partial F}{\partial t}\right)_c - \dot{R}\mathbf{x} \cdot \frac{(\nabla_c F)}{R} \tag{E.8}$$

where the velocity of the stationary frame is $-\dot{R}\mathbf{x}$ with respect to the co-moving frame. With these definitions, the continuity equation (E.1) will read[1]

$$\left[\frac{\partial}{\partial t} - \left(\frac{\dot{R}}{R}\right)\mathbf{x} \cdot \nabla_c\right]\rho(1 + \delta) + \frac{\rho}{R}\nabla_c \cdot [(1 + \delta)(\dot{R}\mathbf{x} + \mathbf{v})] = 0 \tag{E.9}$$

In evaluating this expression, recall that $\nabla\rho = 0$ in a homogeneous universe. Further, if the pressure is small, that is, we are dealing with non-relativistic matter in our cosmic fluid, $\rho \propto 1/R^3$, so $\partial\rho/\partial t = -3\rho\dot{R}/R$. The quantity

---

[1] The following relations are useful in evaluating (E.3) and (E.5):

$$\nabla \cdot \mathbf{r} = \left(\mathbf{i}\frac{\partial}{\partial x} + \mathbf{j}\frac{\partial}{\partial y} + \mathbf{k}\frac{\partial}{\partial z}\right)$$
$$\times (\mathbf{i}x + \mathbf{j}y + \mathbf{k}z) = 3$$
$$(\mathbf{v} \cdot \nabla)\mathbf{x} = (\mathbf{i}v_x + \mathbf{j}v_y + \mathbf{k}v_z)$$
$$\times \left(\mathbf{i}\frac{\partial}{\partial x} + \mathbf{j}\frac{\partial}{\partial y} + \mathbf{k}\frac{\partial}{\partial z}\right)$$
$$\times [\mathbf{i}x + \mathbf{j}y + \mathbf{k}z]$$
$$= \left(v_x\frac{\partial}{\partial x} + v_y\frac{\partial}{\partial y} + v_z\frac{\partial}{\partial z}\right)$$
$$\times [\mathbf{i}x + \mathbf{j}y + \mathbf{k}z]$$
$$= (\mathbf{i}v_x + \mathbf{j}v_y + \mathbf{k}v_z) = \mathbf{v}$$

$\nabla_c \cdot \mathbf{x} = 3$, so that $(\rho/R)\nabla_c \cdot R\mathbf{x} = +3\rho\dot{R}/R$. Finally, second-order terms such as the product $v\delta$ can be neglected. The equation then reads

$$\left(\frac{\partial\delta}{\partial t}\right) + \frac{\nabla_c \cdot \mathbf{v}}{R} = 0 \tag{E.10}$$

The Euler equation (E.2) becomes

$$\left[\frac{\partial}{\partial t} - \left(\frac{\dot{R}}{R}\right)\mathbf{x} \cdot \nabla_c\right](R\mathbf{x} + \mathbf{v}) + (R\mathbf{x} + \mathbf{v}) \cdot \nabla_c \frac{(R\mathbf{x} + \mathbf{v})}{R}$$
$$= -\frac{[\nabla_c \Phi + (\partial P/\partial \rho)\nabla_c(1 + \delta)]}{R} \tag{E.11}$$

Subtracting the equation for the unperturbed system and again neglecting second-order perturbative terms such as $\mathbf{v} \cdot \nabla\mathbf{v}$ gives

$$\frac{\partial\mathbf{v}}{\partial t} + \mathbf{v}\frac{\dot{R}}{R} + \frac{\nabla_c\phi}{R} + v_s^2\frac{\nabla_c\delta}{R} = 0 \tag{E.12}$$

where ϕ (assumed to be $\ll \Phi$) is the gravitational potential due to the perturbation and $\partial P/\partial\rho = v_s^2$ determines the speed of sound in the fluid. The Poisson equation gives $\nabla^2\phi = 4\pi G\rho\delta$, so that subtracting the time derivative of (E.10) from the divergence of (E.12) gives

$$\frac{\partial^2\delta}{\partial t^2} + 2\frac{\dot{R}}{R}\left(\frac{\partial\delta}{\partial t}\right) - 4\pi G\delta\rho - \frac{v_s^2(\nabla^2\delta)}{R^2} = 0 \tag{E.13}$$

The final step is to express the spatial dependence of the pressure and density perturbation as a superposition of plane waves of wavenumbers k, that is, of the form

$$\delta(x, t) = \sum \delta_k(t)\exp(i\mathbf{k} \cdot \mathbf{x}) \tag{E.14}$$

so that for a particular wavenumber k, (E.13) becomes

$$\frac{d^2\delta}{dt^2} + 2\left(\frac{\dot{R}}{R}\right)\frac{d\delta}{dt} = \left(4\pi G\rho - \frac{k^2 v_s^2}{R^2}\right)\delta \tag{E.15}$$

The terms on the right vanish for a value of k corresponding to the Jeans length:

$$\lambda_J = \frac{2\pi R}{k} = v_s\left(\frac{\pi}{G\rho}\right)^{1/2} \tag{E.16}$$

First, we note that, if the expansion of the universe is neglected, that is $\dot{R}(t) = 0$, the solution of (E.15) is either periodic or exponential, according to the following two possibilities:

1. $\lambda \gg \lambda_J$: if the response time for the pressure wave is large compared with the gravitational infall time, the density contrast **grows exponentially**:

$$\delta \propto \exp\left(\frac{t}{\tau}\right) \quad \text{where } \tau = \left[\frac{1}{(4\pi G\rho)}\right]^{1/2} \tag{E.17}$$

2. $\lambda \ll \lambda_J$: in this case the solution to (E.15) is of the form

$$\delta \propto \exp(i\omega t) \quad \text{where } \omega = \frac{2\pi v_s}{\lambda} \tag{E.18}$$

so that the density contrast **oscillates as a sound wave**.

E.1   Growth in the matter-dominated era

In the early stages of the Big Bang, the universe is radiation dominated and in that case the velocity of sound is relativistic, with a value $v_s = c/\sqrt{3}$—see Table 2.2. This means that, using equation (2.36) with $\rho_r c^2 = (3c^2/32\pi G)/t^2$, the Jeans length is

$$\lambda_J = c \left[\frac{\pi}{3G\rho_r} \right]^{1/2} = ct \left(\frac{32\pi}{9} \right)^{1/2} \tag{E.19}$$

In this case the Jeans length and the horizon distance are both of the order of ct, where $t = 1/H$ is the Hubble time (i.e. the time since the start of the Big Bang). Thus, growth in this stage of the radiation era may appear less likely (and it is also true that our assumption of classical Newtonian mechanics in Euclidean space might not be valid on such large length scales).

After radiation and matter decouple, that is, at a temperature of $kT \sim 0.3$ eV (see equation (2.56)) when $t \approx 3 \times 10^5$ yr, the electrons and protons combine to form hydrogen atoms and the velocity of sound and hence **the Jeans length will decrease abruptly**, so that growth of inhomogeneities becomes possible. At the above decoupling temperature, $v_s \sim 5 \times 10^3$ m s^{-1} only and thus the Jeans length has decreased by over 10^4 times.

Let us take the simple case of a matter-dominated universe with $\rho = \rho_c$ and $\Omega = 1$, usually referred to as an Einstein–de Sitter universe. Then, from (2.12) and (2.23),

$$\rho = \frac{3H^2}{8\pi G} \quad \text{and} \quad H = \frac{\dot{R}}{R} = \frac{2}{3t}$$

so that (E.15) becomes, assuming $4\pi G\rho \gg k^2 v_s^2/R^2$,

$$\frac{d^2\delta}{dt^2} + \frac{4}{3t}\frac{d\delta}{dt} - \frac{2}{3t^2}\delta = 0 \tag{E.20}$$

which has a power law solution of the form

$$\delta = At^{2/3} + Bt^{-1} \tag{E.21}$$

where A and B are constants. The second term describes a contracting mode and is of no interest. The first term describes a mode in which the density contrast grows as a power law. Thus, the effect of taking into account the expansion of the universe is to change an exponential growth as in (E.17) to a power law dependence. We note from Table 2.2 that

$$\frac{\delta_0}{\delta_{dec}} = \left(\frac{t_0}{t_{dec}} \right)^{2/3} = \frac{R(0)}{R(dec)} = (1 + z_{dec}) \approx 1100 \tag{E.22}$$

where the brackets '0' and 'dec' refer to quantities today and at the time of decoupling of matter and radiation, that is, when electrons and protons started to combine to form hydrogen atoms. Since today $kT_0 = 0.23$ meV, and at the time of decoupling $kT_{dec} \approx 0.3$ eV, $(1 + z_{dec}) = 1100$, as in (2.57). The above equation is a result based on the assumption of small perturbations, and since we started out with a density contrast of order 10^{-5}, such a large extrapolation may be questionable. Nevertheless, this analysis shows that any small anisotropies at the time of decoupling of matter and radiation will increase in proportion to the scale parameter $R(t)$.

Answers to problems

Answers are given to all the problems. Fully worked solutions are given for those problems marked with an asterisk.

Chapter 1

(1.1) $(GM^2/R)/(2Mc^2)$: 2.5×10^{-14}; 8×10^{-21}.

(1.2) In terms of quarks the reaction is written as follows:

$$d\bar{u} + u\,du \rightarrow u\,ds + d\bar{s}$$

The strong interactions have a range of order $r_0 = 1$ fm, hence a typical cross section of $\sigma = \pi r_0^2 = 31$ mb($= 3.1 \times 10^{-26}$ cm^2). The characteristic time is $r_0/c = 3 \times 10^{-24}$ s. Thus, a value of $\sigma = 1$ mb corresponds to a strong-interaction time in this case of 10^{-23} s. Hence, the ratio of weak coupling to strong coupling is $(10^{-23}/10^{-10})^{1/2} \sim 10^{-6}$.

(1.3) 29.8 MeV: 10 GeV: 5.7 GeV.

(1.4) 5.5×10^{-24} s; 134 fm.

(1.5) (a) yes; (b) and (c) no, $\Delta S = 2$ forbidden to first order in coupling; (d) no, because of energy conservation.

(1.6) If all final state lepton masses are neglected, the rate is proportional to Q^5. In nuclear beta decay, this is known as the **Sargent Rule**. The decay rate W in (1.15) has dimensions E^{-1}. The formula includes a factor E^{-4} from the weak coupling G_F^2 as in (1.27). Hence, the other factors in the expression for W must have dimensions E^5, that is, vary as Q^5 as Q is the important energy in the problem. The values of W/Q^5 in MeV^{-5} s^{-1} are as follows: (a) 3.5×10^{-5} (b) 3.6×10^{-5} (c) 3.4×10^{-4} (d) 2.7×10^{-4} (e) 3.9×10^{-3}. The extreme relativistic approximation for the electron secondary does not hold for processes (c), (d), and (e) and W/Q^5 shows an increase with decreasing Q.

(1.7)
$$\frac{\tau_p}{\tau_\mu} \sim \left(\frac{M_X}{M_W}\right)^4 \times \left(\frac{m_\mu}{m_p}\right)^5 \cdot M_X \sim 2 \times 10^{14} \text{ GeV}$$

(1.8) We start by considering a massless neutrino of very high energy E, momentum p colliding with a nucleon of mass M at rest. The square of the energy in the CMS of the collision will be

$$s = (E + M)^2 - (p + 0)^2 = 2ME + M^2 \sim 2ME$$

If the quark carries a fraction x of the nucleon mass the result in the quark–neutrino CMS will be $s = 2xME$ and the cross section from

(1.27) will be

$$\sigma = \frac{G_F^2 s}{\pi} = \frac{2G_F^2 x M E}{\pi}$$

Inserting the values $G_F = 1.17 \times 10^{-5}$ GeV^{-2}, $M = 0.94$ GeV, 1 GeV$^{-1} = 0.1975 \times 10^{-13}$ cm (see Table 1.1) one obtains $\sigma = 3.2 \times 10^{-38} x E$ cm^2 where E is in GeV, or $\sigma = 0.8 \times 10^{-38} E$ cm^2 for $x = 0.25$. The actual high energy total neutrino cross section per nucleon is $\sigma = 0.74 E \times 10^{-38}$ cm^2.

(1.9) From (1.9) and (1.22) we can write for the differential cross section

$$\frac{d\sigma}{dq^2} = \frac{g_w^4}{[\pi(-q^2 + M_W^2)^2]}$$

where the four-momentum transfer squared has a maximum value $-q^2(\text{max}) = s$, the square of the CMS energy. Hence, the total cross section, integrating from $q^2(\text{min}) = 0$ to $q^2(\text{max})$ becomes

$$\sigma = \frac{g_w^4}{\pi} \int \frac{dq^2}{[-q^2 + M_W^2]^2}$$

$$= \frac{g_w^4 s}{\pi M_W^2 (s + M_W^2)} \quad \rightarrow \quad \frac{G_F^2 s}{\pi} \quad \text{for } s \ll M_W^2$$

$$\rightarrow \quad \frac{G_F^2 M_W^2}{\pi} \quad \text{for } s \gg M_W^2$$

Inserting the values of the constants, the asymptotic cross section equals 0.11 nb. The cross section reaches half the asymptotic value when $s = M_W^2$, that is $E = M_W^2/2m_e = 6.3 \times 10^6$ GeV.

(1.10) 6.3×10^6 GeV.

(1.11) 4×10^{-13} s.

(1.12) The three decays are identified with electromagnetic, weak, and strong interactions, respectively. If we set the strong coupling equal to unity, then, from the data in the table, that for the electromagnetic interactions will be of the order of 1.6×10^{-2}, and that for weak interactions, of the order of 5×10^{-7}, taking the square roots of decay rates as proportional to the couplings.

(1.13) The diagrams are as follows:

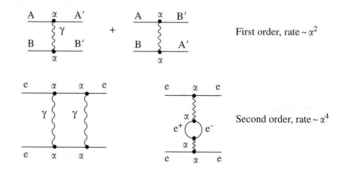

In the first-order process of electron–electron scattering via single photon exchange, there are two diagrams depending on how one labels the final-state particles as A or B. Since all that one observes is the scattered electron and not the vertices, both diagrams should be included.

The second-order diagrams contain factors α^2 in amplitude or α^4 in rate, compared with α^2 for the first-order process, so they are relatively suppressed by a factor $\alpha^2 \sim 10^{-4}$.

(1.14) (a) and (b) are weak processes, (c) is electromagnetic, and (d) is strong. Setting the strong coupling equal to unity, the weak and electromagnetic couplings are $\sim 10^{-8}$ and 10^{-3}, respectively.

(1.15) The ratio $R = 3\sum(Q_i/e)^2$ where the factor 3 is for the number of possible quark colours and the sum is over the charges Q_i of all relevant quark flavours. As a function of the CMS energy $\sqrt{s}$ the quark–antiquark flavours and values of R are as follows:

Quarks	$\sqrt{s}$ (GeV)	R	
$u\bar{u}$, $d\bar{d}$	>0.7	$3\left[\left(\frac{1}{3}\right)^2 + \left(\frac{2}{3}\right)^2\right]$	$= \frac{5}{3}$
$u\bar{u}$, $d\bar{d}$, $s\bar{s}$	>1.0	$3\left[\left(\frac{1}{3}\right)^2 + \left(\frac{2}{3}\right)^2 + \left(\frac{1}{3}\right)^2\right]$	$= \frac{6}{3}$
$u\bar{u}$, $d\bar{d}$, $s\bar{s}$, $c\bar{c}$	>3.5	$3\left[\left(\frac{1}{3}\right)^2 + \left(\frac{2}{3}\right)^2 + \left(\frac{1}{3}\right)^2 + \left(\frac{2}{3}\right)^2\right]$	$= \frac{10}{3}$
$u\bar{u}$, $d\bar{d}$, $s\bar{s}$, $c\bar{c}$, $b\bar{b}$	>10	$3\left[\left(\frac{1}{3}\right)^2 + \left(\frac{2}{3}\right)^2 + \left(\frac{1}{3}\right)^2 + \left(\frac{2}{3}\right)^2 + \left(\frac{1}{3}\right)^2\right] = \frac{11}{3}$	

The diagram for $e^+e^- \to \pi^+ + \pi^- + \pi^0$ is shown below. G represents a (strong) gluon exchange.

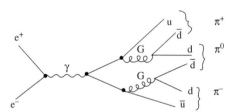

(1.16) The CMS energy of the pion in the decay $\Delta \to \pi + p$ is given by a little calculation in relativistic kinematics, as $E_\pi = (M_\Delta^2 + m_\pi^2 - M_p^2)/2M_\Delta = 0.267$ GeV (at the resonance peak). The corresponding pion momentum is $p_\pi = 0.228$ GeV/c. The CMS wavelength is then $\lambdabar_\pi = \hbar c/p_\pi c = 8.6 \times 10^{-14}$ cm. Inserting $J = \frac{3}{2}$, $s_\pi = 0$, $s_p = \frac{1}{2}$, $\Gamma_\gamma/\Gamma_{total} = 0.0055$, one finds $\sigma = 1.03$ mb. This is the cross section for the reaction $\gamma + p \to \Delta$ at the resonance peak. In the head-on collision of a proton of high energy E_p with a photon of energy E_γ, the CMS energy squared will be $s = M_p^2 + 4E_\gamma E_p = M_\Delta^2$ if the collision excites the Δ resonance. The microwave radiation at $T = 2.73$ K has a mean energy of $2.7\,kT$, and this corresponds to a quantum energy $E_\gamma = 6.3 \times 10^{-4}$ eV. Inserting this in the above expression, one obtains $E_p \sim 10^{21}$ eV. The mean free path of these protons through the microwave radiation will be $\lambda = 1/\rho\sigma$, where $\rho = 400$ cm^{-3} is the density of the microwave photons (see Chapter 2). Inserting the above value for the cross section, one obtains for the mean free path the value

$$\lambda = 2.5 \times 10^{22} \text{ m} \sim 0.8 \text{ Mpc}$$

(For further details, see Section 6.6.)

Chapter 2

(2.1) Binding energy $\sim 10^{69}$ J. Mass energy $\sim 10^{70}$ J.

(2.2) $v^2 > 8\pi G \rho r^2/3$. Inserting $v = Hr$, the limit on the density is just the critical density (2.23).

(2.4) $(1 + z) = 107$. $T = 12 \times 10^6$ yr, assuming matter domination for $z < 107$.

(2.5) 5%.

(2.6) 5750 K.

(2.7) $\varepsilon > 5 \times 10^{-19}$.

(2.8) For a radiation-dominated universe, $\rho = (3/32\pi G)/t^2$ from (2.36) while from (2.11) for matter domination $R = (6G\pi\rho R^3)^{1/3}t^{2/3}$. After integration this gives for the time elapsed to reach a density ρ

$$ t_{\text{rad}} = \left(\frac{3}{32\pi G\rho} \right)^{1/2} \qquad t_{\text{mat}} = \left(\frac{1}{6\pi G\rho} \right)^{1/2} $$

which can be compared with the free-fall time of collapse of a body of density ρ from rest, (see (5.26)):

$$ t_{\text{freefall}} = \left(\frac{3\pi}{32G\rho} \right)^{1/2} $$

(2.9) As indicated in the text (see also Problem 2.14) the freeze out of neutrons and protons from equilibrium occurs when the interaction rate W in (2.58), varying as T^5, falls below the expansion rate H in (2.47), varying as $T^2 g^{*1/2}$. Thus, the freeze-out temperature $T \propto g^{*1/6}$, where $g^* = (22 + 7 N_v)/4$ is the number of states of photons, electrons, positrons, and neutrinos/antineutrinos and N_v is the number of neutrino families (see Section (2.8)). For $N_v = 3$, $g^* = 43/4$, $kT = 0.8$ MeV, so that kT for other values of N_v is easily found. Inserting in (2.61), the initial and final neutron/proton ratios and hence the helium mass fraction can be calculated as a function of the assumed number of neutrino families.

For a neutron–proton mass difference of 1.4 MeV and three neutrino flavours, the initial neutron/proton density ratio in (2.62) becomes 0.174, leading to a helium mass fraction of 0.21.

(2.10) Let the light signal start off at $t = t_1$, to reach us at $t = t_0$. Consider the time interval dt' where $t_1 < t' < t_0$. In this time interval the light signal covers a distance $c\, dt'$, but by the time $t = t_0$, this will have expanded to $c\, dt' R(0)/R(t')$, where $R(t')$ is the expansion factor at $t = t'$. From (2.11) we know that in a matter-dominated universe $R(0)/R(t') = (t_0/t')^{2/3}$. Hence, the total distance travelled by the light signal will be

$$ L = R(0) \int \frac{c\, dt'}{R(t')} = ct_0^{2/3} \int \frac{dt'}{t'^{2/3}} = 3ct_0 \left[1 - \left(\frac{t_1}{t_0} \right)^{1/3} \right] $$

The redshift is given by $(1 + z) = R(0)/R(t_1) = (t_0/t_1)^{2/3}$. Hence, the time elapsed is

$$ t_{\text{elapsed}} = 3t_0 \left[1 - \frac{1}{(1 + z)^{1/2}} \right] = 0.85t_0 $$

(2.11) Referring to equation (1.18) the cross section for the reaction $v_e + n \rightarrow e^- + p$ is given by the formula

$$\sigma = \frac{4}{\pi \hbar^4 c^4} G_F^2 (p_f c)^2$$

where we have taken $v_i = v_f = c$. With $G_F = 1.17 \times 10^{-5} (\hbar c)^3 \text{ GeV}^{-2}$ this gives $\sigma = 1.7 (p_f c)^2 \times 10^{-44} \text{ cm}^2$ with the final state momentum $p_f c$ in MeV. The width or rate for the reaction per neutron target is found from (1.14) as $W = \sigma \phi$ where $\phi = nc$ is the flux of incident neutrinos and n is their number density as given by (2.40) and (2.45), that is, $\phi = 2 \times 10^{42} (kT)^3$ in units cm^{-2} s^{-1}, with kT measured in MeV. This gives a value of $W = 0.05 (kT)^3 (p_f c)^2$ s^{-1} to be compared with $H = 0.7 (kT)^2$ s^{-1} from (2.47) for three neutrino flavours. The value of the final state momentum in the reaction will be $p_f c \sim kT + Q$, where $Q = 1.29$ MeV. Setting $W/H = 1$ for the freeze-out condition gives $kT \sim 1.5$ MeV, as can be found by trial and error. This is an overestimate of the freeze-out temperature for several reasons. First, in assuming equation (1.18) we have ignored the effects of spin, which will increase the cross section; second, we have ignored the electron mass, which, with a value of 0.51 MeV, is comparable to kT; third, an integration should be made over the thermal spectrum and over the isotropic angular distribution of the colliding particles. When all such effects are accounted for, the freeze-out temperature is found to be approximately 0.8 MeV.

(2.12) $\Omega < \frac{2}{3}$, or $\Omega_v > \frac{1}{3}$.

(2.14) Referring to equation (2.33) and Example 2.3, the expression for the age t_0 will be given by the integral

$$H_0 t_0 = \int \frac{dz}{(1+z)[\Omega(1+z)^3 + (1-\Omega)(1+z)^2]^{1/2}}$$

where the integral runs from $z = 0$ to $z = \infty$, and the $(1+z)^3$ and $(1+z)^2$ terms refer to matter and curvature contributions, respectively, and $\Omega_k = 1 - \Omega$ where $\Omega \equiv \Omega_m$. To perform this integral, first make the substitution $(1+z) = [(1-\Omega)/\Omega] \tan^2 \theta$, when it reduces to

$$A \left\{ \int d\theta \left[\frac{1}{\sin^3 \theta} - \frac{1}{\sin \theta} \right] \right\} = -\frac{1}{2} A \left\{ \left(\frac{\cos \theta}{\sin^2 \theta} \right) + \ln \left[\tan \left(\frac{\theta}{2} \right) \right] \right\}$$

where $A = 2\Omega/(1-\Omega)^{3/2}$. The limits of integration are from $z = 0$, when $\tan^2 \theta = \Omega/(1-\Omega)$, to $z = \infty$, when $\tan^2 \theta = \infty$ and $\theta = \pi/2$. Inserting the numerical value $\Omega = 0.35$, one obtains $H_0 t_0 = 0.793$ or $t_0 = 10.9$ Gyr.

Chapter 3

(3.2) Positive and negative pions are particle and antiparticle. Positive and negative sigma baryons are not.

(3.3) The ρ-meson has $C = P = -1$; f-meson has $C = P = +1$. The process $\rho \rightarrow \pi^0 \gamma$ is allowed as an electromagnetic decay, with a branching ratio $\sim\alpha$ (actually 0.07%). The corresponding decay for the f-meson is forbidden by C-invariance.

(3.5) 30 mrad per year. 1.5×10^{20} yr.

(3.6) If $\mathbf{p}_e$, E_e, m_e, and $\mathbf{p}_p$, E_p, and M_p refer to the three-momenta, total energies, and masses of the electron and proton, respectively, then the square of the total four-momentum, equal to the CMS energy squared, is (see Appendix B)

$$s = (E_e + E_p)^2 - (\mathbf{p}_e + \mathbf{p}_p)^2 = m_e^2 + m_p^2 + 2(E_e E_p - \mathbf{p}_e \cdot \mathbf{p}_p) \approx 4E_e E_p$$

where in the final step we have used the fact that both particles are extreme relativistic, so that masses can be neglected, and the fact that the electron and proton momenta are in opposite directions.

(a) Inserting numbers, the value of $s = 98\,400$ GeV^2.
(b) The CMS energy squared of the electron–quark system is $s/4$.
(c) The cross section for the electromagnetic interaction is given by (1.23), which assumes that $q^2 \ll q_{max}^2 = s$ and q^2 represents the absolute value $|q^2|$. In this approximation

$$\left(\frac{d\sigma}{dq^2}\right)_{em} = \frac{4\pi\alpha^2 |Q/e|^2}{q^4} \tag{i}$$

where $|Q/e| = \frac{2}{3}$ is the u-quark charge. The cross section for the weak charged current interaction is given by (1.27b) which, after allowing for the W propagator at high q^2 assumes the form

$$\left(\frac{d\sigma}{dq^2}\right)_{wk} = \frac{G_F^2}{[\pi(1 + q^2/M_W^2)^2]} \tag{ii}$$

If we make the substitution $x = 3G_F M_W^2/4\pi\alpha$ and set $\gamma = q^2/M_W^2$, then, equating the above cross sections, gives the quadratic

$$\gamma^2(x^2 - 1) - 2\gamma - 1 = 0$$

with solution $\gamma = (1 + x)/(x^2 - 1)$. Inserting numbers (see Appendix A) one obtains $x = 2.45$ and $\gamma = 0.69$, so that the cross sections become equal at $q^2 = 4400$ GeV^2. Above this value, the charged weak current cross section exceeds the electromagnetic cross section.

[Note: the cross section (i) has been stated in simplified form. At large q^2 it should be multiplied by a factor $[1 + (1 - q^2/q_{max}^2)^2]/2$, but since the appropriate value of $q^2 \ll s$ this correction is small.]

(d) At large momentum transfers, neutral current (Z exchange) as well as photon exchange in the process $e + p \rightarrow e + $ hadrons will become important.

(3.7) (a) Under interchange of space and spin coordinates, the wavefunction acquires a factor $(-1)^{L+S}$, that is, $(-1)^S$ for a system with $L = 0$, and $S = 0$ or 1. But interchange of spatial and spin coordinates of electron and positron is equivalent to interchange of positive and negative charges, so that $C = (-1)^S$. If the positronium decays to two photons, it must have $C = +1$ so that this is the singlet state of $S = 0$, while decay to three photons implies $C = -1$ and $S = 1$. On account of the opposite parity of particle and antiparticle, the parity is $P = (-1)^{L+1} = -1$. Hence, the quantum numbers are $J^{PC} = 0^{-+}$ for the two-photon decay and 1^{--} for the three-photon decay.

(b) The energy levels are $E_n = -\alpha^2 mc^2/(4n^2) = 6.806/n^2$ eV. The $n = 2 \to n = 1$ transition energy is $0.75 \times 6.806 = 5.1$ eV.

(c) The annihilation process needs the overlap of the electron and positron wavefunctions inside the volume they occupy, which is of the order of the cube of the Bohr radius $a = 2h/(mc\alpha)$. So, for either decay, a factor $(m\alpha)^3$ enters the rate. The two-photon decay involves two lepton–photon vertices, hence a factor α^2, giving an overall factor $m^3\alpha^5$. A rate or width has dimensions of energy, hence dividing by m^2 to get the correct dimensions we can guess $\Gamma(2\gamma) \sim m\alpha^5$. In fact, the true width is just half this, $m\alpha^5/2$. The three-photon decay clearly involves a third vertex and hence another factor of α. The full calculation yields $\Gamma(3\gamma) = [2(\pi^2-9)/9\pi]m\alpha^6$.

(3.8) $J^{PC} = 1^{--}$. $\alpha_s \sim 0.7$. (A more sophisticated analysis of upsilon levels gives $\alpha_s \sim 0.2$.)

(3.9) The transformations are as follows:

	T	P
$\mathbf{r}$	$\mathbf{r}$	$-\mathbf{r}$
$\mathbf{p}$	$-\mathbf{p}$	$-\mathbf{p}$
σ	$-\sigma$	σ
$\mathbf{E}$	$\mathbf{E}$	$-\mathbf{E}$
$\mathbf{B}$	$-\mathbf{B}$	$\mathbf{B}$
$\sigma \cdot \mathbf{E}$	$-\sigma \cdot \mathbf{E}$	$-\sigma \cdot \mathbf{E}$
$\sigma \cdot \mathbf{B}$	$\sigma \cdot \mathbf{B}$	$\sigma \cdot \mathbf{B}$
$\sigma \cdot \mathbf{p}$	$\sigma \cdot \mathbf{p}$	$-\sigma \cdot \mathbf{p}$

It is seen that an electric dipole moment for the neutron would violate both P and T invariance. So we can write for the dipole moment

$$\text{EDM} = \text{charge } (|e|) \times \text{length}$$
$$\times \, P\text{-violating parameter} \times T\text{-violating parameter}$$

The fact that P is violated means that we must introduce the weak coupling, with magnitude $G_F = 1.17 \times 10^{-5}$ GeV^{-2}. We can get from this a characteristic length, which has dimensions 1/energy, by introducing a mass, which can be taken as the neutron mass. Thus, with 1 GeV$^{-1} = 1.97 \times 10^{-14}$ cm, we find for the length $l = G_F M_n \sim 2 \times 10^{-19}$ cm. For the T-violating parameter we assume the *CPT* theorem and take the equivalent *CP* violation rate from neutral kaon decay. The direct *CP*-violating rate is $\varepsilon' \sim 10^{-7}$, giving as our guess for the neutron electric dipole moment EDM $\sim 10^{-26}e$ cm. It is a pure accident that this is also the present (2003) experimental upper limit to the dipole moment. A full calculation with the Standard Model yields an estimate of $10^{-31}e$ cm, but other theories of 'physics beyond the Standard Model' yield values as high as $10^{-26}e$ cm.

A polarization asymmetry in proton–proton scattering, that is, a dependence of the scattering cross section on the sign of the beam helicity, would be a sign of parity violation. The expected level will be of the order of the ratio of weak to strong coupling amplitudes, that is of the order of 10^{-7}.

(3.10) All the decays are allowed, except for

$\rho^0 \to \pi^0 + \pi^0$ (forbidden by Bose symmetry, for which J must be even)

$\rho^0 \to \pi^0 + \eta$ ($C = -1 \to C = +1$ transition forbidden in e.m. interaction)

$\eta \to e^+ + e^-$ ($C = +1 \to C = -1$ transition forbidden in e.m. interaction)

The rate for $\pi^0 \to \gamma + e^+ + e^-$ is suppressed by a factor α relative to the two-photon decay.

Chapter 4

(4.1) $\Delta\theta = \dfrac{2\,GM}{bc^2}$

[Note: This is too small by a factor 2, as compared with the value given by the general theory of relativity. One reason for the discrepancy is that, in the Newtonian approach, only the spatial coordinates of the photon are considered, while in fact the gravitational field also affects the time coordinate, as explained in Appendix B. This introduces a time delay (called the Shapiro delay), which must introduce an extra deflection (as is familiar in classical optics, when the speed of light changes as it travels from a less dense to a more dense medium).]

(4.2) (1) Inserting the value of the Fermi constant, and with s in MeV^2, the cross section is found to be $\sigma = 2.82 \times 10^{-45}$ s cm^2.

(2) If the momentum and energy of the electron and positron are $\mathbf{p}_1(= E_1)$ and $\mathbf{p}_2(= E_2)$, then $s = (E_1 + E_2)^2 - (\mathbf{p}_1 + \mathbf{p}_2)^2 = 2E_1E_2(1 - \cos\theta)$ where θ is the angle between the two momenta. These are isotropically distributed, so that $\langle\cos\theta\rangle = 0$ and $\langle s\rangle = 2\langle E\rangle^2 = 2 \times (3.15kT)^2$.

(3) Again referring to (2.40), the density of electrons or positrons is $N_e = (\frac{3}{4}) \times 2.404 \times (kT)^3/(\pi^2\hbar^3c^3) = 2.39 \times 10^{31}(kT)^3$ cm^{-3} (with kT in MeV). Setting the relative velocity $v \sim c$, the value of $1/W = 1/\langle\sigma N_e v\rangle \sim 25/(kT)^5$ s.

(4) From (2.38) and (2.46) one obtains $t = 0.74/(kT)^2$ s for $g^* = \frac{43}{4}$. Equating t with $1/W$ gives for the freeze-out temperature $kT \sim 3$ MeV.

[Note: This is only an approximate value, for several reasons. First, the calculation of both cross sections and particle densities assumes all particles are extreme relativistic, and the corrections for electron mass will reduce the cross section, the particle density, and the relative velocity and thus increase the freeze-out temperature. Second, the calculation of the cross section ignores the effects of Z^0 exchange (neutral weak currents), which will increase it by about 15% and hence reduce the critical temperature.]

(4.3) $v^2 = GM/R.$

$v = 170 \text{ km s}^{-1}$

Optical depth $\tau = v^2/c^2 = 10^{-6}$.

(4.4) $E_R = [4M_D M_R/(M_D + M_R)^2]E_D \cos^2 \theta$

E_R maximum when $\theta = 0$.

$E_R(\text{max}) = E_D$ when $M_R = M_D$.

For $M_R \ll M_D, E_R = 4E_D(M_R/M_D)$

For $M_D \ll M_R, E_R = 4E_D(M_D/M_R)$.

$E_R \sim 70 \text{ keV}$ in numerical problem.

(4.5) $1.03 \times 10^{-11} \text{ m s}^{-2} \sim 10^{-12} \text{ g}$.

Chapter 5

(5.1) $\tau = 2\pi/\omega = (3\pi/G\rho)^{1/2}$

(5.2) $v_s = 360 \text{ m s}^{-1}$

$\lambda_J = 6.88 \times 10^7 \text{ m}$

$M_J = \pi\rho\lambda_J^3/6 = 2.2 \times 10^{23} \text{ kg (cf. Earth mass} = 6 \times 10^{24} \text{ kg)}$

(5.3) The Thomson cross section is given in (1.26) as

$$\sigma = \frac{8\pi\alpha^2\hbar^2c^2}{(3m_e^2c^4)} = 0.67 \text{ b.}$$

The present baryon density from (2.28) is

$$\rho_b = 4.5 \times 10^{-28} \text{ kg m}^{-3}$$

corresponding to a proton and electron number density of the order of $N_p = N_e = \rho_b/M_p = 0.27$ electrons or protons m^{-3}. The mean free path for Thomson scattering of the radiation by free electrons is then

$$\lambda \sim \frac{1}{(F\sigma N_e)} \sim 5 \times 10^{28}/F \text{ m}$$

where F is the fraction of intergalactic matter that is ionized and therefore subject to scattering, $\gamma + e \rightarrow \gamma + e$. So the mean free path is large compared with the horizon distance, of the order of 10^{26} m, and scattering is not likely to be important. Note that, since matter is conserved, this result does not depend on details of the expansion of the universe.

(5.5) $t = 3\lambda^2/(c^2 t_i)$

$\lambda = 1 \text{ cm}; t = 10^4 \text{ yr}; M = 10^{16}M_\odot \sim \text{cluster mass}$

$\lambda = 1 \text{ mm}; t = 10^2 \text{ yr}; M = 10^{12}M_\odot \sim \text{galaxy mass}$

Chapter 6

(6.1) $E^{-2.22}$

(6.2) 110 m.

(6.3) 2.4 km

(6.4) The probability that a pion is produced in an element of depth dx at depth x gm cm^{-2} in the atmosphere by a primary proton is $\exp(-x/\lambda)\,dx/\lambda$, where the interaction length $\lambda \sim 100$ gm cm^{-2}. The probability that it will then survive to depth y without interaction is $\exp[-(x-y)/\lambda]$, where to keep the problem simple we assume the same interaction length for pions and protons.

The connection between the depth x in gm cm^{-2} and height h is $h = H \ln(X/x)$, where $X = 1030$ gm cm^{-2} is the total atmospheric depth, and again for simplicity we have assumed an isothermal, exponential atmosphere (strictly only true in the top third of the atmosphere) with $H = 6.5$ km.

In traversing the interval x to y the pion covers a distance $s = H \ln(y/x)$, and the probability that it does this without decaying, and subsequently decays in an element ds is

$$dP = \exp\left(-\frac{s}{\gamma c \tau}\right)\frac{ds}{\gamma c \tau} = \frac{H}{\gamma c \tau}\frac{dy}{y}$$

$$\times \exp\left[-\frac{H}{\gamma c \tau}\ln\left(\frac{y}{x}\right)\right] = \alpha\left(\frac{x}{y}\right)^{\alpha}\frac{dy}{y}$$

where $\alpha = H/\gamma c\tau = E_0/E$. Here, E is the pion energy, $\gamma = E/mc^2$, and m and τ are the pion mass and lifetime. Thus, the joint probability that a pion is created in dx and survives to decay in the depth interval dy is

$$P(x,y)\,dx\,dy = \frac{dx}{\lambda}\exp\left(-\frac{x}{\lambda}\right)\exp\left(-\frac{x-y}{\lambda}\right)\alpha\left(\frac{x}{y}\right)^{\alpha}\frac{dy}{y}$$

and the probability that the pion is produced at **any** value of $x < y$ and decays in dy is found by straightforward integration to be

$$P(y)\,dy = \frac{\alpha}{\alpha+1}\exp\left(-\frac{y}{\lambda}\right)\frac{dy}{\lambda}$$

The overall probability that the pion decays anywhere in the atmosphere is found by integrating from $y = 0$ to $y = X$. Since $X \gg \lambda$ the y integration just gives unity. So, the overall pion decay probability is

$$P_{\text{decay}} = \frac{\alpha}{\alpha+1} = \frac{E_0}{E+E_0}$$

where $E_0 = Mmc^2/c\tau = 117$ GeV.

If the pion is produced at a zenith angle θ, the depths remain the same, but all distances are multiplied by the secant of this angle, so that the energy E_0 is simply replaced by $E_0 \sec \theta$.

(6.5) In the pion rest-frame, let the muon and neutrino have three-momenta p^* and $-p^*$. Then, conservation of energy gives, assuming zero neutrino mass (units $c = \hbar = 1$)

$$m_\pi = \sqrt{p^{*2} + m_\mu^2} + p^*$$

Hence, the momentum and energy of the muon in the pion rest-frame are

$$p^* = \frac{(m_\pi^2 - m_\mu^2)}{2m_\pi} \qquad\qquad E^* = \frac{(m_\pi^2 + m_\mu^2)}{2m_\pi}$$

In the laboratory system, the muon energy is therefore (see Appendix B, equation (B.12))

$$E_\mu = \gamma(E^* + \beta p^* \cos\theta^*)$$

where $\gamma = E_\pi/m_\pi$, $\beta = (1 - 1/\gamma^2)^{1/2}$, and θ^* is the angle of emission of the muon relative to the pion line-of-flight, measured in the pion rest-frame. Since the pion has spin zero, the muon will be emitted isotropically so the distribution in $\cos\theta^*$ will be flat and the lab energy distribution of the muon, in the approximation that $\beta \approx 1$, will also be a flat distribution extending over the range $E_\mu = \gamma(E^* \pm p^*)$, that is, from $E_\mu = E_\pi$ to $E_\mu = (1 - a)E_\pi$, where $a = (1 - m_\mu^2/m_\pi^2) = 0.42$.

If the pion energy spectrum has the form $E^{-3} dE$ it follows that pions in the energy interval dE will produce the following laboratory muon spectrum at sea-level (neglecting ionization energy losses in the atmosphere):

$$dN = \left(\frac{dE_\mu}{aE}\right)\left(\frac{dE}{E^3}\right) P(E)$$

where $P(E) = E_0/(E_0 + E)$ is the pion decay probability from Problem (6.4). Setting $x = E/E_0$ and integrating over pion energy we obtain for the muon spectrum

$$\frac{dN}{dE_\mu} = \frac{1}{aE_0^3} \int \frac{dx}{x^4(1 + x)}$$

where the integral runs from $x = E_\mu/E_0$ to $x = E_\mu/aE_0$.

We can express the integrand in the form

$$\frac{1}{x^n(1 + x)} = \frac{1 - x + x^2 - \cdots (-1)^{n-1}x^{n-1}}{x^n} + \frac{(-1)^n}{(1 + x)}$$

The integration is straightforward and yields the result

$$\frac{dN}{dE_\mu} = \left(\frac{1}{aE_\mu^3}\right)\left[\frac{(1 - b^3)}{3} - (1 - b^2)\left(\frac{E_\mu}{2E_0}\right)\right.$$
$$\left. + (1 - b)\left(\frac{E_\mu}{E_0}\right)^2 + \left(\frac{E_\mu}{E_0}\right)^3 \ln\left[\frac{(E_\mu/E_0 + b)}{(E_\mu/E_0 + 1)}\right]\right]$$

where $b = 1 - a = 0.58$. Typical values for the ratio $R(E)$ of muons at sea-level with energy E to pions created in the atmosphere with energy

E are as follows:

E/E_0	~0	0.1	1.0	5.0
R		0.64	0.56	0.29	0.09

(6.6) No *CP*-violating effects are possible with only two flavours in vacuum, since a *CP*-violating phase requires at least three flavours and a 3×3 mixing matrix. If matter effects in the Earth are taken into account, induced *CP*-violating effects, that is a difference in the oscillation amplitudes for neutrinos and antineutrinos are possible with only two neutrino flavours, because the Earth is not *CP*-symmetric, being made of matter without antimatter.

(6.7) $\Delta m^2 < 0.064$ eV^2.

(6.8) Applying the conservation of energy and the conservation of momentum, parallel and perpendicular to the direction of the incident neutrino, allows one to eliminate the energy and angle of emission of the recoil electron, and there results a relation between the angle of emission θ of the scattered neutrino and its energy, E', in terms of the incident energy E:

$$\cos\theta = 1 - m\left(\frac{1}{E'} - \frac{1}{E}\right)$$

where m is the electron mass, and the neutrino is assumed massless. Since $E \gg m$, we can expand $\cos\theta \approx 1 - \theta^2/2$, whence we obtain the relation

$$\theta = \sqrt{2m\left(\frac{1}{E'} - \frac{1}{E}\right)}$$

(6.9) From the solution to Problem (6.5) we know the energy and momentum, E^* and p^*, of the muon in the pion restframe:

$$E^* = \frac{m_\pi^2 + m_\mu^2}{2m_\pi}; \quad p^* = \frac{m_\pi^2 - m_\mu^2}{2m_\pi}$$

The laboratory energy of the muon from the decay of a relativistic pion of Lorentz factor $\gamma = E_\pi/m_\pi$ and $\beta \approx 1$ is (see Appendix B for examples of Lorentz transformations)

$$E_\mu = \gamma(E^* + p^* \cos\theta^*)$$

where θ^* is the angle of emission of the muon in the pion rest-frame. As explained in the answer to Problem 6.5, this angular distribution is isotropic, and the muon energy in the laboratory therefore extends from $(m_\mu^2/m_\pi^2)E_\pi = 0.58E_\pi$ to E_π, with a mean value of $0.79E_\pi$, so the neutrino receives an average energy of $0.21E_\pi$. In its subsequent decay, the muon transforms into a positron, an electron–neutrino and a muon-antineutrino, each receiving approximately one-third of the muon energy, that is, about $0.26E_\pi$. To summarize, therefore, the average energies of the various neutrinos are as follows:

$$\pi^+ \to \mu^+ + \nu_\mu \qquad \langle E(\nu_\mu) \rangle = 0.21 E_\pi$$

$$\mu^+ \to e^+ + \nu_e + \nu_\mu \qquad \langle E(\nu_e) \rangle = \langle E(\nu_\mu) \rangle = 0.26 E_\pi$$

The numbers here ignore the effects of spin polarization of the muons from pion decay, which can affect the mean energies by several per cent.

(6.10) $P = 5\,\mu\text{W}$

(6.12) 3.8×10^{19} m ~ 1 kpc. No.

(6.13) 10^{-7}.

(6.16) Suppose the jet of relativistic particles emits a light signal at time t_0 and a second signal at time $(t_0 + \Delta t)$. Taking the x-axis as the line of sight to Earth and the y-axis in the transverse direction, the actual transverse velocity of the jet is $\Delta y / \Delta t = v \sin \theta$, but this is not the value observed at the Earth. Since the jet is moving towards the Earth with velocity $v \cos \theta$, the time on the Earth between the two signals is

$$\Delta t_E = \Delta t - \frac{\Delta x}{c} = \Delta t \left[1 - \frac{v}{c} \cos \theta \right]$$

Hence, the apparent transverse velocity measured on the Earth is

$$\frac{u_{\text{trans}}}{c} = \frac{\Delta y}{c \Delta t_E} = \frac{\beta \sin \theta}{1 - \beta \cos \theta}$$

where $\beta = v/c$ and $\gamma = 1/\sqrt{1 - \beta^2}$. Differentiation shows that u_{trans}/c has a maximum value of $\gamma \beta$ when $\sin \theta = 1/\gamma \beta$, and therefore exceeds unity when $\beta > 1/\sqrt{2}$. On the contrary, when $\theta > \pi/2$, the 'away jet' will be observed to have a transverse velocity less than $\beta \sin \theta$.

(6.17) We refer to (B.15) in Appendix B. Assume the electron travels along the x-axis, and set $p_z = 0$ for convenience, so the transverse momentum is p_y. The angle of emission in the electron rest-frame is given by

$$\tan \theta^* = \frac{p_y^*}{p_x^*} = \frac{p_y}{\gamma (p_x - \beta E/c)}$$

where symbols with an asterisk refer to the electron rest-frame, and those without to the laboratory system, and $p_y^* = p_y$. With $p_y = p \sin \theta$, $p_x = p \cos \theta$, and $E = pc$ for a photon one obtains

$$\tan \theta^* = \frac{\sin \theta}{\gamma (\cos \theta - \beta)}$$

In the electron rest-frame, half of the photons will have $\theta^* < \pi/2$, or $\cos \theta > \beta$, or $\sin \theta < 1/\gamma$. For ultra-relativistic particles, the half-width of the beam of emitted photons is therefore $\theta \sim 1/\gamma$.

Chapter 7

(7.1) $\omega = 0.63$ rad s^{-1}.

(7.2) Five billion years.

(7.3) If the mass M is to be supported by degeneracy pressure, the density is given by (7.29)

$$\rho_{\text{deg}} = \left(\frac{4m_e^3}{\hbar^6}\right)\left(\frac{Am_P}{Z}\right)^5\left(\frac{4\pi}{3}\right)^3 M^2$$

If M is small enough, ρ_{deg} will fall below normal solid matter densities and atomic (electromagnetic) forces will then prevent gravitational collapse. So the maximum mass not dependent on electron degeneracy for stability is found by setting $\rho_{\text{deg}} = \rho_{\text{matter}} = 10^4$ kg m^{-3}. Inserting the various constants yields $M \sim 5 \times 10^{27}$ kg or about 0.25% of the solar mass. The largest planet in the solar system is Jupiter with $M = 0.001M_\odot$, for which the increase in central density due to electron degeneracy would be only about 10%.

(7.4) $t < 660$ yr, inconsistent with its origin in AD 1054.

(7.5) The particle horizon distance is nct_0 where $n = 2$ for a radiation-dominated universe and $n = 3$ for matter domination. If the mass is M, then,

$$R_{\text{Schw}} = \frac{2GM}{c^2} = nct_0$$

or

$$M_{\text{BH}} = \frac{nc^3t_0}{2G}$$

The mass of the universe of age t_0 and critical density is $(4\pi/3)\rho_c(nct_0)^3$. Inserting the values in SI units of $\rho_c = 9 \times 10^{-27}$ and $G = 6.7 \times 10^{-11}$, this is equal to the above mass when $t_0 = 1.5 \times 10^{10}/n$ yr.

(7.6) $R \sim 10^{-15}$ m (about equal to the radius of a proton).
$M \sim 10^{12}$ kg (about equal to the mass of a typical mountain).

(7.8) To answer this question, consult a text on celestial mechanics. The period $\tau = 15$ yr $= 4.7 \times 10^8$ s and perigee $s = 17$ light-hours $= 1.84 \times 10^{13}$ m. The period relation is $(M + m) = 4\pi^2a^3/(G\tau^2)$, called Kepler's Law, where the semi-major axis of the elliptic orbit is $a = s/(1 - \varepsilon)$, where $\varepsilon = 0.87$ is the eccentricity, and m and M are the masses of the star and the black hole respectively. Then $a = 1.42 \times 10^{14}$ m, yielding

$$M = 3.65 \times 10^6 M_{\text{sun}}$$

The velocity v at radius vector r is given by $v^2 = GM[(2/r) - 1/a]$ which at perigee gives $v^2 = GM(1 + \varepsilon)/s$ and $v = 7170$ km per sec.

References

Adler, R., Bazin, M., and Schiffer, M. (1965). *Introduction to General Relativity*. McGraw-Hill, New York.

Alcock, C. *et al.* (1993). *Nature*, **365**, 621.

Anderson, C.D. (1933). *Phys. Rev.*, **43**, 491.

Anderson, H.L. *et al.* (1952). *Phys. Rev.*, **85**, 934.

Arnison, G. *et al.* (1983). *Phys. Lett.*, **122B**, 103.

Bahcall, J.N. (1989). *Neutrino Astrophysics*. Cambridge University Press, Cambridge.

Bahcall, J.N., Pinsonneault, H.M., and Basu, S. (2001). *Astrophys. J.*, **555**, 990.

Barrow, J.D. (1988). *Quart. J. Roy. Astron. Soc.*, **29**, 101.

Bathow, G. *et al.* (1970). *Nucl. Phys.*, **B20**, 592.

Benoit, A. *et al.* (2002). astro-ph/0206271.

Bernabei, R. *et al.* (2002). *Phys. Lett.*, **B480**, 23.

Bionta R.M. *et al.* (1987) *Phys. Rev. Lett.* **58**, 1494.

Broeils, A.H. (1992). *Astron. Astrophys.*, **256**, 19.

Buks, E. and Roukes, M.L. (2002). *Nature*, **419**, 119.

Cabbibo, N. (1963). *Phys. Rev. Lett.*, **10**, 531.

Casimir, H.B.G. (1948). *Proc. Kon. Ned. Akad.*, **51**, 793.

Chaboyer, B. (1996). *Nucl. Phys.*, **B51**, 11.

Chandrasekhar, S. (1931). *Astrophys. J.*, **74**, 81.

Christenson, J.H. *et al.* (1964). *Phys. Rev. Lett.*, **13**, 138.

Cronin, J.W. (1999). *Rev. Mod. Phys.*, Supplement p. S166.

de Bernardis, P. *et al.* (2002). *Astrophys. J.*, **564**, 539.

de Lapparent, *et al.* (1986). *Astrophys. J. Lett.*, **302**, L1.

Dirac, P.A.M. (1931). *Proc. Roy. Soc.*, **A133**, 60.

Enge, H.A. (1972). *Introduction to Nuclear Physics*. Addison-Wesley.

Friedmann, H.A. (1922). *Zeit. Physik*, **10**, 377.

Freedman, J.L. *et al.* (2001). *Astrophys. J.*, **553**, 47.

Fukuda, Y. *et al.* (1998). *Phys. Lett.*, **436**, 33; **433**, 9.

Gell-Mann, M. (1964). *Phys. Lett.*, **8**, 214.

Georgi, H. and Glashow, S.L. (1974). *Phys. Rev. Lett.*, **32**, 438.

Glashow, S.L. (1961). *Nucl. Phys.*, **22**, 579.

Guth, A.H. (1981). *Phys. Rev.*, **D23**, 347; (2000). *Phys. Rep.*, **333**, 555.

Halverson, N.W. *et al.* (2001). astro-ph/0104489.

Harun-or-Rashid, S.M. and Roos, M. (2001). *Astron. Astrophys.*, **373**, 369.

Hawking, S.W. (1974). *Nature*, **248**, 30.

Hewish, A. *et al.* (1968). *Nature*, **217**, 709.

Higgs, P.W. (1964). *Phys. Lett.*, **12**, 132; (1966). *Phys. Rev.*, **145**, 1156.

Hirata K.S. *et al.* (1987) *Phys. Rev. Lett.* **58**, 1490.

Hoyle, F. (1954). *Astrophys. J. Suppl.*, **1**, 121.

Hulse, R.A. and Taylor, J.H. (1974). *Astrophys. J.*, **191**, L59; (1975). **201**, L55.

Itzykson, C. and Zuber, J.B. (1985). *Quantum Field Theory.* McGraw-Hill, New York.

Kamionkowski, M. and Kosowski, A. (1999). *Ann. Rev. Nucl. Part. Sci.*, **49**, 77.

Kobayashi, M. and Maskawa, K. (1972). *Prog. Theor. Phys.*, **49**, 282.

Koks, F.W.J. and Van Klinken, J. (1976). *Nucl. Phys.*, **A272**, 61.

Kolb, E.W. (1998). Proc. 29th Int. Conf. High En. Phys. (Vancouver, ed. A. Astbury *et al.*).

Lamoreaux, S.K. (1997). *Phys. Rev. Lett.*, **78**, 5.

Lee, A.T. *et al.* (2001). astro-ph/0104459.

Lehraus, I. *et al.* (1978). *Nucl. Instrum. Meth.*, **153**, 347.

Linde, A.D. (1982). *Phys. Rev. Lett.*, **B108**, 389; (1984). *Rep. Prog. Phys.*, **47**, 925.

Mikhaev, S.P. and Smirnov, A.Y. (1986). *Nuov. Cim.*, **9C**, 17.

Milne, E.A. (1934). *Quart. J. Math.* (Oxford), **5**, 64.

Ong, R.A. (1998). *Phys. Rep.* **305**, 93.

Peccei, R. and Quinn, H. (1977). *Phys. Rev. Lett.*, **38**, 1440.

Penzias, A.A. and Wilson, R.W. (1965). *Astrophys. J.*, **142**, 419.

Perlmutter, S. *et al.* (1999). *Astrophys. J.*, **517**, 565.

Pound R.V. and Snider J.L. (1965) *Phys. Rev.* **140B**, 788.

Riess, A.G. *et al.* (2000). *Astrophys. J.*, **536**, 62.

Rolfs, C.E. and Rodney, W.S. (1988). *Cauldrons in the Cosmos.* University of Chicago Press, Chicago.

Roy, A. *et al.* (1999). *Phys. Rev.*, **D60**, 111101.

Sakharov, A. (1967). *JETP Lett.*, **5**, 241.

Salam, A. (1967). *Elementary Particle Theory.* Almquist and Wiksell, Stockholm.

Schramm, D.N. and Turner, M.S. (1998). *Rev. Mod. Phys.*, **70**, 303.

Simpson, J.A. (1983). *Ann. Rev. Nucl. Part. Sci.*, **33**, 326.

Smith N.J.T. (2002). (UK DMC report at Dark Matter Conference, York, July).

Sparnaay, M.J. (1958). *Physica (Utrecht)*, **24**, 751.

Straumann, N. (2002). astro-ph/0203330.

Suntzeff, N.B. *et al.* (1992). *Astrophys. J. Lett.*, **384**, L33.

Surdej, J. *et al.* (1987). *Nature*, **329**, 695.

Webber, W.R. (1958). *Nuov. Cim. Suppl.* II, **8**, 532.

Weekes, T.C. (1998). *Phys. Rep.*, **160**, 1.

Weinberg, S. (1967). *Phys. Rev. Lett.*, **19**, 1264.

Wolfenstein, L. (1978). *Phys. Rev.*, **D17**, 2369.

Wu, C.S. and Shaknov, I. (1950). *Phys. Rev.*, **77**, 136.

Yukawa, H. (1935). *Proc. Math. Soc. Japan*, **17**, 48.

Zweig, R. (1964). CERN Report 8419/Th. 412.

Bibliography

Books on astrophysics and cosmology at a similar level:

Cauldrons in the Cosmos, C.E. Rolfs and W.S. Rodney (University of Chicago Press, Chicago 1988)

The Big Bang, J. Silk (W.H. Freeman and Co, New York 1989)

Cosmology and Particle Astrophysics, L. Bergstrom and A. Goobar (John Wiley and Sons, Praxis Publishing, Chichester 1999)

Cosmology, M. Rowan-Robinson (Clarendon Press, Oxford 1996)

The Dynamic Cosmos, M.S. Madsen (Chapman and Hall, London 1995)

Particle Astrophysics, H.V. Klapdor-Kleingrothaus and K. Zuber (IOP Publishing Ltd, Bristol 2000)

Books on astrophysics and cosmology at a more advanced level:

The Early Universe, E.W. Kolb and M.S. Turner (Addison-Wesley, New York 1990)

Physical Cosmology, P.J.E. Peebles (Princeton University Press 1971)

The Large Scale Structure of the Universe, P.J.E. Peebles (Princeton University Press 1980)

Principles of Physical Cosmology, P.J.E. Peebles (Princeton University Press 1993)

Modern Cosmology, D.W. Sciama (Cambridge University Press, Cambridge 1971)

Introduction to Cosmology, J.V. Narlikar (Jones and Bartlett Inc, Boston 1983)

Cosmological Physics, J.A. Peacock (Cambridge University Press 1999)

Books and articles on more specialist topics:

Neutrino Astrophysics, J. Bahcall (Cambridge University Press 1989)

The Physics of Stars, A.C. Phillips (Manchester Physics Series, John Wiley and Sons, Chichester 1994)

Cosmic Rays and Particle Physics, T.K. Gaisser (Cambridge University Press, Cambridge 1990)

Very high energy gamma ray astronomy, R.A. Ong, Phys. Rep., **305**, 93–202 (1998)

Very high energy gamma ray astronomy, T.C. Weekes, Phys. Rep., **160**, 1–121 (1988)

Review of gravitational wave detectors, F. Ricci and A. Brillet, Ann. Rev. Nucl. Part. Sci., **47**, 111 (1997)

Extensive air showers associated with discrete astrophysical sources, D.E. Nagle, T.K. Gaisser, and R.J. Protheroe Ann. Rev. Nucl. Part. Sci., **38**, 609 (1988)

Nuclear reactions in stars, B.W. Filippone *Ann. Rev. Nucl. Part. Sci.*, **36**, 717 (1986)

Neutrinos from supernova explosions, A. Burrows *Ann. Rev. Nucl. Part. Sci.*, **40**, 181 (1990)

Search for discrete astrophysical sources of energetic gamma radiation, J.W. Cronin, K.G. Gibbs, and T.C. Weekes, *Ann. Rev. Nucl. Part. Sci.*, **43**, 687 (1993)

Gamma-Ray Bursts: Ligo/Virgo sources of gravitational radiation, M. Van Putten *Phys. Rep.*, **345**, 1 (2001)

Cosmic Rays at the Highest Energies, A.A. Watson, *Contemp. Phys.* **43**, 181 (2002)

Cosmic Microwave Background Anisotropies, W. Hu and S. Dodelson, *Ann. Rev. Astron. Astroph.*, **40**, 171 (2002)

- Special Centenary Issue of Reviews of Modern Physics, **71** (1999) S145–197, has a number of short articles covering cosmology, cosmic microwave radiation, cosmic rays, black holes, dark matter, etc.
- Physics Reports, **333** (2000) dedicated to the late David Schramm contains several reports on microwave background anisotropies, high energy neutrino astronomy, inflation, dark matter, the Hubble constant, globular clusters, etc.

Index